# Molecular Diagnostics

Current Research and Applications

Edited by

Jim F. Huggett

Molecular and Cell Biology
LGC
Teddington
UK

and

Justin O'Grady

Norwich Medical School
University of East Anglia
Norwich
UK

Caister Academic Press

Caister Academic Press
Norfolk, UK

www.caister.com

British Library Cataloguing-in-Publication Data
A catalogue record for this book is available from the British Library

ISBN: 978-1-908230-41-6 (hardback)
ISBN: 978-1-908230-64-5 (ebook)

Description or mention of instrumentation, software, or other products in this book does not imply endorsement by the author or publisher. The author and publisher do not assume responsibility for the validity of any products or procedures mentioned or described in this book or for the consequences of their use.



Cover design by Lucy Baker, LGC.

# Contents

# Contributors

**Robert Anderson**
National Institute of Biological Standards and Control (NIBSC)
Centre of the Medicines and Healthcare products Regulatory Agency (MHRA)
Division of Virology
Potters Bar
UK

rob.anderson@nibsc.org

**Thomas Barry**
Department of Microbiology
School of Natural Sciences
National University of Ireland
Galway
Ireland

thomas.barry@nuigalway.ie

**David S. Boyle**
Program for Appropriate Technologies in Health
Seattle, WA
USA

dboyle@path.org

**Stephen A. Bustin**
Postgraduate Medical Institute
Faculty of Health, Social Care and Education
Anglia Ruskin University
Chelmsford
UK

stephen.bustin@anglia.ac.uk

**Bambos M. Charalambous**
UCL Centre for Clinical Microbiology
Royal Free Campus
University College London
London
UK

b.charalambous@ucl.ac.uk

**Eoin Clancy**
Molecular Diagnostics Research Group
NCBES; and
Department of Microbiology
School of Natural Sciences
National University of Ireland
Galway
Ireland

eoin.clancy@nuigalway.ie

**Sophie Collot-Teixeira**
National Institute of Biological Standards and Control (NIBSC)
Centre of the Medicines and Healthcare products Regulatory Agency (MHRA)
Division of Virology
Potters Bar
UK

s.collot@retroscreen.com

**Siobhan Dorai-Raj**
Department of Microbiology
School of Natural Sciences
National University of Ireland
Galway
Ireland

siobhandorairaj@gmail.com

**Phillip Evans**
Department of Research and Applied Markets
GE Healthcare Lifesciences
Cardiff
UK

phil.evans@ge.com

**Agnieszka M. Falinska**
Imperial College Hospital NHS Trust
Hammersmith Hospital
London
UK

afalinska@doctors.org.uk

**Tanya M. Ferguson**
Claremont BioSolutions
Upland, CA
USA

tferguson@claremontbiosolutions.com

**Jeremy A. Garson**
Division of Infection and Immunity
University College London
London
UK

j.garson@ucl.ac.uk

**Rebecca L. Gorton**
UCL Centre for Clinical Microbiology
Royal Free Campus
University College London
London
UK

rebecca.gorton@nhs.net

**Jim F. Huggett**
Molecular and Cell Biology
LGC
Teddington
UK

jim.huggett@lgcgroup.com

**Diane Lee**
Porton Consulting Research Limited
Salisbury
UK

diane@portonconsultingresearch.co.uk

**Martin Lee**
Porton Consulting Research Limited
Salisbury
UK

martin@portonconsultingresearch.co.uk

**Timothy D. McHugh**
UCL Centre for Clinical Microbiology
Royal Free Campus
University College London
London
UK

t.mchugh@ucl.ac.uk

**Philip Minor**
National Institute of Biological Standards and Control (NIBSC)
Centre of the Medicines and Healthcare products Regulatory Agency (MHRA)
Division of Virology
Potters Bar
UK

philip.minor@nibsc.org

**Jamie Murphy**
Division of Colon and Rectal Surgery
Mayo Clinic
Scottsdale, AZ
USA

jamie.murphy@qmul.ac.uk

**Angelika Niemz**
Keck Graduate Institute of Applied Life Sciences
Claremont, CA
USA

aniemz@kgi.edu

**Justin O'Grady**
Norwich Medical School
University of East Anglia
Norwich
UK

justin.ogrady@uea.ac.uk

**Kate Reddington**
Department of Microbiology
School of Natural Sciences
National University of Ireland
Galway
Ireland

kate.reddington@nuigalway.ie

**Joachim L. Schultze**
Department for Genomics and Immunoregulation
LIMES (Life and Medical Sciences Bonn)
University of Bonn
Bonn
Germany

j.schultze@uni-bonn.de

**G. Michael Taylor**
Division of Microbial Sciences
Faculty of Health and Medical Sciences
University of Surrey
Guildford
UK

gm.taylor@surrey.ac.uk

**Gemma L. Vanstone**
UCL Centre for Clinical Microbiology
Royal Free Campus
University College London
London
UK

gemma.vanstone@nhs.net

**Tomasz K. Wojdacz**
Karolinska Institutet
Institute of Environmental Medicine (IMM)
Stockholm
Sweden

tomasz.wojdacz@ki.se

# Current books of interest

| | |
|---|---|
| Microarrays: Current Technology, Innovations and Applications | 2014 |
| Metagenomics of the Microbial Nitrogen Cycle: Theory, Methods and Applications | 2014 |
| Pathogenic *Neisseria*: Genomics, Molecular Biology and Disease Intervention | 2014 |
| Proteomics: Targeted Technology, Innovations and Applications | 2014 |
| Biofuels: From Microbes to Molecules | 2014 |
| Human Pathogenic Fungi: Molecular Biology and Pathogenic Mechanisms | 2014 |
| Applied RNAi: From Fundamental Research to Therapeutic Applications | 2014 |
| Halophiles: Genetics and Genomes | 2014 |
| Phage Therapy: Current Research and Applications | 2014 |
| Bioinformatics and Data Analysis in Microbiology | 2014 |
| Pathogenic *Escherichia coli*: Molecular and Cellular Microbiology | 2014 |
| *Campylobacter* Ecology and Evolution | 2014 |
| *Burkholderia*: From Genomes to Function | 2014 |
| Myxobacteria: Genomics, Cellular and Molecular Biology | 2014 |
| Next-generation Sequencing: Current Technologies and Applications | 2014 |
| Omics in Soil Science | 2014 |
| Applications of Molecular Microbiological Methods | 2014 |
| *Mollicutes*: Molecular Biology and Pathogenesis | 2014 |
| Genome Analysis: Current Procedures and Applications | 2014 |
| Bacterial Membranes: Structural and Molecular Biology | 2014 |
| Bacterial Toxins: Genetics, Cellular Biology and Practical Applications | 2013 |
| Cold-Adapted Microorganisms | 2013 |
| *Fusarium*: Genomics, Molecular and Cellular Biology | 2013 |
| Prions: Current Progress in Advanced Research | 2013 |
| RNA Editing: Current Research and Future Trends | 2013 |
| Real-Time PCR: Advanced Technologies and Applications | 2013 |
| Microbial Efflux Pumps: Current Research | 2013 |
| Cytomegaloviruses: From Molecular Pathogenesis to Intervention | 2013 |
| Oral Microbial Ecology: Current Research and New Perspectives | 2013 |
| Bionanotechnology: Biological Self-assembly and its Applications | 2013 |
| Real-Time PCR in Food Science: Current Technology and Applications | 2013 |
| Bacterial Gene Regulation and Transcriptional Networks | 2013 |
| Bioremediation of Mercury: Current Research and Industrial Applications | 2013 |
| *Neurospora*: Genomics and Molecular Biology | 2013 |

Full details at www.caister.com

# Foreword

This book comes at an extraordinary time for those of us in the Clinical Front Line. The scale and pace of diagnostics development is ever accelerating, matched only by the rising expectations of our patients and our Governments. There must be more answers, more correct answers, ever faster, ever cheaper.

Until recently molecular diagnostics only achieved real clinical traction in Virology Departments, where culture has all but disappeared. Virologists were the first to develop these techniques for real (and real-time!) clinical benefit, more recently extending from monoplex PCR-based diagnostics for single pathogens to a more syndromic, multiplex diagnostic approach. In addition, our ability to rapidly sequence PCR products has in addition allowed for rapid anti viral resistance analysis for HIV.

Virologists certainly led the charge; and clinical bacteriologists are at last becoming more comfortable with molecular diagnostics, which do not use a rather slow biological PCR-based phenomenon referred to as "colony formation" using solidified essence of Japanese seaweed, sometimes complemented with other dubious natural products such as Horse Blood. Perhaps some die-hards will bemoan the lack of smell in the Bacteriology laboratory of the future.

Joking apart – the development of molecular diagnostics in bacteriology is now beginning to release the potential offered by speed – "same day" bacteriology was hitherto impossible due to the seemingly universal requirement for culture to the point where enough material was available for direct visualization and biochemical profiling. Whilst modern automation without doubt made the bacteriology laboratory's work more efficient and less "hands on" – the fact that pathogens had to be allowed to grow first remained an absolute barrier to rapid diagnostics. Not so with molecular tests, which are ever faster, currently easily achievable in the same half-day, and perhaps soon within the hour. These tests can furthermore be brought to bear not only on questions of identification, but also antibiotic resistance. The latter remains a very significant contributor to the necessary delay incurred by the need to not only grow the organism, but to then subsequently re-grow it (or not) in the presence of a range of antibiotics. Delay times of 72 hours continue to be accepted – if not acceptable. Similarly, gone are the bad old days where initial forays into microbe detection were constantly marred by issues of cross contamination and false positive results due to PCR product contamination at the front end of the process, although false conclusions can still be made for such reasons of exquisite sensitivity, as illustrated in one of this books' chapters.

These technologies are now also being brought to bear on myriad other matters of Human Biology as this relates to our Genome, both in health and disease, inherited and acquired. There are parallels here with pathogen diagnostics, where initially monoplex PCR-based systems are being replaced by several other far more advanced search strategies – both in terms of "wet chemistry" but also *in silico*.

Accordingly, the time is rapidly approaching where the technology will make it simpler and cheaper to openly sequence the tumour or even the whole individual, and then subsequently ask relevant questions of sequence *in silico*.

Many challenges remain: can, where and why should this technology be applied near to the patient? The much-vaunted and possibly rather ill-defined "Point of Care Testing" paradigm is nowhere near being resolved – due not so much to the power of the technology to deliver in such a format, but more because such instruments have to *contribute to delivering a clinical solution,* rather than merely a test result. Knowing that a sequence of interest, either human or pathogen, is, or is not there – is pointless if you cannot bring this to bear on an improved clinical outcome.

Similarly, how to reduce the vast amount of data which the more recent Next Generation technologies yield for Human Genomes into a framework of "normality"? This last question can only be answered when our data layers are sufficiently populated with Human sequence information in order to intelligently interpret any particular sequence variant. This will take time, although thankfully it appears that Cloud Based computing can stay one step ahead of the vast computational needs of such problems.

I suspect that the rapidly emerging field of Human transcriptomics will also very heavily rely on such high dimensional computing power.

In conclusion, what only a few years ago remained the property of very few, very well funded academic centres, with publications rather than patient benefit as output, is set to step into the clinical arena – providing not just test results, but solutions for far more complex problems.

And yet – we remain several years away from seeing the every day application of such powerful technologies truly embedded for patient benefit.

Nevertheless, molecular diagnostics have arrived and are here to stay. I'm not sure where "here" is, or will end up. But I do look forward to watching the "here" emerge and take its place at the centre of the New Medicine.

Dr Vanya Gant
The Department of Microbiology
UCLH NHS Foundation Trust
London
UK

# Biographies

**Jim F. Huggett**

I gained my BSc in Genetics at Liverpool University and moved to Cardiff University to study my PhD where I focused upon studying mechanically regulated genes in bone remodelling. After completing my PhD I made a subject jump by taking up a Senior Fellowship at University College London (UCL) where I developed an interest in infectious diseases. During my seven years at UCL I was fortunate to be able to work on a range of subjects ranging from molecular diagnostics to immunology and developed a considerable interest in molecular measurement and standardization. In 2009 I moved to LGC (while maintaining an honorary status at UCL) to lead the diagnostics and genomics research. At LGC my fledging interests in standardization developed into a strong enthusiasm for metrology (the science of measurement) which I apply to a wide range of subjects including cancer/fetal genotyping and RNA biomarker analysis and, of course, a substantial amount of my work has focused upon microbial analysis ranging from trace detection, metagenomics to diagnostics. A wide remit of my research deals with methodological efficacy of both established methods, like real time quantitative polymerase chain reaction (PCR), as well as newer methodologies such as digital/high throughput PCR and next generation sequencing.

**Justin O' Grady**

I gained my BSc in Microbiology, my MSc (Res) in infectious diseases molecular diagnostics and my PhD in molecular diagnosis of bacterial pathogens in food all at the National University of Ireland Galway (NUIG). My PhD research focused on real-time polymerase chain reaction (PCR)-based detection of *Listeria monocytogenes* and related species in multiple food types using a novel bacterial diagnostic target. I continued my research in the food microbiology sector, with a two-year post-doc at NUIG, but with a focus on lab-on-a-chip technology and isothermal amplification based diagnostics. This was followed by a two-year stint in industry (Beckman Coulter) developing real-time PCR-based molecular diagnostics assays for infectious diseases including *Mycobacterium tuberculosis*. During my time at Beckman Coulter, I developed a strong interest in tuberculosis (TB) research, which inspired me to move back into academia and take up a post-doc position at University College London on TB diagnostics. I was lucky enough to spend time in southern Africa while at UCL, where I developed a passion for infectious diseases focused on Global Health research. In January 2013, I was appointed Lecturer in Medical Microbiology at University of East Anglia (UEA) where I continue my research on microbial pathogen molecular diagnostics with the aim of translating this research broadly, in different sectors and diseases, to maximize community/patient benefit.

# Molecular Diagnostics: An Introduction

1

Jim F. Huggett, Siobhan Dorai-Raj, Agnieszka M. Falinska and Justin O'Grady

The clinical diagnostics sector is the younger sibling to the pharmaceuticals industry in the development of solutions to clinical problems. With diagnostics representing only a small percentage of clinical research, it is, perhaps, ironic when one considers that a correct diagnosis is usually key to good patient management. Nonetheless, the diagnostic sector is growing, increasingly seen as an untapped market by the biotechnological and pharmaceutical industries. The proportion of the diagnostic market held by molecular diagnostics is small but increasing.

The development of the polymerase chain reaction (PCR) in 1983 catapulted the field of molecular biology into mainstream biological and clinical research. Molecular methods are a quintessential component in fields ranging from vaccine development to toxicology. The simplicity and versatility of molecular techniques has meant that there are few fields of biological and clinical research that do not rely, in some way, on molecular biology.

Molecular diagnostics is defined here as diagnostic/prognostic approaches that employ the measurement of nucleic acids (DNA and/or RNA) in clinical situations. The potential application of PCR in molecular diagnostics was identified shortly after its discovery and was expected to revolutionize clinical diagnostics. However, three decades later, molecular diagnostics has failed to have as big a clinical impact as expected. While there are some examples of molecular diagnostics almost fully replacing traditional diagnostics methods, such as in virology, uptake of molecular methods in other areas has been slower than expected. This may be due to cost, lack of available expertise and the poor performance of some molecular methods relative to traditional methods. It is fair to say, however, that there has been a reluctance in certain fields to embrace molecular diagnostics. A case in point is described below where the contrast in the available diagnostic tests for different potential diagnoses is highlighted.

A 10-day-old patient (Fig. 1.1) is suffering from neonatal shock. She was born without complication at 37 weeks. Five days after birth she stopped feeding. There were no other clinical signs until day 7, when she started to fit and progressed to neonatal collapse. The differential diagnosis broadly consisted of four potential causes: infection, metabolic disorder, congenital defect or trauma. The technology to evaluate the last two differentials, such as echocardiography or magnetic resonance imaging, was capable of ruling them out within hours of the patient's arrival to the tertiary hospital. However, the diagnosis of

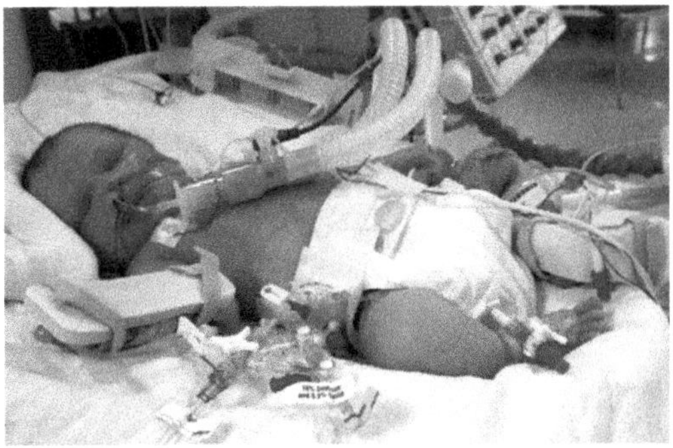

**Figure 1.1** A 10-day-old patient in intensive care with neonatal shock.

a potential metabolic disorder or infection effectively represented a waiting and guessing game, whereby the correct test that would be able to identify the problem had to be chosen from a long list of possibilities.

The patient was treated empirically with a range of antibiotics and antivirals and recovered without complications. The diagnosis was made ten days after first symptoms and turned out to be a rare enterovirus (Parechovirus) infection associated with neonatal collapse (Harvala *et al.*, 2010). It is worth noting that it is modern intensive care medicine that saved the patient. Diagnostic methods, while rapidly ruling out two of the differentials, failed to identify the causative pathogen or rule out a metabolic disorder until the patient had started to recover. Parechovirus has no specific cure, however, the patient could have been managed slightly differently had the diagnosis been known. There are molecular methods capable of testing for genetic metabolic disorders and infection. Unfortunately, results are not available within hours due to the requirement of specialized equipment located in in centralized laboratories. The ultimate challenge that the molecular diagnostics sector faces is the translation of such approaches to frontline routine application in a time frame that allows for appropriate clinical intervention.

Solving this challenge is far from simple. Papers describing the application of molecular methods to diagnostics always follow shortly after the introduction of such molecular methods. The advancement of molecular diagnostics from 'bench to bedside' can be compared to a daring and risky military campaign breaking through the enemy lines, which, once successful, requires considerable logistical support to develop, strengthen and maintain. The initial and frequently hyped successes of molecular diagnostics technologies are dwarfed in terms of effort when compared to that required to ensure the technologies are sufficiently robust and reliable for routine clinical application. The frequently disparate findings seen when applying molecular methods such as micro-arrays, quantitative PCR and next generation sequencing (NGS) to areas that could, if reproducible, be translated into diagnostics has led to calls for standardization. This has resulted in the publication of technical guidelines such as MIAME (Brazma *et al.*, 2001), the MIQE series (Bustin *et al.*, 2009; Huggett *et al.*, 2013) and MIGS (Field *et al.*, 2008) which facilitate technical evaluation of experimental design and data interpretation and will ultimately speed up and simplify the adoption of molecular methods in clinical practice.

A major, and often unrecognized, challenge in molecular diagnostics is the transition from conducting a molecular test to identify the presence of a molecular species (examples ranging from the identification of cancer point mutations to detection of pathogen genomes) to quantifying its abundance – qualitative vs quantitative clinical measurement. Where the latter is performed routinely (e.g. viral load monitoring or quantification of BCR-ABL transcripts in chronic myeloid leukaemia) considerable efforts have and are being made to support interlaboratory standardization (Fryer *et al.*, 2008; White and Hughes, 2012). The frequently wide measurement disparity in the absence of quantitative reference materials (Fryer *et al.*, 2008) indicates that methods like qPCR can be very precise but are often very biased; much work remains to be done to ensure that analysts are aware of inherent biases in certain technologies when developing molecular diagnostic methods.

Ensuring potential tests are sensitive, specific and reproducible, while fundamental for efficacy, are also at the mercy of the disease in question. Carriage of potentially pathogenic microorganisms as commensals can be a problem because the analytical sensitivity of a molecular method may facilitate the detection of the organism. This may not tally with the diagnostic sensitivity leading to a reduced specificity of the test unless factors like load are considered (Huggett *et al.*, 2008). Hence, quantitative diagnostics methods may be preferred in certain cases. The difficulty lies, however, in determining the cut-off point between carriage and infection. This is where sample quality plays a major role, having a direct impact on both qualitative and quantitative measurements. While some attempts have been made to control for the differences in sample quality and to define the cut-off between carriage and disease, these are very difficult parameters to standardize and are likely to be determined by a large variety of factors.

Epidemiological factors, such as disease prevalence, also play a crucial role in how a specific result will be analysed and this can be a particular problem when using very analytically sensitive molecular methods as diagnostic tools. False positive results are a major consideration when disease prevalence is low. For example, a diagnostic test that is 99% sensitive (correctly identifies 99 out of 100 positive patients as positive) and 99% specific (correctly identifies 99 out of 100 negative patients as negative) gives the impression of being very accurate. However, if the disease being diagnosed occurs in 1 out of 100 of the patients being tested, then for every patient that is correctly diagnosed one will be incorrectly diagnosed. Hence, when diagnosing diseases with a low prevalence, it can be difficult to maintain a high positive predictive value (PPV).

In clinical practice, where thousands of patients may be screened for a given disease, a low prevalence can lead to low PPV resulting in more patients being treated who do not have the disease than those that do. To counter this, physicians rarely treat based on a single test, frequently applying a range of different tests and empirical observations to better stratify and diagnose patients. The danger occurs when a new method is (a) heralded as a 'magic bullet' and clinicians depend less on alternative tests and/or (b) designed to diagnose a disease with a very low prevalence, increasing the challenge of PPV.

Preclinical research to identify potential molecular prognostic and diagnostic markers should consider the factors discussed above as central to their experimental design. Diagnostic sensitivity and predictive values need to be representative of the population being measured and not a polarized, selected group. Findings that cannot be reproduced should be considered as much a responsibility of the discovering laboratory as it is for those who attempt to repeat the work. Without such considerations, preclinical findings will remain exactly that and will never be translated to clinical practice.

In this book, we have chosen a series of chapters to represent two key molecular diagnostic areas; namely cancer and infectious diseases. The cancer section deals with the challenges associated with identifying genetic, epigenetic and transcriptomic biomarkers. The infectious disease section describes the current clinical applications of molecular diagnostics to detect viral, bacterial and fungal pathogens as well as an example of the use of molecular diagnostics outside the clinic – for infectious disease diagnosis in an archaeological context. A cautionary tale describing what can go wrong when molecular methods are applied incorrectly is also provided and makes fascinating reading.

We have also chosen to dedicate a substantial component of this book to the process of translating a preclinical test to the bedside and take a detailed look on how that is progressing in the near patient point-of-care molecular diagnostics market. As mention above, this is a fundamental consideration for successful translation of diagnostics tests from bench to bedside and is crucial if molecular diagnostics research is to have the impact on patient care that was hyped after the invention of PCR. In the final chapter we take a look at the future and offer some prediction as to where molecular techniques may take us, using the example of infectious diseases.

## References

Brazma, A., Hingamp, P., Quackenbush, J., Sherlock, G., Spellman, P., Stoeckert, C., Aach, J., Ansorge, W., Ball, C.A., Causton, H.C., *et al.* (2001). Minimum information about a microarray experiment (MIAME)-toward standards for microarray data. Nat. Genet. *29*, 365–371.

Bustin, S.A., Benes, V., Garson, J.A., Hellemans, J., Huggett, J., Kubista, M., Mueller, R., Nolan, T., Pfaffl, M.W., Shipley, G.L., *et al.* (2009). The MIQE guidelines: minimum information for publication of quantitative real-time PCR experiments. Clin. Chem. *55*, 611–622.

Field, D., Garrity, G., Gray, T., Morrison, N., Selengut, J., Sterk, P., Tatusova, T., Thomson, N., Allen, M.J., Angiuoli, S.V., *et al.* (2008). The minimum information about a genome sequence (MIGS) specification. Nat. Biotechnol. *26*, 541–547.

Fryer, J.F., Baylis, S.A., Gottlieb, A.L., Ferguson, M., Vincini, G.A., Bevan, V.M., Carman, W.F., and Minor, P.D. (2008). Development of working reference materials for clinical virology. J. Clin. Virol. *43*, 367–371.

Harvala, H., Wolthers, K.C., and Simmonds, P. (2010). Parechoviruses in children: understanding a new infection. Curr. Opin. Infect. Dis. *23*, 224–230.

Huggett, J.F., Taylor, M.S., Kocjan, G., Evans, H.E., Morris-Jones, S., Gant, V., Novak, T., Costello, A.M., Zumla, A., and Miller, R.F. (2008). Development and evaluation of a real-time PCR assay for detection of Pneumocystis jirovecii DNA in bronchoalveolar lavage fluid of HIV-infected patients. Thorax *63*, 154–159.

Huggett, J.F., Foy, C.A., Benes, V., Emslie, K., Garson, J.A., Haynes, R., Hellemans, J., Kubista, M., Mueller, R.D., Nolan, T., *et al.* (2013). The digital MIQE guidelines: minimum information for publication of quantitative digital PCR experiments. Clin. Chem. *59*, 892–902.

White, D.L., and Hughes, T.P. (2012). Classification of patients with chronic myeloid leukaemia on basis of BCR-ABL transcript level at 3 months fails to identify patients with low organic cation transporter-1 activity destined to have poor imatinib response. J. Clin. Oncol. *30*, 1144–1145; author reply 1145–1146.

# Part I

# Molecular Diagnostics in Cancer: Research and Development of Biomarkers

# Transcriptome-based Biomarkers: A Road Map Exemplified for Peripheral Blood-based Biomarker Discovery, Development and Clinical Use

2

Joachim L. Schultze

## Abstract

By using high-throughput genomic technologies an enormous number of novel biomarker candidates have been suggested for all the major diseases including cancer, metabolic disorders, autoimmune diseases and infections. Biomarkers are a prerequisite for developing truly personalized medicine. Yet the enormous efforts of biomarker discovery are not paralleled by similarly efficient biomarker development. In fact, only a very small number of biomarkers have made it into clinical practice in recent years. There are many reasons for the lack of translation of discovery into development and clinical practice. Using blood transcriptomics as an example, recent and current development, but also pitfalls and potential solutions for biomarker development are outlined here.

## Introduction

The genomic revolution enables the assessment of thousands of data points from thousands of patients in a rather short time frame. Initially, using array-based technology, genomic, transcriptomic, or epigenomic information has been accumulated from patients with all major diseases including cancer, metabolic disorders, autoimmune diseases, chronic inflammation and infections (Hoffman, 2004). With the advent of next-generation sequencing technologies, the information available is increasing by several orders of magnitude (Barabasi *et al.*, 2011). Therefore, now and certainly in the future, there is no shortage of information to enable the discovery of biomarkers for many important diseases (Nevins and Potti, 2007). Moreover, innovative new technologies for isolating specific cell types from injured or diseased tissues allows for sophisticated analyses even down to the single cell level. The resolution of information available from diseased tissue has been improved massively over the last decade. This is not only true for nucleotide-based analysis (genomics, transcriptomics, epigenomics) but also for metabolomics (Griffin *et al.*, 2011) and even proteomics (Mallick and Kuster, 2010). Yet the number of clinical biomarker tests solely based on high-throughput data assessment, either array-based or sequencing-based, is rather disappointing (Poste, 2011). One might argue that the development phase of such biomarkers takes time and therefore it is too early to expect a large number of assays being applicable to clinical practice. Yet, there is ample evidence that reasons other than time seem to play an important role for the lack of clinically applicable biomarkers based on genomic technology (Poste, 2011). For example, in 1999 Golub and colleagues were first to postulate that transcriptomics could be utilized to distinguish patients with acute myeloid leukaemia from patients with acute lymphatic leukaemia, a differential diagnosis of two devastating types of leukaemia that require immediate medical attention and different therapy regimens (Golub *et al.*, 1999). While this initial small study prompted hundreds of investigators to apply the same technology to other patients with cancer and many different diseases, leading to numerous published studies (Hoffman, 2004), more than a decade later, we still do not have a transcriptome-based test for the differential diagnosis of acute leukaemias. In fact, no transcriptome-based test for any clinical question (diagnosis, differential diagnosis or prognosis) concerning leukaemias has been approved for clinical use. Only very few products for very specialized clinical questions have made it to the clinic. One such area is outcome prediction of a subset of patients with breast cancer (van 't Veer *et al.*, 2002; van de Vijver *et al.*, 2002). Three products [MammaPrint® assay, Onco*type* DX™, and the Breast Cancer Profiling (BCP)] are utilized to stratify patients with an increased risk score to more aggressive therapy (Oakman *et al.*, 2010). While these transcriptome-based tests are approved for clinical use, many questions concerning their utilization in clinical practice remain (Marchionni *et al.*, 2008). This is mainly due to the difficult goal of these tests, namely risk and therapy outcome prediction. It is surprising that technologies such as transcriptome analysis have not been utilized sufficiently for the development of biomarkers addressing more straightforward clinical questions, such as diagnosis and differential diagnosis of common diseases, thereby replacing old and outdated technologies with new high-resolution and high-throughput genomic technologies. In contrast, the majority of transcriptome studies in the past addressed much more difficult clinical questions such as risk assessment, prognosis and therapy outcome prediction. From studies evaluating clinical predictors it is very clear that biomarkers targeting such difficult clinical questions need to be much more advanced than those that can be utilized for diagnosis, differential diagnosis and disease subclassification. It is therefore not entirely surprising that high-throughput technologies, such as transcriptome analyses, for biomarker discovery and development have seen a significant decline after a very intense phase of discovery in the few years after the millennium. The combination of immature technology, the infancy of bioinformatics, instability in sample handling, insufficient study sizes and overly challenging clinical questions can be judged retrospectively as a deadly combination for this promising field of research (Allison *et al.*, 2006). Yet, there is hope. Since almost all the pitfalls have been successfully addressed over the last years, we should not count out these technologies for future biomarker discovery, development, and clinical use (Allison *et al.*, 2006).

## Defining biomarker classes as a prerequisite for successful biomarker development

Prior to discussing some important methodological improvements necessary for biomarker development, we need to define the term biomarker in context of genomic technologies. Biomarkers can be defined as 'a characteristic that is objectively measured and evaluated as an indicator of normal biologic processes, pathogenic processes, or pharmacologic responses to a therapeutic intervention' (Biomarkers, 2001). In context of transcriptomics, a biomarker could be defined as specific alteration of gene expression that occurs during the course of a particular disease, e.g. in cancer tissue (van de Vijver *et al.*, 2002; Raz and Jablons, 2007; Merritt *et al.*, 2008). Such expression changes can be utilized for diagnosis, differential diagnosis, or subclassification of disease. Indeed, numerous studies have addressed these questions for cancer diagnosis, however; very few reached the clinic. One of the first tests approved by the FDA in 2008 is the Pathwork Diagnostic Tissue of Origin Test, developed to identify the origin of cancer tissues of unknown primary (Dumur *et al.*, 2008). Similar to the gene signature tests applied in breast cancer patients (see above), the impact of the test on patient's outcome is still not known and awaits the results of ongoing large prospective randomized clinical trials.

More recently, gene signatures have been suggested to facilitate early diagnosis of disease prior clinical symptoms (Mutti, 2008; Zander *et al.*, 2011). This is a particularly important area of research in cancer patients, since presentation with progressive disease is usually associated with a poor prognosis compared to those who are diagnosed at early stages of the disease. Whether blood transcriptomics might play a role in this area of research will be discussed later.

Other biomarkers include markers for prediction, prognosis, pharmacodynamics and risk assessment, which have been reviewed elsewhere (Staratschek-Jox and Schultze, 2010). Biomarkers predicting therapy outcome will be critical for the successful development of personalized medicine. However, whether transcriptome-based assays will play a role is under intense investigation.

The Human Immunology Profiling Consortium (HIPC) is very active in this area studying the role of vaccination on gene signatures in peripheral blood. The first promising results from this group have already been reported (Querec *et al.*, 2009; Nakaya *et al.*, 2011). Similarly, the impact of corticosteroid therapy, in patients with systemic lupus erythematodes, on blood gene signatures also suggested that transcriptome analysis could be used for therapy outcome prediction (Bennett *et al.*, 2003; Allantaz *et al.*, 2007). These examples emphasize the need for development of predictive biomarkers as a prerequisite for successful development of molecularly defined treatments. Early implementation of biomarker discovery and development during the drug development process is urgently needed (Singer, 2005; Phillips *et al.*, 2006; Tan *et al.*, 2009). A prerequisite for successful predictive biomarkers will be strong collaborations between basic research and clinical research in academia together with the pharmaceutical industry (Phillips *et al.*, 2006). Such collaborative efforts are exemplified by the Biomarker Consortium founded by the National Institutes of Health, the Food and Drug Administration (FDA) and the Pharmaceutical Research and Manufacturers of America group as a coordinated cross-sector public–private partnership (http://www.biomarkersconsortium.org). The goal is to develop drugs and corresponding predictive biomarkers in parallel to more precisely assess the therapeutic benefit during the entire development process of a new treatment or compound (Phillips *et al.*, 2006; Kuhlmann, 2007).

In contrast to predictive biomarkers, prognostic biomarkers predict the natural course of a given disease irrespective of any treatment. Prognostic biomarkers can be strongly influenced in their clinical values when novel therapy regimens are introduced that can change the natural progression of disease. A prognostic biomarker might be useful as a predictive biomarker or vice versa (Kim *et al.*, 2009). One of the most prominent gene signature-based prognostic biomarker tests is the MammaPrint® test, based on a 70-transcript signature identifying lymph node negative breast cancer patient at high risk for disease progression (van de Vijver *et al.*, 2002). Several validation studies supported the prognostic value of this signature (van de Vijver *et al.*, 2002; Glas *et al.*, 2006; Bueno-de-Mesquita *et al.*, 2007) and in 2007 the FDA approved the MammaPrint® test for clinical use. Another example is the Onco*type* DX™ signature, a quantitative RT-PCR-based test assessing a 21-transcript signature. This signature identifies lymph node-negative, oestrogen receptor-positive patients with early-stage invasive breast cancer who might benefit from hormonal therapy (Cobleigh, 2003; Esteban, 2003; Paik *et al.*, 2004). Ongoing research on large prospective randomized clinical trials with several thousand patients is currently assessing open questions concerning the use of this test for therapy decision-making and outcome (Sparano, 2006; Mook *et al.*, 2007; Cardoso *et al.*, 2008; Marchionni *et al.*, 2008; Zujewski and Kamin, 2008). As is the nature of predictive and prognostic biomarkers, definitive proof for their clinical utility can only be reached after an extended period of time. It is, therefore, not surprising that similar gene signatures for other cancers such as lung cancer (Beer *et al.*, 2002; Chen *et al.*, 2007) and colon cancer (Garman *et al.*, 2008) have been established; yet await prospective randomized clinical trials.

Pharmacodynamic (PD) biomarkers are developed to assess a biological effect of a biologically active compound. Important endpoints are safety, individual dosage and mechanism of action (Colburn and Lee, 2003). An early introduction of a PD biomarker into the drug development pipeline will allow earlier determination of clinical efficacy(Kola and Landis, 2004; Phillips *et al.*, 2006; Kummar *et al.*, 2007). The Investigational New Drug Guidance issued by the FDA in 2006 acknowledges this need for co-development of PD biomarkers during pre-clinical and early clinical drug development (US FDA, 2006). The result of this FDA initiative, in conjunction with the National Cancer Institute, was the establishment of so-called phase 0 clinical trials (Kummar *et al.*, 2007). These trials are intended to determine pharmacokinetic (PK) and PD endpoints by introducing novel biomarkers. For such PK and PD biomarkers, surrogate tissues such as blood or urine are promising since they are easily accessible and mirror ongoing processes within the body. Moreover, relevant biomarkers can be discovered in peripheral blood *in vitro* in the presence of the respective compound under study using, for example, blood transcriptomics (Staratschek-Jox *et al.*, 2009). In a second step such biomarkers are then tested *in vivo* in preclinical animal models before they are further tested in phase 0 and then phase 1 clinical trials. Examples for such an approach are the monitoring of an anti-TGFβR1 inhibitor in cancer patients (Classen *et al.*, 2010) and the monitoring of an anti-IFN-α antibody therapy in systemic lupus erythematodes (SLE) (Yao *et al.*, 2009). While transcriptome-based biomarkers are still a valid approach for the above-mentioned types of biomarkers, it is rather unlikely that transcriptomics will play a role in biomarker development for risk assessment (Pharoah *et al.*, 2008; Staratschek-Jox and Schultze, 2010). Overall, biomarkers should enable clinicians to take the most appropriate preventative or therapeutic strategy for the patient, thereby improving disease outcome. Each of the respective biomarker applications requires a tailored development strategy in order to avoid misleading interpretation.

## Focus on peripheral blood-based biomarkers as an example

The field of biomarker development is broad and complex with different settings requiring different strategies. There will be no one successful strategy for the development of biomarkers for all diseases, specimens to be analysed, or clinical questions to be addressed. Therefore, this chapter will focus on a specific specimen, in this case peripheral blood, a specific analytical approach, in this case transcriptome analysis, and a set of well-defined clinical questions, namely the development of biomarkers that could be applied to early detection, diagnosis, differential diagnosis, or sub-classification of disease.

## Blood transcriptomics started in leukaemia research

The first studies of transcriptome profiling in peripheral blood were focused on identifying genes that might be associated with the pathophysiology of, mainly, leukaemias. In 1999, Golub and colleagues laid the foundations for blood transcriptomics by demonstrating that the signatures of patients with acute myeloid leukaemia (AML) and those with acute lymphatic leukaemia (ALL) differed significantly, suggesting that blood transcriptomics could be used for differential diagnosis of leukaemias (Golub *et al.*, 1999). All the early studies were rather small, error prone, and assessed only a limited number of transcripts due to the array technology available at that time. A surprising development within this field of research was the early focus on rather difficult clinical questions such as disease sub-classification (Yeoh *et al.*, 2002; Ross *et al.*, 2003; Bullinger *et al.*, 2004; Haslinger *et al.*, 2004; Valk *et al.*, 2004; Hoffmann *et al.*, 2006), prognosis (Bullinger *et al.*, 2004; Verhaak *et al.*, 2005; Wilson *et al.*, 2006) or therapy outcome prediction (Metzeler *et al.*, 2008) rather than applying the new technology to more accessible endpoints such as primary diagnosis and differential diagnosis (Golub *et al.*, 1999; Haferlach *et al.*, 2005). In this context, blood transcriptomics could be introduced as a substitution technology (substituting for older technologies) rather than as an add-on technology for new clinical questions, which introduce additional hurdles before clinical application. In case of the leukaemias, this opportunity has been missed so far. The study by Golub *et al.* clearly indicated that blood transcriptomics is a valid option for differential diagnosis (Golub *et al.*, 1999), yet, no clinical blood transcriptomics test for the diagnosis and differential diagnosis of the acute leukaemias is available. Current primary diagnosis for the acute leukaemias is based on clinical assessment and light microscopy. Moreover, molecular diagnosis still depends on a myriad of time-consuming, labour-intensive, and expensive diagnostic tests. Other similar examples include autoimmune diseases such as rheumatoid arthritis or lupus erythematosus. Overall, the number of studies with a sufficiently large patient population (e.g. $n$ >100) addressing the role of blood transcriptomics and questions of diagnosis, differential diagnosis, or disease sub-classification is still limited.

A major early focus of blood transcriptomics was on the subclassification of AML (Yeoh *et al.*, 2002; Ross *et al.*, 2003; Bullinger *et al.*, 2004; Haslinger *et al.*, 2004; Valk *et al.*, 2004; Hoffmann *et al.*, 2006). According to previous clinical and morphological classification systems, AML is a heterogeneous disease with several subtypes that are characterized by different biological and clinical outcomes. Introduced by Bennett in 1976, the French–American–British classification (FAB) of AML was based on cytomorphology and cytochemistry only. Further updates of AML classification have introduced additional biological hallmarks such as characteristic cytogenetic features, molecular genetic changes and immunophenotypic variations (Jaffe *et al.*, 1999). So far, only few studies addressing differential diagnosis – but not primary diagnosis – of acute leukaemias have been reported (van Delft *et al.*, 2005; Haferlach *et al.*, 2005; Andersson *et al.*, 2007). Validation studies of these initial trials have not been reported as of yet.

Since induction therapy for AML patients has to be applied immediately after initial diagnosis, current diagnostic technologies are too slow to enable tailoring of induction therapy. This is rather disappointing since different prognostic values are related to certain chromosomal aberrations and gene mutations, strongly suggesting that faster molecular technologies are urgently needed to develop novel and tailored therapies for AML patients. In 2004, two studies independently reported transcriptome-based sub-classifications of AML (Bullinger *et al.*, 2004; Valk *et al.*, 2004) (Table 2.1). In the first, utilizing samples from 285 AML patients, Valk *et al.* (2004) described 16 different AML subtypes that – in some cases – were directly linked to the presence of specific chromosomal aberrations. More importantly, transcriptome analysis revealed new subtypes in a group of patients previously described as cytogenetically normal AML (CN-AML). In the second study, Bullinger *et al.* (2004) reported on the analysis of 116 AML patients including 46

**Table 2.1** Important studies of blood transcriptomics

| Author | Area of study | Reference |
|---|---|---|
| Zander | Early detection of lung cancer | Zander *et al.* (2011) |
| Querec | Gene signature change after yellow fever vaccine | Querec *et al.* (2009) |
| Nakaya | Gene signature change after influenza vaccine | Nakaya *et al.* (2011) |
| Haferlach | Differential diagnosis of acute leukaemias | Haferlach *et al.* (2005) |
| van Delft | Differential diagnosis of acute leukaemias | van Delft *et al.* (2005) |
| Anderson | Differential diagnosis of acute leukaemias | Andersson *et al.* (2007) |
| Valk | AML subclassification | Valk *et al.* (2004) |
| Bullinger | AML subclassification, prognosis | Bullinger *et al.* (2004) |
| Verhaak | AML subclassification, prognosis | Verhaak *et al.* (2005) |
| Wilson | AML subclassification, prognosis | Wilson *et al.* (2006) |
| Metzeler | Build predictor for course of AML | Metzeler *et al.* (2008) |
| Yeoh | ALL subclassification | Yeoh *et al.* (2002) |
| Ross | ALL subclassification, independent validation | Ross *et al.* (2003) |
| Haslinger | B-CLL subclassification | Haslinger *et al.* (2004) |
| Ramilo | (Differential) diagnosis of infectious disease | Ramilo *et al.* (2007) |
| Chaussabel | (Differential) diagnosis of infectious disease | Chaussabel *et al.* (2005) |
| Berry | Gene signature change in patients with active tuberculosis | Berry *et al.* (2010) |
| Chaussabel | Differential diagnosis of infectious disease and autoimmune disease | Chaussabel *et al.* (2008) |
| Hoffmann | ALL subclassification | Hoffmann *et al.* (2006) |
| Potti | Prediction of VTE in APS | Potti *et al.* (2006) |

AML, acute myeloid leukaemia; ALL, acute lymphoblastic leukaemia; CLL, chronic lymphoblastic leukaemia; APS, anti-phospholipid antibody syndrome; VTE, venous thromboembolism.

cases of CN-AML also defining new subclasses that were linked to the presence of chromosomal aberrations. Moreover, transcriptome analysis revealed two prognostically relevant subtypes among CN-AML. Using supervised class prediction, a clinical outcome predictor was established for AML patients that was further validated in an independent Cancer and Leukaemia Group B (CALCB) cohort consisting of 64 adult AML patients (Radmacher *et al.*, 2006). Transcriptome-based subclassification also served as the basis for the identification of additional mutations, e.g. in the nucleophosmin (NPM) gene (Falini *et al.*, 2005; Verhaak *et al.*, 2005), associated with one of the newly identified subtypes. Additional independent studies assessed signatures for clinical outcome prediction (Metzeler *et al.*, 2008; Wilson *et al.*, 2006) or prediction of cytogenetically defined subtypes of AML (Valk *et al.*, 2004). Subclassification has also been addressed for other leukaemias such as ALL (Yeoh *et al.*, 2002; Ross *et al.*, 2003) and chronic lymphocytic leukaemia (CLL) (Haslinger *et al.*, 2004) but not chronic myeloid leukaemia (CML). While there is evidence that blood transcriptomics can be used for subclassification of leukaemias and potentially for differential diagnosis, the simple question concerning clinical utility of blood transcriptomics for primary diagnosis has not been adequately assessed.

## Is there a role for blood transcriptomics in other diseases?

Developing blood transcriptomics for blood-based diseases such as leukaemias is an obvious goal. However, is it also possible to use blood as a surrogate tissue for diseases that are related to other organs? In fact, numerous smaller studies have addressed transcriptional changes in peripheral blood in diseases ranging from cardiovascular diseases, acute or chronic inflammation, infectious diseases, autoimmune diseases or even metabolic disorders. While many of these studies indeed suggest transcriptional changes in blood, they are generally underpowered to address questions concerning diagnosis, differential diagnosis, or subclassification of disease. However, some studies deserve to be mentioned. A very active field of investigation is the assessment of blood transcriptomics for a better diagnosis of infectious diseases, for which diagnosis is still inadequate (Fauci, 2005). Infections induce significant changes of transcription of immune cells and these cells can be assessed by blood transcriptomics (Medzhitov and Janeway, 2000; Jenner and Young, 2005; Ramilo *et al.*, 2007). One of the earliest studies addressing diagnostic issues in infectious diseases was reported by Chaussabel *et al.* (2005) analysing gene signatures of peripheral blood mononuclear cells (PBMC) from patients suffering from infectious diseases including influenza, *Escherichia coli*, *Staphylococcus aureus* and *Streptococcus pneumoniae* (Chaussabel *et al.*, 2005) (Table 2.1). In addition to defining a common inflammation signature, it was possible to define a signature specific for patients infected with influenza A. A follow-up study suggested that it is also possible to distinguish patients infected with the different bacteria (Ramilo *et al.*, 2007). In a more recent publication, it was shown that patients with active tuberculosis exhibit changes in their blood transcriptome when compared with patients with inactive disease or healthy individuals (Berry *et al.*, 2010). Taken together, bacterial infections caused by different pathogens are associated with specific changes in the transcriptome of blood cells. While certainly promising, further studies are required to estimate the true value of these observations with regard to diagnosis and differential diagnosis of bacterial infection.

Another area of intense investigation in blood transcriptomics is the search for signatures in autoimmune diseases. One of the more recent studies utilized PBMC samples from 239 individuals with different autoimmune disorders including systemic juvenile idiopathic arthritis (SJIA), systemic lupus erythematosus (SLE), type I diabetes, but also metastatic melanoma and acute infections (Chaussabel *et al.*, 2008). While the results obtained from this study were very promising questions concerning its clinical utility were not addressed, as highlighted by an associated commentary (Wang and Marincola, 2008). The authors used a proprietary approach based on biological models instead of well-recognized approaches for classifier generation, therefore the clinical value of the identified signatures for diagnosis, differential diagnosis, sub-classification or even prognostic or predictive questions remains unknown. Many smaller studies concerning SLE have been reviewed elsewhere (Mandel and Achiron, 2006). Similarly, many studies – too small to obtain reliable results for diagnostic purposes – have been reported for other autoimmune diseases including rheumatoid arthritis, juvenile rheumatoid arthritis, ankylosing spondylitis, ulcerative colitis, multiple sclerosis and Crohn's disease.

## Is there a role of blood transcriptomics for early detection of disease?

For many clinical scenarios, particularly malignant diseases, early detection of disease is a very attractive goal since patients recognized at earlier stages of the disease have a significantly better chance of survival. Classical examples are the PAP smear test to detect cervical dysplasia (Naucler *et al.*, 2007) or prophylactic colonoscopy (Lieberman, 2009) to identify colonic polyps. While successful in principle, compliance for prophylactic colonoscopy is usually too low and therefore less invasive screening tests for early detection with higher compliance rates are urgently needed. Blood-based tests seem to be an attractive alternative. Many studies have already reported on the discrimination of tumour patients from healthy individuals based on the analysis of tumour-associated antigens (Clarke-Pearson, 2009) or antibodies (Tan and Zhang, 2008), or assessment of defined sets of mRNA (Staratschek-Jox *et al.*, 2009) or non-coding RNAs (Keller *et al.*, 2009a,b) in serum or plasma. These tests are based on the assumption that even small tumour masses release sufficient amounts of tumour-associated proteins or other molecules (RNA, DNA) that can be detected in serum or plasma within the circulation. Yet, many of these proposed, or even clinically used, tests are characterized by low sensitivity and sometimes even low specificity. Recently published mathematical models strongly suggest that the amount of molecules directly released by small numbers of tumour cells is several logs too low to be detected reliably in the circulation (Brown and Palmer, 2009; Hazelton and Luebeck, 2011; Hori and Gambhir, 2011). Blood transcriptomics might therefore be an alternative approach. Immune cells within the circulation monitor diseased tissue and are therefore exposed to tumour-derived molecules in much higher concentrations. Since immune cells are equipped with a myriad of pattern recognition and other receptors; immune cells are sentinels that react towards signals derived from small numbers of tumour cells. Moreover, the transcriptional changes induced in these immune cells by the tumour-derived signals can amplify these signals. Therefore, blood transcriptomics of blood-derived immune cells can theoretically be used to identify transcriptional changes induced by tumour-derived signals. To test this hypothesis we studied transcriptional changes in early-stage lung cancer patients where tumour size was small and no metastases had yet occurred (Zander *et al.*, 2011). The study resulted in a lung cancer-associated transcriptomic signature that could distinguish lung cancer patients from healthy individuals with both high sensitivity and specificity (Zander *et al.*, 2011). In unpublished work (Schultze, unpublished data) we have extended these findings to individuals recruited on the European Prospective Investigation into Cancer and Nutrition (EPIC) study. Blood transcriptomes derived from individuals diagnosed within 2 years of initial blood sampling were compared with individuals reported to be tumour-free for a period of more than 10 years. Again, using blood transcriptomics, we were able to distinguish individuals with later clinical diagnoses of lung cancer from tumour-free individuals. This suggests that blood transcriptomics may indeed be an attractive alternative to current serum and plasma-based tests, which perform poorly in the early stages of disease. Other cancer types might also be recognized by transcriptional changes in peripheral blood. In fact, the German biotech company Signature Diagnostics is currently developing a blood transcriptome-based test for early detection of colon cancer (Signature Diagnostics, 2010). Nevertheless, for blood transcriptomics to become successful in the early detection setting, a large number of healthy individuals have to be screened and followed up until a minority of those will develop the disease under investigation. A prerequisite for such an endeavour is the integration of additional blood sampling, suitable for transcriptomics, into the large population-based cohorts already existing in several countries (Wichmann *et al.*, 2005; Omenn, 2007; Vineis and Riboli, 2009). Within these cohorts, biomarkers for early detection and diagnosis could be retrospectively developed and validated, however, prospective testing would also be possible (Schrohl *et al.*, 2008). As for all early detection tests, sensitivity and specificity of the test need to be primary endpoints since they are most informative in terms of diagnostic test development.

## General guidance towards successful biomarker development

As outlined above, blood transcriptomics is still an attractive area of biomarker research with undiscovered potential. However, as also mentioned, despite numerous reported studies, much more work is necessary to make this interesting technology applicable to routine clinical use. The first and most important issue to be resolved is to ensure that investigators always ask an appropriate question of significant clinical relevance (Table 2.2). The next step would then be the discovery of an appropriate biomarker that might be suitable for answering the proposed clinical question. Most of the aforementioned studies are discovery studies. Even when small additional validation studies were performed, they were insufficiently powered for use as a validation step towards clinical use. However, these small validation studies are an important step towards risk minimization, a prerequisite for the initiation of the biomarker development phase.

One aspect to be carefully considered when using high-throughput genomic technologies, such as blood transcriptomics, for biomarker discovery is whether the chosen technology is appropriate for biomarker development. For example, is it possible to develop reliable quantitative multiplex RT-PCR assays that target identified biomarkers rather than using the high-throughput technology used to identify the biomarkers. It should be considered, however, that every platform change requires additional validation trials. Certainly, an important consideration for the biomarker development phase is technical robustness of the chosen analysis method. For example, can the test be performed at multiple sites with comparable results or is it necessary to provide a central service to ensure high quality data. With modern transportation capabilities both options are of value. Another important consideration during the biomarker development phase is the accurate prediction of the number of clinical validation trials needed and the number of patients to be tested. One needs to ensure that the clinical trials designed to answer the initial clinical question are not underpowered (Sargent *et al.*, 2005). In most cases, there will be a requirement to perform prospective randomized multicentre clinical trials prior to approval by the regulatory authorities. Institutions like the U.S. FDA Office of In Vitro Diagnostic evaluation and Safety (OIVD) are providing Guidance Documents on In Vitro Diagnostic Multivariate Index Assays (IVDMIAs) that will be the basis for future development of multi-parameter assays in the USA. These guidelines also cover issues such as robustness, cost effectiveness and feasibility of usage at multiple sites of any new biomarker test. In addition, laboratories that plan to offer such assays – as any other laboratory providing

**Table 2.2** Milestones of biomarker development

| | |
|---|---|
| 1 | Definition of medical needs with significant clinical relevance |
| 2 | Discovery phase of an appropriate biomarker |
| 3 | Design robust biomarker assay, potential platform change |
| 4 | Initiate development phase with proof of principle clinical trials |
| 5 | Further development by validation in independent cohorts at multiple sites |
| 6 | Approval of biomarker assay for routine clinical application |

diagnostic services – must adhere to the Clinical Laboratory Improvement Act (CLIA). Considering such regulatory issues, it becomes more and more clear that while biomarker discovery can be performed in both academia and industry, the development phase is best suited to the private sector.

The field of subclassification of leukaemias and lymphomas during the last decade can be used as an example to illustrate how regulation is necessary for successful biomarker development. Initial biomarker discovery studies showed very promising results. However, since very different gene expression platforms were used, comparison was very difficult. Moreover, most of the early discovery studies used small sample sizes excluding the possibility of drawing conclusions with significant statistical power (Staratschek-Jox *et al.*, 2009). To overcome these limitations, it was necessary to (1) initiate larger prospective clinical trials with higher statistical power to capture sufficiently large test and validation cohorts, (2) introduce strict standard operating procedures for sample handling, microarray experiments and bioinformatic analysis, and (3) ensure sufficient reference diagnostics were used based on previous technology for the biomarker development phase. Following these lines, a large international consortium initiated the Microarray Innovations in Leukemia (MILE) study within the European LeukemiaNet (ELN). More than 3000 patients with leukaemias were included for transcriptome analysis in this large study (Kohlmann *et al.*, 2008).

## Importance of technical and organizational aspects during biomarker development

Clinical utility and validity – and therefore acceptance – of a biomarker test heavily relies on its reproducibility and performance. High accuracy is an absolute prerequisite for successful implementation into the clinical decision tree, which is necessary to truly improve disease outcome. To achieve high accuracy, technical and organizational aspects of the test have to be addressed from biomarker discovery to late development phases and approval. Moreover, sample handling has to be highly standardized to achieve valid and reliable results (Sargent and Allegra, 2002; Mandrekar *et al.*, 2005). Even after approval and during widespread clinical use, biomarker tests need to be tightly controlled to avoid invalid or incorrect test results. An inglorious example is the immunohistochemistry test to assess expression of the human epidermal growth factor receptor 2 (HER2) in breast cancer. In a well-recognized systematic review of the literature by the College of American Pathologists and by the American Society of Clinical Oncology in 2007, an unexpectedly high rate (about 20%) of incorrect HER2 assessments by immunohistochemistry was uncovered (Wolff *et al.*, 2007). These devastating results required the development of guidelines to reduce assay variations. Issues that were addressed include test interpretation, assay protocol standardization, variability of test reagents, sample heterogeneity, sample collection, sample storage, sample handling and sample fixation (Wolff *et al.*, 2007). Such guidelines are now in place for future clinical trials assessing other biomarkers in other cancer entities, e.g. for monitoring EGFR inhibitory therapy in lung cancer (Eberhard *et al.*, 2008). Clearly, such guidelines are required for any kind of biomarker development and should be established prior clinical testing even prior to the discovery phase. Unfortunately, most of the previously published studies in blood transcriptomics either do not report the use of such guidelines or obviously did not follow such guidelines. We, and some of our collaborators, have spent tremendous effort in developing such guidelines for blood transcriptomics (Debey *et al.*, 2004, 2006; Cobb *et al.*, 2005; Debey-Pascher *et al.*, 2009, 2011, 2012; Classen *et al.*, 2010; Gaarz *et al.*, 2010; http://www.spidia.eu/). In the case of blood transcriptomics, all biomarkers rely on an 'instable' biomaterial, since RNA content and distribution within blood cells heavily depends on exogenous signals and even sample handling can quickly change RNA content and distribution. Therefore standard operating procedures for sample stabilization, handling and storage as well as quality control of the biomaterial prior assay application had to be established (Debey *et al.*, 2004, 2006; Debey-Pascher *et al.*, 2009, 2011, 2012; Gaarz *et al.*, 2010). A landmark study was the Microarray

Quality Control (MAQC) study I – clearly demonstrating that standardized sample handling and procurement can lead to very comparable results between different laboratories (Shi *et al.*, 2006). For blood transcriptomics, the European Union recognized the need for standardization of sampling by funding the SPIDIA project through the Seventh Research Framework Program (http://www.spidia.eu/) SPIDIA is a consortium of 16 academic institutions, life sciences companies, and international organizations together with the European Committee for Standardization (CEN). The goal of SPIDIA is to improve tools for in-vitro diagnostics including high-throughput technologies such as array-based transcriptome analyses. In addition, SPIDIA will provide quality assurance schemes and guidelines for biomaterial sampling within Europe. Another aspect to be taken into account is the comparability of laboratory processes between different laboratories. Certification of laboratories testing biomarkers based on quality scores (Delost *et al.*, 2009) and implementation of ring trials for inter-laboratory comparison (Dowsett *et al.*, 2007; Franciotta and Lolli, 2007) are necessary even at the biomarker development phase.

There are additional requirements for blood transcriptomics to be utilized efficiently for biomarker discovery and development (Table 2.3). Especially important will be the availability of reagents and technology platforms over an extended period of time. Since biomarker development can easily take a decade from discovery to approval, rapid changes in genomic technologies are an unexpected challenge to the biomarker developer (Eggle *et al.*, 2009). One positive example is the availability of the Affymetrix U133A platform for more than a decade allowing for continuous use of a single platform technology. An alternative to this technology in the future might be transcriptome sequencing, however, sequencing is still not cost-effective for lengthy biomarker discovery and development projects. Moreover, RNA-seq is a still evolving technology and there is accumulating evidence that comparability of RNA-seq data between different platforms or even within one platform with different chemistry might not be as high as initially suggested (Schultze, unpublished results).

## Requirements for bioinformatics analysis during biomarker development

Blood transcriptomics is heavily dependent on bioinformatics analysis. During the last 10 years there has been enormous improvement in algorithm development and overall workflows for the development of disease classifiers based on high-throughput data (Radmacher *et al.*, 2002; Glas

**Table 2.3** Requirements for future biomarker discovery and development utilizing blood transcriptomics

| | |
|---|---|
| 1 | Ask important clinical questions focusing on unmet medical needs |
| 2 | Define precise endpoint of studies, major areas should be primary diagnosis, differential diagnosis, subclassification. More critical endpoints are therapy outcome, prognosis or prediction of disease |
| 3 | Follow standard operating procedures for sample handling, storage and preparation (e.g. as outlined by MAQC-I) |
| 4 | Utilize standardized study design. Use large enough patient populations during discovery and development phases. Avoid underpowered studies |
| 5 | RNA stabilization is a prerequisite in blood transcriptomics during all phases of discovery and development, particularly in later phase multi-centre studies |
| 6 | Standardized operating procedures for the generation of cDNA and/or cRNA |
| 7 | Utilize standards of bioinformatics analysis established, e.g.by MAQC-II |
| 8 | Development of bioinformatics approaches to reduce risk for further development after initial promising results in biomarker discovery |
| 9 | Long-term commitment of biotech industry to provide comparable array products for many years or affordable sequencing technologies |
| 10 | Widely apply recently established guidelines for transcriptome analysis in clinical studies |

*et al.*, 2006; Reme *et al.*, 2008; Liu *et al.*, 2009; Parry *et al.*, 2010). However, there have been concerns about the validity of transcriptome-based gene signatures (Michiels *et al.*, 2005; Dupuy and Simon, 2007; Ioannidis *et al.*, 2009). These concerns were diminished when the results of the large MAQC-II study, a large consortium effort to improve transcriptome data analysis for classifier development, were revealed (Shi *et al.*, 2010). MAQC-II clearly established that the performance of classifier model prediction is largely dependent on the biological endpoint. Moreover, the MAQC-II study also showed that many classifiers with similar performance can be developed from any given data set. While these findings are very reassuring, they do not address some important continuing concerns over biomarker discovery and development. There is an urgent need to better estimate the risk of biomarker development, once biomarkers are discovered. Handing over a biomarker project from academia-based discovery to industry-based development is most often hampered by an unacceptably high risk for further development. While academia usually cannot provide sufficiently large studies as a risk minimizing strategy, industry will not take the lead in developing rather preliminary discovery data established in small patient cohorts. This is certainly a promising new area of mathematical, statistical and bioinformatics research, developing strategies and algorithms that allow better estimation of statistical performance of new biomarkers during the development phase and beyond. Questions that need to be addressed include the estimation of (1) informational value of pilot trials performed during the discovery phase, (2) mean and maximum statistical performance to be expected in larger validation trials, (3) the size of validation trials required during further development phases using existing data, and (4) the most promising features to be selected for further biomarker development.

## The next decade of blood transcriptomics

After almost a decade of hype in genomics and genomic technologies we have finally entered a time of realism. Not surprisingly, it is a long road from the first pilot trials inducing the early hype to something we can apply with high efficiency and reliability to daily clinic practice. Based on solid research and development in genomics, bioinformatics, quality assurance, and clinical trial research during the last decade we are finally enabled to perform biomarker discovery and development at the level necessary to bring blood transcriptomics to the clinic. An unexpected hurdle introduced by the biotech industry is the very short half-life of genomic technology, making transcriptome-based biomarker discovery and development cumbersome. This hurdle is an important one that we need to overcome quickly (Table 2.4). Other hurdles for blood transcriptomics in the past

**Table 2.4** Future issues that need to be addressed when developing blood transcriptomics-based biomarkers

| | |
|---|---|
| 1 | So far, no test utilizing blood transcriptomics has been approved for diagnosis, differential diagnosis, subclassification, prediction or prognosis. Industry-based biomarker development projects are under way for the subclassification of certain leukaemias and early detection of colon cancer. Additional projects need to be moved from discovery to development phase and towards approval |
| 2 | Future studies need to be focused on clinically relevant questions, particularly in areas where there are urgent medical needs but reliable diagnostic tests are not existing |
| 3 | Where applicable novel blood transcriptomics-based biomarkers should substitute for assays based on older technologies rather than being introduced as novel add-on technologies. In light of capped health care budgets world-wide this is a very critical issue |
| 4 | Future studies need to be sufficiently larger to avoid underpowered studies |
| 5 | Standardization along the complete experimental and bioinformatics workflows of the biomarker assay need to be reached by strictly following well-established standard operating procures |
| 6 | The biotech industry providing genomic technologies needs to provide technology platforms that are available at least for a decade. Moreover, costs still need to drop significantly, particularly when utilizing next generation sequencing technologies |

(too much hype for too little studies, insufficient standardization, comparability, and reliability) certainly can be overcome. However, if we want to make blood transcriptomics clinically useful, we need to cross these hurdles within the next 10 years. Otherwise, blood transcriptomics might never become part of clinical practice.

## References

Allantaz, F., Chaussabel, D., Stichweh, D., Bennett, L., Allman, W., Mejias, A., Ardura, M., Chung, W., Smith, E., Wise, C., *et al.* (2007). Blood leukocyte microarrays to diagnose systemic onset juvenile idiopathic arthritis and follow the response to IL-1 blockade. J. Exp. Med. *204*, 2131–2144.

Allison, D.B., Cui, X., Page, G.P., and Sabripour, M. (2006). Microarray data analysis: from disarray to consolidation and consensus. Nat. Rev. Genet. *7*, 55–65.

Andersson, A., Ritz, C., Lindgren, D., Eden, P., Lassen, C., Heldrup, J., Olofsson, T., Rade, J., Fontes, M., Porwit-Macdonald, A., *et al.* (2007). Microarray-based classification of a consecutive series of 121 childhood acute leukaemias: prediction of leukemic and genetic subtype as well as of minimal residual disease status. Leukaemia *21*, 1198–1203.

Barabasi, A.L., Gulbahce, N., and Loscalzo, J. (2011). Network medicine: a network-based approach to human disease. Nat. Rev. Genet. *12*, 56–68.

Beer, D.G., Kardia, S.L., Huang, C.C., Giordano, T.J., Levin, A.M., Misek, D.E., Lin, L., Chen, G., Gharib, T.G., Thomas, D.G., *et al.* (2002). Gene-expression profiles predict survival of patients with lung adenocarcinoma. Nat. Med. *8*, 816–824.

Bennett, L., Palucka, A.K., Arce, E., Cantrell, V., Borvak, J., Banchereau, J., and Pascual, V. (2003). Interferon and granulopoiesis signatures in systemic lupus erythematosus blood. J. Exp. Med. *197*, 711–723.

Berry, M.P., Graham, C.M., McNab, F.W., Xu, Z., Bloch, S.A., Oni, T., Wilkinson, K.A., Banchereau, R., Skinner, J., Wilkinson, R.J., *et al.* (2010). An interferon-inducible neutrophil-driven blood transcriptional signature in human tuberculosis. Nature *466*, 973–977.

Biomarkers Definitions Working Group (2001). Biomarkers and surrogate endpoints: preferred definitions and conceptual framework. Clin. Pharmacol. Ther. *69*, 89–95.

Brown, P.O., and Palmer, C. (2009). The preclinical natural history of serous ovarian cancer: defining the target for early detection. PLoS Med. *6*, e1000114.

Bueno-de-Mesquita, J.M., van Harten, W.H., Retel, V.P., van't Veer, L.J., van Dam, F.S., Karsenberg, K., Douma, K.F., van Tinteren, H., Peterse, J.L., Wesseling, J., *et al.* (2007). Use of 70-gene signature to predict prognosis of patients with node-negative breast cancer: a prospective community-based feasibility study (RASTER). Lancet Oncol. *8*, 1079–1087.

Bullinger, L., Dohner, K., Bair, E., Frohling, S., Schlenk, R.F., Tibshirani, R., Dohner, H., and Pollack, J.R. (2004). Use of gene-expression profiling to identify prognostic subclasses in adult acute myeloid leukaemia. N. Engl. J. Med. *350*, 1605–1616.

Cardoso, F., Van't Veer, L., Rutgers, E., Loi, S., Mook, S., and Piccart-Gebhart, M.J. (2008). Clinical application of the 70-gene profile: the MINDACT trial. J. Clin. Oncol. *26*, 729–735.

Chaussabel, D., Allman, W., Mejias, A., Chung, W., Bennett, L., Ramilo, O., Pascual, V., Palucka, A.K., and Banchereau, J. (2005). Analysis of significance patterns identifies ubiquitous and disease-specific gene-expression signatures in patient peripheral blood leukocytes. Ann. N.Y. Acad. Sci. *1062*, 146–154.

Chaussabel, D., Quinn, C., Shen, J., Patel, P., Glaser, C., Baldwin, N., Stichweh, D., Blankenship, D., Li, L., Munagala, I., *et al.* (2008). A modular analysis framework for blood genomics studies: application to systemic lupus erythematosus. Immunity *29*, 150–164.

Chen, H.Y., Yu, S.L., Chen, C.H., Chang, G.C., Chen, C.Y., Yuan, A., Cheng, C.L., Wang, C.H., Terng, H.J., Kao, S.F., *et al.* (2007). A five-gene signature and clinical outcome in non-small-cell lung cancer. N. Engl. J. Med. *356*, 11–20.

Clarke-Pearson, D.L. (2009). Clinical practice. Screening for ovarian cancer. N. Engl. J. Med. *361*, 170–177.

Classen, S., Muth, C., Debey-Pascher, S., Eggle, D., Beyer, M., Mallmann, M.R., Rudlowski, C., Zander, T., Polcher, M., Kuhn, W., *et al.* (2010). Application of T cell-based transcriptomics to identify three candidate biomarkers for monitoring anti-TGFbetaR therapy. Pharmacogenet. Genomics *20*, 147–156.

Cobb, J.P., Mindrinos, M.N., Miller-Graziano, C., Calvano, S.E., Baker, H.V., Xiao, W., Laudanski, K., Brownstein, B.H., Elson, C.M., Hayden, D.L., *et al.* (2005). Application of genome-wide expression analysis to human health and disease. Proc. Natl. Acad. Sci. U.S.A. *102*, 4801–4806.

Cobleigh, M.B., P; Baker, J; Cronin, M; Liu, M-L; Borchik, R; Tabesh, B; Mosquera, J-M; Walker, MG; Shak, S. (2003). Tumor gene expression predicts distant disease-free survival (DDFS) in breast cancer patients with 10 or more positive nodes: high throughout RT-PCR assay of paraffin-embedded tumor tissues [abstract]. Prog. Proc. Am. Soc. Clin. Oncol. *22*, 850.

Colburn, W.A., and Lee, J.W. (2003). Biomarkers, validation and pharmacokinetic–pharmacodynamic modelling. Clin. Pharmacokinet. *42*, 997–1022.

Debey, S., Schoenbeck, U., Hellmich, M., Gathof, B.S., Pillai, R., Zander, T., and Schultze, J.L. (2004). Comparison of different isolation techniques prior gene expression profiling of blood derived cells: impact on physiological responses, on overall expression and the role of different cell types. Pharmacogenomics J. *4*, 193–207.

Debey, S., Zander, T., Brors, B., Popov, A., Eils, R., and Schultze, J.L. (2006). A highly standardized, robust, and cost-effective method for genome-wide transcriptome analysis of peripheral blood applicable to large-scale clinical trials. Genomics *87*, 653–664.

Debey-Pascher, S., Eggle, D., and Schultze, J.L. (2009). RNA stabilization of peripheral blood and profiling by bead chip analysis. Methods Mol. Biol. *496*, 175–210.

Debey-Pascher, S., Hofmann, A., Kreusch, F., Schuler, G., Schuler-Thurner, B., Schultze, J.L., and Staratschek-Jox, A. (2011). RNA-stabilized whole blood samples but not peripheral blood mononuclear cells can be stored for prolonged time periods prior to transcriptome analysis. J. Mol. Diagn. *13*, 452–460.

Debey-Pascher, S., Chen, J., Voss, T., and Staratschek-Jox, A. (2012). Blood-Based miRNA preparation for non-invasive biomarker development. Methods Mol. Biol. *822*, 307–338.

van Delft, F.W., Bellotti, T., Luo, Z., Jones, L.K., Patel, N., Yiannikouris, O., Hill, A.S., Hubank, M., Kempski, H., Fletcher, D., *et al.* (2005). Prospective gene expression analysis accurately subtypes acute leukaemia in children and establishes a commonality between hyperdiploidy and t(12;21) in acute lymphoblastic leukaemia. Br. J. Haematol. *130*, 26–35.

Delost, M.D., Miller, W.G., Chang, G.A., Korzun, W.J., and Nadder, T.S. (2009). Influence of credentials of clinical laboratory professionals on proficiency testing performance. Am. J. Clin. Pathol. *132*, 550–554.

Dowsett, M., Hanna, W.M., Kockx, M., Penault-Llorca, F., Ruschoff, J., Gutjahr, T., Habben, K., and van de Vijver, M.J. (2007). Standardization of HER2 testing: results of an international proficiency-testing ring study. Mod. Pathol. *20*, 584–591.

Dumur, C.I., Lyons-Weiler, M., Sciulli, C., Garrett, C.T., Schrijver, I., Holley, T.K., Rodriguez-Paris, J., Pollack, J.R., Zehnder, J.L., Price, M., *et al.* (2008). Interlaboratory performance of a microarray-based gene expression test to determine tissue of origin in poorly differentiated and undifferentiated cancers. J. Mol. Diagn. *10*, 67–77.

Dupuy, A., and Simon, R.M. (2007). Critical review of published microarray studies for cancer outcome and guidelines on statistical analysis and reporting. J. Natl. Cancer Inst. *99*, 147–157.

Eberhard, D.A., Giaccone, G., and Johnson, B.E. (2008). Biomarkers of response to epidermal growth factor receptor inhibitors in Non-Small-Cell Lung Cancer Working Group: standardization for use in the clinical trial setting. J. Clin. Oncol. *26*, 983–994.

Eggle, D., Debey-Pascher, S., Beyer, M., and Schultze, J.L. (2009). The development of a comparison approach for Illumina bead chips unravels unexpected challenges applying newest generation microarrays. BMC Bioinformatics *10*, 186.

Esteban, J.B. Jr., Cronin, M., Liu, M.-L., Llamas, M.G., Walker, M.G., Mena, R., and Shak, S. (2003). Tumor gene expression and prognosis in breast cancer: multi-gene RT-PCR assay of paraffin-embedded tissue [abstract]. Prog. Proc. Am. Soc. Clin. Oncol. *22*, 850.

Falini, B., Mecucci, C., Tiacci, E., Alcalay, M., Rosati, R., Pasqualucci, L., La Starza, R., Diverio, D., Colombo, E., Santucci, A., *et al.* (2005). Cytoplasmic nucleophosmin in acute myelogenous leukaemia with a normal karyotype. N. Engl. J. Med. 352, 254–266.

Franciotta, D., and Lolli, F. (2007). Interlaboratory reproducibility of isoelectric focusing in oligoclonal band detection. Clin. Chem. 53, 1557–1558.

Fauci, A.S. (2005). The global challenge of infectious diseases: the evolving role of the National Institutes of Health in basic and clinical research. Nat. Immunol. *6*, 743–747.

Gaarz, A., Debey-Pascher, S., Classen, S., Eggle, D., Gathof, B., Chen, J., Fan, J.B., Voss, T., Schultze, J.L., and Staratschek-Jox, A. (2010). Bead array-based microrna expression profiling of peripheral blood and the impact of different RNA isolation approaches. J. Mol. Diagn. *12*, 335–344.

Garman, K.S., Acharya, C.R., Edelman, E., Grade, M., Gaedcke, J., Sud, S., Barry, W., Diehl, A.M., Provenzale, D., Ginsburg, G.S., *et al.* (2008). A genomic approach to colon cancer risk stratification yields biologic insights into therapeutic opportunities. Proc. Natl. Acad. Sci. U.S.A. *105*, 19432–19437.

Glas, A.M., Floore, A., Delahaye, L.J., Witteveen, A.T., Pover, R.C., Bakx, N., Lahti-Domenici, J.S., Bruinsma, T.J., Warmoes, M.O., Bernards, R., *et al.* (2006). Converting a breast cancer microarray signature into a high-throughput diagnostic test. BMC Genomics *7*, 278.

Golub, T.R., Slonim, D.K., Tamayo, P., Huard, C., Gaasenbeek, M., Mesirov, J.P., Coller, H., Loh, M.L., Downing, J.R., Caligiuri, M.A., *et al.* (1999). Molecular classification of cancer: class discovery and class prediction by gene expression monitoring. Science *286*, 531–537.

Griffin, J.L., Atherton, H., Shockcor, J., and Atzori, L. (2011). Metabolomics as a tool for cardiac research. Nat. Rev. Cardiol. *8*, 630–643.

Haferlach, T., Kohlmann, A., Schnittger, S., Dugas, M., Hiddemann, W., Kern, W., and Schoch, C. (2005). Global approach to the diagnosis of leukaemia using gene expression profiling. Blood *106*, 1189–1198.

Haslinger, C., Schweifer, N., Stilgenbauer, S., Dohner, H., Lichter, P., Kraut, N., Stratowa, C., and Abseher, R. (2004). Microarray gene expression profiling of B-cell chronic lymphocytic leukaemia subgroups defined by genomic aberrations and VH mutation status. J. Clin. Oncol. 22, 3937–3949.

Hazelton, W.D., and Luebeck, E.G. (2011). Biomarker-based early cancer detection: is it achievable? Sci. Transl. Med. 3, 109fs109.

Hoffman, E.P. (2004). Expression profiling--best practices for data generation and interpretation in clinical trials. Nat. Rev. Genet. *5*, 229–237.

Hoffmann, K., Firth, M.J., Beesley, A.H., de Klerk, N.H., and Kees, U.R. (2006). Translating microarray data for diagnostic testing in childhood leukaemia. BMC Cancer 6, 229.

Hori, S.S., and Gambhir, S.S. (2011). Mathematical model identifies blood biomarker-based early cancer detection strategies and limitations. Sci. Transl. Med. 3, 109ra116.

SPIDIA (2013). Standardisation and improvement of generic pre-analytical tools and procedures for in-vitro diagnostics. Available at: http://www.spidia.eu/.

Ioannidis, J.P., Allison, D.B., Ball, C.A., Coulibaly, I., Cui, X., Culhane, A.C., Falchi, M., Furlanello, C., Game, L., Jurman, G., *et al.* (2009). Repeatability of published microarray gene expression analyses. Nat. Genet. *41*, 149–155.

Jaffe, E.S., Harris, N.L., Diebold, J., and Muller-Hermelink, H.K. (1999). World Health Organization classification of neoplastic diseases of the hematopoietic and lymphoid tissues. A progress report. Am. J. Clin. Pathol. *111*, S8–12.

Jenner, R.G., and Young, R.A. (2005). Insights into host responses against pathogens from transcriptional profiling. Nat. Rev. Microbiol. 3, 281–294.

Keller, A., Leidinger, P., Borries, A., Wendschlag, A., Wucherpfennig, F., Scheffler, M., Huwer, H., Lenhof, H.P., and Meese, E. (2009a). miRNAs in lung cancer – studying complex fingerprints in patient's blood cells by microarray experiments. BMC Cancer *9*, 353.

Keller, A., Leidinger, P., Lange, J., Borries, A., Schroers, H., Scheffler, M., Lenhof, H.P., Ruprecht, K., and Meese, E. (2009b). Multiple sclerosis: microRNA expression profiles accurately differentiate patients with relapsing–remitting disease from healthy controls. PLoS One *4*, e7440.

Kim, C., Taniyama, Y., and Paik, S. (2009). Gene expression-based prognostic and predictive markers for breast cancer: a primer for practicing pathologists. Arch. Pathol. Lab. Med. *133*, 855–859.

Kohlmann, A., Kipps, T.J., Rassenti, L.Z., Downing, J.R., Shurtleff, S.A., Mills, K.I., Gilkes, A.F., Hofmann, W.K., Basso, G., Dell'orto, M.C., *et al.* (2008). An international standardization programme towards the application of gene expression profiling in routine leukaemia diagnostics: the Microarray Innovations in Leukaemia study prephase. Br. J. Haematol. *142*, 802–807.

Kola, I., and Landis, J. (2004). Can the pharmaceutical industry reduce attrition rates? Nat. Rev. Drug Discov. *3*, 711–715.

Kuhlmann, J. (2007). The applications of biomarkers in early clinical drug development to improve decision-making processes. Ernst Schering Res. Found. Workshop, 29–45.

Kummar, S., Kinders, R., Rubinstein, L., Parchment, R.E., Murgo, A.J., Collins, J., Pickeral, O., Low, J., Steinberg, S.M., Gutierrez, M., *et al.* (2007). Compressing drug development timelines in oncology using phase '0' trials. Nat. Rev. Cancer 7, 131–139.

Lieberman, D.A. (2009). Clinical practice. Screening for colorectal cancer. N. Engl. J. Med. *361*, 1179–1187.

Liu, Q., Sung, A.H., Chen, Z., Liu, J., Huang, X., and Deng, Y. (2009). Feature selection and classification of MAQC-II breast cancer and multiple myeloma microarray gene expression data. PLoS One *4*, e8250.

Mallick, P., and Kuster, B. (2010). Proteomics: a pragmatic perspective. Nat. Biotechnol. *28*, 695–709.

Mandel, M., and Achiron, A. (2006). Gene expression studies in systemic lupus erythematosus. Lupus *15*, 451–456.

Mandrekar, S.J., Grothey, A., Goetz, M.P., and Sargent, D.J. (2005). Clinical trial designs for prospective validation of biomarkers. Am. J. Pharmacogenomics *5*, 317–325.

Marchionni, L., Wilson, R.F., and Marinopoulos, S.S. (2008). Impact of Gene Expression Profiling Tests on Breast Cancer Outcomes. In Evidence Reports/Technology Assessments, No. 160, A.f.H.R.a.Q. (US), ed. (Agency for Healthcare Research and Quality (US), Rockville, MD, USA).

Medzhitov, R., and Janeway, C., Jr. (2000). Innate immune recognition: mechanisms and pathways. Immunol. Rev. *173*, 89–97.

Merritt, W.M., Lin, Y.G., Han, L.Y., Kamat, A.A., Spannuth, W.A., Schmandt, R., Urbauer, D., Pennacchio, L.A., Cheng, J.F., Nick, A.M., *et al.* (2008). Dicer, Drosha, and outcomes in patients with ovarian cancer. N. Engl. J. Med. *359*, 2641–2650.

Metzeler, K.H., Hummel, M., Bloomfield, C.D., Spiekermann, K., Braess, J., Sauerland, M.C., Heinecke, A., Radmacher, M., Marcucci, G., Whitman, S.P., *et al.* (2008). An 86-probe-set gene-expression signature predicts survival in cytogenetically normal acute myeloid leukaemia. Blood *112*, 4193–4201.

Michiels, S., Koscielny, S., and Hill, C. (2005). Prediction of cancer outcome with microarrays: a multiple random validation strategy. Lancet *365*, 488–492.

Mook, S., Van't Veer, L.J., Rutgers, E.J., Piccart-Gebhart, M.J., and Cardoso, F. (2007). Individualization of therapy using Mammaprint: from development to the MINDACT Trial. Cancer Genomics Proteomics *4*, 147–155.

Mutti, A. (2008). Molecular diagnosis of lung cancer: an overview of recent developments. Acta Biomed. *79 Suppl 1*, 11–23.

Nakaya, H.I., Wrammert, J., Lee, E.K., Racioppi, L., Marie-Kunze, S., Haining, W.N., Means, A.R., Kasturi, S.P., Khan, N., Li, G.M., *et al.* (2011). Systems biology of vaccination for seasonal influenza in humans. Nat. Immunol. *12*, 786–795.

Naucler, P., Ryd, W., Tornberg, S., Strand, A., Wadell, G., Elfgren, K., Radberg, T., Strander, B., Johansson, B., Forslund, O., *et al.* (2007). Human papillomavirus and Papanicolaou tests to screen for cervical cancer. N. Engl. J. Med. *357*, 1589–1597.

Nevins, J.R., and Potti, A. (2007). Mining gene expression profiles: expression signatures as cancer phenotypes. Nat. Rev. Genet. *8*, 601–609.

Oakman, C., Santarpia, L., and Di Leo, A. (2010). Breast cancer assessment tools and optimizing adjuvant therapy. Nat. Rev. Clin. Oncol. 7, 725–732.

Omenn, G.S. (2007). Chemoprevention of lung cancers: lessons from CARET, the beta-carotene and retinol efficacy trial, and prospects for the future. Eur. J. Cancer Prev. *16*, 184–191.

Paik, S., Shak, S., Tang, G., Kim, C., Baker, J., Cronin, M., Baehner, F.L., Walker, M.G., Watson, D., Park, T., *et al.* (2004). A multigene assay to predict recurrence of tamoxifen-treated, node-negative breast cancer. N. Engl. J. Med. *351*, 2817–2826.

Parry, R.M., Jones, W., Stokes, T.H., Phan, J.H., Moffitt, R.A., Fang, H., Shi, L., Oberthuer, A., Fischer, M., Tong, W., *et al.* (2010). k-Nearest neighbor models for microarray gene expression analysis and clinical outcome prediction. Pharmacogenomics J. *10*, 292–309.

Pharoah, P.D., Antoniou, A.C., Easton, D.F., and Ponder, B.A. (2008). Polygenes, risk prediction, and targeted prevention of breast cancer. N. Engl. J. Med. *358*, 2796–2803.

Phillips, K.A., Van Bebber, S., and Issa, A.M. (2006). Diagnostics and biomarker development: priming the pipeline. Nat. Rev. Drug Discov. *5*, 463–469.

Poste, G. (2011). Bring on the biomarkers. Nature *469*, 156–157.

Potti, A., Bild, A., Dressman, H.K., Lewis, D.A., Nevins, J.R., and Ortel, T.L. (2006). Gene-expression patterns predict phenotypes of immune-mediated thrombosis. Blood *107*, 1391–1396.

Querec, T.D., Akondy, R.S., Lee, E.K., Cao, W., Nakaya, H.I., Teuwen, D., Pirani, A., Gernert, K., Deng, J., Marzolf, B., *et al.* (2009). Systems biology approach predicts immunogenicity of the yellow fever vaccine in humans. Nat. Immunol. *10*, 116–125.

Radmacher, M.D., McShane, L.M., and Simon, R. (2002). A paradigm for class prediction using gene expression profiles. J. Comput. Biol. *9*, 505–511.

Radmacher, M.D., Marcucci, G., Ruppert, A.S., Mrozek, K., Whitman, S.P., Vardiman, J.W., Paschka, P., Vukosavljevic, T., Baldus, C.D., Kolitz, J.E., *et al.* (2006). Independent confirmation of a prognostic gene-expression signature in adult acute myeloid leukaemia with a normal karyotype: a Cancer and Leukaemia Group B study. Blood *108*, 1677–1683.

Ramilo, O., Allman, W., Chung, W., Mejias, A., Ardura, M., Glaser, C., Wittkowski, K.M., Piqueras, B., Banchereau, J., Palucka, A.K., *et al.* (2007). Gene expression patterns in blood leukocytes discriminate patients with acute infections. Blood *109*, 2066–2077.

Raz, D.J., and Jablons, D.M. (2007). Five-gene signature in non-small-cell lung cancer. N. Engl. J. Med. *356*, 1582; author reply 1583.

Reme, T., Hose, D., De Vos, J., Vassal, A., Poulain, P.O., Pantesco, V., Goldschmidt, H., and Klein, B. (2008). A new method for class prediction based on signed-rank algorithms applied to Affymetrix microarray experiments. BMC Bioinformatics *9*, 16.

Ross, M.E., Zhou, X., Song, G., Shurtleff, S.A., Girtman, K., Williams, W.K., Liu, H.C., Mahfouz, R., Raimondi, S.C., Lenny, N., *et al.* (2003). Classification of pediatric acute lymphoblastic leukaemia by gene expression profiling. Blood *102*, 2951–2959.

Sargent, D., and Allegra, C. (2002). Issues in clinical trial design for tumor marker studies. Semin. Oncol. *29*, 222–230.

Sargent, D.J., Conley, B.A., Allegra, C., and Collette, L. (2005). Clinical trial designs for predictive marker validation in cancer treatment trials. J. Clin. Oncol. *23*, 2020–2027.

Schrohl, A.S., Wurtz, S., Kohn, E., Banks, R.E., Nielsen, H.J., Sweep, F.C., and Brunner, N. (2008). Banking of biological fluids for studies of disease-associated protein biomarkers. Mol. Cell. Proteomics *7*, 2061–2066.

Shi, L., Reid, L.H., Jones, W.D., Shippy, R., Warrington, J.A., Baker, S.C., Collins, P.J., de Longueville, F., Kawasaki, E.S., Lee, K.Y., *et al.* (2006). The MicroArray Quality Control (MAQC) project shows inter- and intraplatform reproducibility of gene expression measurements. Nat. Biotechnol. *24*, 1151–1161.

Shi, L., Campbell, G., Jones, W.D., Campagne, F., Wen, Z., Walker, S.J., Su, Z., Chu, T.M., Goodsaid, F.M., Pusztai, L., *et al.* (2010). The MicroArray Quality Control (MAQC)-II study of common practices for the development and validation of microarray-based predictive models. Nat. Biotechnol. *28*, 827–838.

Signature Diagnostics (2010). Dector C. Available at: http://www.signature-diagnostics.de/detectorc.htm (accessed 9 September 2013).

Singer, E. (2005). Personalized medicine prompts push to redesign clinical trials. Nat. Med. *11*, 462.

Sparano, J.A. (2006). TAILORx: trial assigning individualized options for treatment (Rx). Clin. Breast Cancer *7*, 347–350.

Staratschek-Jox, A., and Schultze, J.L. (2010). Re-overcoming barriers in translating biomarkers to clinical practice. Expert Opin. Med. Diagn. *4*, 103–112.

Staratschek-Jox, A., Classen, S., Gaarz, A., Debey-Pascher, S., and Schultze, J.L. (2009). Blood-based transcriptomics: leukaemias and beyond. Expert Rev. Mol. Diagn. *9*, 271–280.

Tan, D.S., Thomas, G.V., Garrett, M.D., Banerji, U., de Bono, J.S., Kaye, S.B., and Workman, P. (2009). Biomarker-driven early clinical trials in oncology: a paradigm shift in drug development. Cancer J. *15*, 406–420.

Tan, E.M., and Zhang, J. (2008). Autoantibodies to tumor-associated antigens: reporters from the immune system. Immunol. Rev. *222*, 328–340.

US Food and Drug Administration (2006). Guidance for Industry, Investigators, and Reviewers, Exploratory IND Studies. US Food and Drug Administration. Available at: http://wwwfdagov/cder/guidance/7086fnlpdf.

Valk, P.J., Verhaak, R.G., Beijen, M.A., Erpelinck, C.A., Barjesteh van Waalwijk van Doorn-Khosrovani, S., Boer, J.M., Beverloo, H.B., Moorhouse, M.J., van der Spek, P.J., Lowenberg, B., *et al.* (2004). Prognostically useful gene-expression profiles in acute myeloid leukaemia. N. Engl. J. Med. *350*, 1617–1628.

van 't Veer, L.J., Dai, H., van de Vijver, M.J., He, Y.D., Hart, A.A., Mao, M., Peterse, H.L., van der Kooy, K., Marton, M.J., Witteveen, A.T., *et al.* (2002). Gene expression profiling predicts clinical outcome of breast cancer. Nature *415*, 530–536.

van de Vijver, M.J., He, Y.D., van't Veer, L.J., Dai, H., Hart, A.A., Voskuil, D.W., Schreiber, G.J., Peterse, J.L., Roberts, C., Marton, M.J., *et al.* (2002). A gene-expression signature as a predictor of survival in breast cancer. N. Engl. J. Med. *347*, 1999–2009.

Verhaak, R.G., Goudswaard, C.S., van Putten, W., Bijl, M.A., Sanders, M.A., Hugens, W., Uitterlinden, A.G., Erpelinck, C.A., Delwel, R., Lowenberg, B., *et al.* (2005). Mutations in nucleophosmin (NPM1) in acute myeloid leukaemia (AML): association with other gene abnormalities and previously established gene expression signatures and their favorable prognostic significance. Blood *106*, 3747–3754.

Vineis, P., and Riboli, E. (2009). The EPIC study: an update. Recent Results Cancer Res. *181*, 63–70.

Wang, E., and Marincola, F.M. (2008). Bottom up: a modular view of immunology. Immunity *29*, 9–11.
Wichmann, H.E., Gieger, C., and Illig, T. (2005). KORA-gen--resource for population genetics, controls and a broad spectrum of disease phenotypes. Gesundheitswesen *67* (Suppl. 1), S26–S30.
Wilson, C.S., Davidson, G.S., Martin, S.B., Andries, E., Potter, J., Harvey, R., Ar, K., Xu, Y., Kopecky, K.J., Ankerst, D.P., *et al.* (2006). Gene expression profiling of adult acute myeloid leukaemia identifies novel biologic clusters for risk classification and outcome prediction. Blood *108*, 685–696.
Wolff, A.C., Hammond, M.E., Schwartz, J.N., Hagerty, K.L., Allred, D.C., Cote, R.J., Dowsett, M., Fitzgibbons, P.L., Hanna, W.M., Langer, A., *et al.* (2007). American Society of Clinical Oncology/College of American Pathologists guideline recommendations for human epidermal growth factor receptor 2 testing in breast cancer. Arch. Pathol. Lab. Med. *131*, 18–43.
Yao, Y., Higgs, B.W., Morehouse, C., de Los Reyes, M., Trigona, W., Brohawn, P., White, W., Zhang, J., White, B., Coyle, A.J., *et al.* (2009). Development of potential pharmacodynamic and diagnostic markers for anti-IFN-*α* monoclonal antibody trials in systemic lupus erythematosus. Hum. Genomics Proteomics *2009*, article ID 374312.
Yeoh, E.J., Ross, M.E., Shurtleff, S.A., Williams, W.K., Patel, D., Mahfouz, R., Behm, F.G., Raimondi, S.C., Relling, M.V., Patel, A., *et al.* (2002). Classification, subtype discovery, and prediction of outcome in pediatric acute lymphoblastic leukaemia by gene expression profiling. Cancer Cell *1*, 133–143.
Zander, T., Hofmann, A., Staratschek-Jox, A., Classen, S., Debey-Pascher, S., Maisel, D., Ansen, S., Hahn, M., Beyer, M., Thomas, R.K., *et al.* (2011). Blood-based gene expression signatures in non-small cell lung cancer. Clin. Cancer Res. *17*, 3360–3367.
Zujewski, J.A., and Kamin, L. (2008). Trial assessing individualized options for treatment for breast cancer: the TAILORx trial. Future Oncol. *4*, 603–610.

# Development of Methylation Biomarkers for Clinical Applications and Methylation-sensitive High-resolution Melting Technology

3

Tomasz K. Wojdacz

## Abstract

Aberrations of locus-specific methylation are a well-established hallmark of cancer. These epigenetic changes, in principle, cause or contribute to the pathological process, but at the same time can be utilized as biomarkers in clinical disease management. Within clinical disease management there are four fields of biomarker application: prevention, diagnosis, prognosis/prediction and treatment management. Methylation biomarkers fulfil criteria of applicability for each of these fields. However, current use of methylation biomarkers in clinical practice is very limited. The inadequate translation of research findings into clinical use in the field of methylation biomarkers can be mainly attributed to poor standardization of the technologies for methylation biomarker discovery and screening as well as a paucity of large-scale clinical validation studies of potential biomarkers. Methylation-sensitive high-resolution melting (MS-HRM) is an established method enabling in-tube cost and labour efficient locus specific methylation screening. The method has been successfully utilized in clinical methylation biomarker validation studies, hence the potential application of this method in diagnostic settings is very promising. This chapter addresses the challenges of the methylation biomarker development process and describes the use of MS-HRM for methylation biomarker development and diagnostic applications.

## DNA methylation

DNA Methylation is one of the most studied epigenetic mechanisms of gene expression regulation. In humans, DNA methylation is generally referred to as the enzymatic addition of a methyl group to the carbon 5 of cytosine (5-mC) within a 5′-CpG-3′ dinucleotide and is catalysed by DNA methyl transferases (DNMT). CpG sites are non-randomly distributed in the human genome with regions of higher than expected density of CpG sites referred to as CpG islands (CGI). CGIs group to promoters and first exons of protein coding genes. Methylation of a single CpG site within promoter CGI can potentially alter binding of the transcription factors and methylation of a number of CpGs across a CGI can lead to changes in conformation of the methylated loci, from opened and accessible to transcription factors, euchromatin, to closed and transcriptionally inactive heterochromatin (Esteller, 2008).

Deregulation of the normal pattern of methylation leads to aberrant expression of the genes, which in turn initiates or contributes to the disease phenotype. Thus, measurement of locus specific methylation is of high importance from both biological and clinical points of view. Research over the last 30 years has established methylation changes as a hallmark of carcinogenesis (Feinberg and Tycko, 2004). At the same time, broad applicability of the locus specific methylation changes in clinical disease management has been shown. This review focuses on the process of development of methylation biomarkers for diagnostic applications and the use of methylation-sensitive high-resolution melting (MS-HRM) in that process.

## Methylation-based *in vitro* diagnostic tests

There are two components of an *in vitro* diagnostic (IVD) test: the biomarker (analyte), which is an indicator of a specific stage and the detection method (technology), which enables investigating of the presence of the biomarker in specific settings. In practical medical terms, a biomarker is an analyte that is detectable in a specific clinically relevant condition. A condition specific change of methylation status of a locus is defined as a methylation biomarker. Despite the fact that a vast number of loci undergo aberrant methylation in a specific pathological process, it has been shown that methylation change of single locus can provide clinically valuable information. Therefore a single locus methylation aberration can be used as a target for an *in vitro* diagnostic (IVD) test.

## Fields of applications of *in vitro* diagnostic tests

In general there are four fields of biomarkers application in clinical disease management:

1 *Prevention.* Environmental factors like food habits, e.g. folate intake, can have significant impact on the methylation pattern of individuals (Jirtle and Skinner, 2007; Kastrup *et al.*, 2008; Shitara *et al.*, 2010). The strong correlation between environment and an individual's methylation make up has long been shown in animal models where experimental exposure to specific environmental condition(s) lead to locus specific methylation changes. Environmentally induced changes have not only been observed in the directly exposed animals but also been shown to be stably transmitted through the germ line (Ng *et al.*, 2010). In humans, there is still a lack substantial molecular evidence for this phenomenon; however, epidemiological evidence shows that exposure to specific environmental factors predisposes the individual to the development of specific diseases (as reviewed by Jirtle and Skinner (2007)). Controversy still remains with regard to the molecular mechanisms associated with the interaction between the environment and the individual's methylation make-up. Recent studies have shown that the initially proposed mechanisms of interaction require further investigation (Dobrovic and Kristensen, 2009). Nevertheless, epidemiological and molecular evidence suggest that methylation changes can potentially be used for risk group screening, which would have a significant impact on disease prevention.

2 *Diagnosis.* Early detection of disease always has the most significant effect on treatment outcome. Methylation changes are involved in the development of pathological phenotypes from the very early stages of the process. Thus, specific methylation aberrations can be detected in pathologically changed cells long before clinical manifestation of the disease. The fact that tumour DNA (or DNA from pathologically changed tissue) is released and can be extracted from easily accessible body fluids which have had direct contact with tumour site (e.g. blood or urine) opens the possibility for non-invasive early diagnosis of the disease. Recently, *SEPTIN9* methylation testing in plasma DNA was shown be very useful in early colon cancer detection (Warren *et al.*, 2011). Such promising research suggests that there will be significant future developments in the field of methylation biomarkers for early disease detection.

3 *Prognosis/prediction.* The disease phenotype largely depends on the methylation make up of the pathologically altered tissues. Hence, testing of pathologically changed tissue for the specific methylation pattern can provide valuable information with regards to prognosis and prediction of disease outcome. There is substantial research evidence supporting this application of methylation biomarkers, especially in cancer clinical management and more and more studies show similar results for other diseases such as Alzheimer's and Parkinson's diseases (Iraola-Guzman *et al.*, 2011).

4 *Treatment management.* Disease-specific methylation changes can be targeted by chemotherapy. Drugs that restore the function of the genes that have undergone hypermethylation as a consequence of

the disease process are already approved by FDA for the treatment of specific diseases (Kuendgen and Lubbert, 2008). Furthermore, methylation biomarkers can be used in a personalized approach for patient treatment. For example, methylation of the *MGMT* promoter renders glioblastoma tumours non-responsive for alkylating chemotherapy agents. Similarly, methylation of *MDR1* correlates with drug resistance in acute myelogenous leukaemia (Nakayama *et al.*, 1998; Hegi *et al.*, 2005). These findings show that testing for locus-specific aberrant methylation can potentially be used for patient stratification in personalized treatment approaches.

## Challenges of methylation biomarker development process

There is increasing research evidence for potential clinical applicability of methylation biomarkers, however, the current impact of such biomarkers on clinical disease management is very limited. The challenges the filed of methylation biomarker faces can be discussed at three levels: biomarker discovery, technologies for biomarker validation and accessibility of the settings for clinical biomarker validation.

### New methylation biomarker discovery

Over 60% of the protein coding genes in the human genome harbour CGIs in their promoters, therefore, a vast number of loci can potentially undergo aberrant methylation during disease development. Each pathological process seems to require different methylation changes, which is best exemplified by the fact that different cancer types show different methylation profiles. Furthermore, different changes seem to have different levels of significance for the disease process, with some of them driving pathological processes and others occurring randomly without impact on the phenotype of the diseased cells (Zeller *et al.*, 2012). The most challenging part of methylation biomarker discovery is to identify the methylation changes presenting the highest value for the clinical management of specific diseases. There is no doubt that the hunt for disease specific methylation changes has to start at the genome-wide level (Laird, 2010). Currently, two major technologies, microarrays and next-generation sequencing (NGS), enable the investigation of genome-wide methylation patterns and have already been shown to be irreplaceable methylation biomarker discovery tools. Both microarray and NGS approaches allow the identification of condition specific methylation changes at the genome level. However, the data obtained using NGS and microarray approaches require skilful analysis and interpretation due to the high complexity involved. Diagnostically applicable tests/technologies must provide a clear result that cannot be subject to ambiguous interpretation. Therefore, the diagnostic applicability of the above methods is still debatable and, at the current stage of the technological development, the findings of genome-wide approaches have to be validated with PCR-based technologies.

### Methylation biomarker validation technologies

PCR-based technologies allow for investigation of locus specific methylation and are currently the only technologies suitable for immediate IVD test application of methylation biomarkers. A vast number of such technologies have been developed over the years and they can be utilized for both qualitative and quantitative locus specific methylation assessment. Currently, only qualitative assessment of methylation has been shown to provide clinically valuable data. However, diagnostically applicable methods for locus specific methylation assessment should be quantitative, as this would allow clinically significant methylation level cut-off points to be set. The assessment of the cut off point for clinically significant methylation is important from both technological and physiological points of view. From a technological point of view, it reduces the number of false positive results obtained and enables normalization for background noise (which, in most cases, depends on the instrument and chemistry used). From a physiological point of view, the cut off point allows normalization for low levels of methylation, which is increasingly shown to accrue in healthy tissues but does not seem to have pathological consequences (Wojdacz *et al.*, 2011). The low level methylation has to be differentiated from pathologically significant methylation levels when diagnostic screening for a given methylation marker is performed. There is still a lack of consensus in the field with regard to which PCR methods are most suitable for methylation biomarker validation. Even more noticeable is the lack of standardization of the analytical parameters of the methods used. The magnitude of the challenges that the field of methylation biomarkers is facing is exemplified in the conclusion of a multicentre study performed by Hegi *et al.* (2005), which states that: 'the success rate of methylation-specific PCR on paraffin-embedded tumour samples was highly variable and centre-dependent' This, among other studies, demonstrates the urgent need for the standardization of the methodologies in the field. Nevertheless, emerging technologies like MS-HRM should, in the near future, allow for more rigorous standardization of clinical validation studies in the filed.

## Validation of biomarker in clinical settings

Once the technology is accessible, studies aiming to evaluate the power of the biomarker, involving a large number of clinical specimens, are a necessity. Such studies have to show what kind of clinically applicable information can be obtained by testing for given biomarker and what is the impact of that information on the clinical management of the disease. The ideal study addressing such questions should be based on large number of the clinical specimens with detailed epidemiological and clinical records. Detailed clinical information enables the correlation of biomarker testing results with the disease phenotype and the evaluation of the clinical applicability of a given methylation change. It is challenging to say what number of samples is ideal for this process. In general, this number is a trade off between the type of disease feature that a given methylation aberration correlates with and the frequency of the change (prevalence) in a given clinical specimen population. Not many retrospective validation studies have been performed, consequently, the clinical applicability of very few methylation biomarkers have been tested. Hence, the field is in immediate need of well-designed studies aiming to validate clinical applications of methylation biomarkers.

## Methylation biomarker development workflow

The process from discovery of methylation biomarkers (and any other biomarker) to clinical use is a labour, time consuming and costly process. However when properly scheduled the road from discovery to clinical application of the biomarker can be relatively efficient. Overall the biomarker development process can be subdivided in to five steps:

1. Biomarker discovery: in this part of the workflow a genome-wide methylation pattern screening technology is used to discover condition specific methylation changes. The integral part of this step is the validation of the genome-wide methylation screening results via a PCR-based technique. The validation step has to be performed for at least a subset of newly discovered loci to confirm the accuracy of the genome-wide methylation screening results.
2. Initial clinical biomarker validation: this step aims to: 'determine the capacity of the biomarker to distinguish between a healthy and sick individual' (National Cancer Institute, Third Report 2005, USA). In the case of methylation biomarkers this part of the workflow aims to screen for new methylation changes in two groups of clinical samples: pathologically changed specimens and healthy controls. The size of the group depends on the specific methylation change, however, the sample number has to be large enough to show statistically that the biomarkers can distinguish 'healthy' from 'sick' individuals.
3. Retrospective biomarker validation: this part of the biomarker development process investigates which of the biomarker application areas the specific biomarker can be used in. Different study designs are required to

qualify the use of the biomarker in different application areas. If a given biomarker is assessed as an early diagnostic or preventative biomarker, its methylation has to be tested in a large number of samples from sick individuals and healthy controls. The studies aiming to evaluate prognostic, predictive or epipharmacological applications of the biomarkers require a large number of the patient samples with detailed clinical records enabling correlation of the disease phenotype with the methylation status of the biomarker in the specific sample population. Retrospective validation of the biomarkers should evaluate the analytical parameters (Table 3.1) of the test (for detailed description of this process from legislative stand point with MS-HRM as an example, see Wojdacz, 2012b).

4 Prospective biomarker validation: when clear clinical benefit of the use of a given biomarker in disease management is shown, a study should be designed to prospectively evaluate the performance of the test. This part of biomarker development aims not only to show the benefit of the use of the new test over current diagnostic procedures, but also to thoroughly assess the clinical sensitivity and specificity of the test in specific applications, e.g. clinical sensitivity and specificity, with regards to different stages of cancer. Furthermore, the impact of the new testing procedure on diagnostic laboratories as well as the economic impact of the testing procedure on disease management should be evaluated in this step. The design of the studies within this part of the biomarker development process also largely depend on the legislative requirements for a given IVD test to be approved by the appropriate regulatory authorities (Wojdacz, 2012b).

5 Evaluation of the role of the biomarker-based test in clinical disease management: this part of the biomarker development process aims to evaluate the overall impact of the use the IVD test on the population. Thus, the studies addressing this part of biomarker development workflow have to be carried on a population level and over a long period of time. Evaluation of the use of biomarker on population is a natural consequence of the biomarker testing implementation.

## PCR amplification and melting analyses in methylation studies

### DNA melting

DNA melting is a process of dissociation the complementary DNA double helix into single strands with DNA melting temperature is defined as a temperature at which DNA dissociates into single strands. The amount of energy required to dissociate DNA is predominantly determined by the number of hydrogen bonds holding the two strands together, thus the melting temperature is dependent on the base composition of the specific DNA fragment. The detection of the melting

**Table 3.1** The outline of the analytical parameters to be assessed for methods used for biomarker detection

| Feature | Description |
|---|---|
| Specificity | Ability to assess biomarker in clinical sample |
| Accuracy | Agreement between the results of analytical procedure and reference values for the procedure |
| Precision | Agreement between series of measurements by analytical procedure |
| Detection limit | Lowest amount of the analyte detectable by analytical procedure |
| Quantitation limit | Lowest amount of analyte quantitatively detectable by analytical procedure |
| Linearity | Ability to proportionally detect analyte in given sample |
| Range | The interval between lowest and highest amount of analyte detectable by the analytical procedure |
| Robustness | Ability of the procedure to remain unaffected by protocol variation |

temperature of a PCR product can be performed using temperature gradients and fluorescent intercalating dyes. These dyes give high levels of fluorescence when intercalated in to double stranded DNA and relatively no fluorescence in the presence of single stranded DNA. Therefore, if PCR product mixed with intercalating dye is subjected to a temperature gradient, a sharp drop in fluorescence is observed when the DNA strands separate. The temperature at which this occurs is known as the melting temperature and it is specific for a given PCR product (Fig. 3.1).

## PCR amplification in methylation studies

The use of PCR amplification for the specific identification of methylation biomarkers is not straightforward. This is because in normal cell physiology the enzymes from DNA methyltransferase family (DNMT) synthetize methyl groups on newly replicated DNA strand preserving methylation marks in semiconservative fashion and in a regular PCR reaction no DNMTs are present. Therefore genomic template has to be modified to conserve methylation marks before the use of PCR, as PCR would erase methylation marks from template in the first round amplification. The most common procedure enabling the preservation of methylation marks is bisulfite modification. Sodium bisulfite deaminates unmethylated cytosines into uracil and leaves methylated cytosines intact. Hence bisulfite-modified DNA acquires methylation-dependent nucleotide sequence (Fig. 3.1). During PCR, DNA polymerase recognizes deaminated cytosine (uracil) as adnenine and incorporates complementary thymine. Therefore, amplification of bisulfite-modified DNA results in relatively GC-rich PCR products from methylated template and TA-rich PCR products from unmethylated template.

## PCR primers for amplification of bisulfite modified template

Two types of primers can be designed to amplify bisulfite-modified DNA: MSP (methylation specific PCR primers) and MIP (methylation-independent primers) (Clark *et al.*, 1994; Herman *et al.*, 1996; Wojdacz *et al.*, 2008b). MSP primers target only methylated template (another set of primers targeting unmethylated template has to be designed to confirm the unmethylated status of the given locus) and the presence of amplification is indicative of the methylated status of the locus of interest. In the MIP approach, one primer set is designed to amplify the locus of interest regardless of its methylation status. The origin of the PCR product is resolved in post-PCR fashion using, e.g. sequencing or MS-HRM, on the basis of the principle that product originating from an unmethylated allele harbours Ts at CpG sites and PCR product from a methylated allele harbours Cs at CpG sites.

## Development of methylation-sensitive high-resolution melting (MS-HRM)

The melting temperature-based analyses of methylation status is an example of a post-PCR analysis technique that enables the investigation of the methylation status of the locus of interest immediately after PCR amplification. The use of temperature gradient in methylation studies was first described over a decade ago (Worm *et al.*, 2001). Recent technological advancements involving interlacing dyes, instruments with high-precision fluorescence acquisition along the temperature gradient and new data analyses software has led to the development of high-resolution melting technology (HRM) (Wittwer *et al.*, 2003). Initially HRM was utilized for the investigation of nucleotide variations in DNA. More recently, we have applied HRM for locus specific methylation studies (Wojdacz and Dobrovic, 2007).

The protocol that initially described the use of melting temperature for locus-specific methylation screening has been shown to lack analytical sensitivity (Guldberg *et al.*, 2002; Dahl and Guldberg, 2003; Clark *et al.*, 2006) and this did not improve with the development of HRM (Wojdacz *et al.*, 2009). High analytical sensitivity of methylation detection is required in many methylation studies especially when using: degraded clinical material [e.g. formalin-fixed paraffin-embedded (FFPE) samples], samples with very low abundance of the target DNA (e.g. biopsies or free

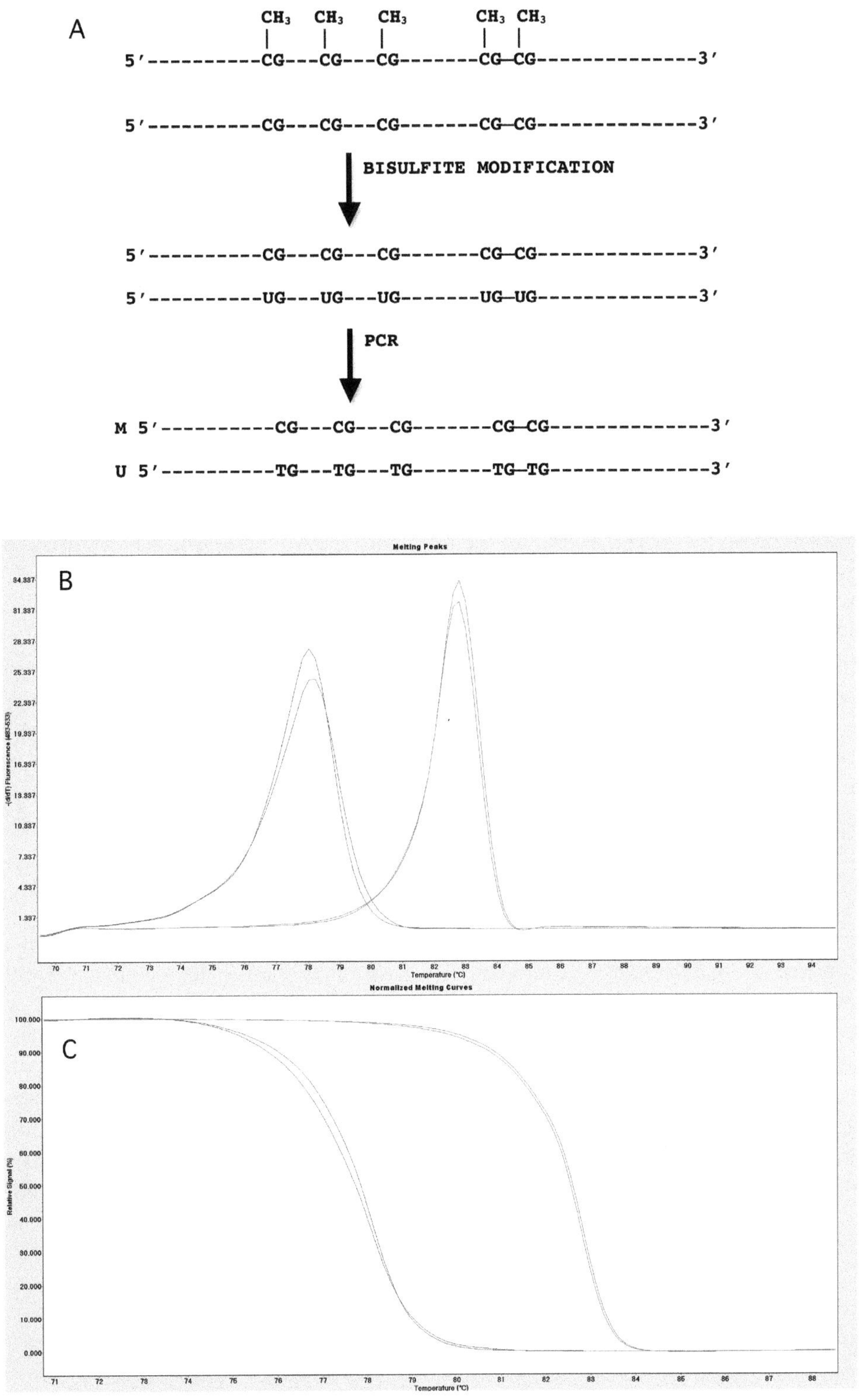


**Figure 3.1** Principles of bisulfite modification and the use of PCR in methylation studies (A) and the melting analyses of methylated (red) and unmethylated (blue) alleles visualized as melting peak (B), and normalized melting curves (C).

tumour circulating DNA extracted from plasma or serum) or highly heterogeneous clinical specimens (e.g. samples with very low tumour fraction). The poor analytical sensitivity of the original melting temperature-based protocol is likely to be attributed to the PCR bias phenomenon (Warnecke *et al.*, 1997). PCR bias is defined as an allele specific efficiency of PCR amplification. In methylation studies, unmethylated alleles have been shown to have higher amplification efficiency than methylated alleles of the same loci when amplified by one primer set in single PCR (Warnecke *et al.*, 1997; Wojdacz and Hansen, 2006). As a consequence of the PCR bias, unmethylated template is amplified in high abundance and can mask the presence of the PCR product amplified form methylated template. PCR bias can potentially make MIP-based amplification unsuitable for the investigation of methylation status of certain loci (Wojdacz *et al.*, 2009). Optimization of the PCR amplification and reaction conditions did not improve the proportional amplification of the methylated and unmethylated templates (Warnecke *et al.*, 1997). During development of our MS-HRM protocol we have addressed the PCR bias issue and proposed new guidelines for primer design for MIP-based amplification (Wojdacz *et al.*, 2008b). Primers designed according to the new guidelines in combination with empirical optimization of the PCR conditions have been shown to eliminate PCR bias, improve the amplification of methylated template and ensure the analytical sensitivity of the MS-HRM method is equivalent to the gold standard methods used in the field (Wojdacz *et al.*, 2009).

## Characteristics of the MS-HRM protocol

### Qualitative and quantitative aspects of MS-HRM results

Currently, in the studies aiming to demonstrate clinical applicability of methylation biomarkers, methylation screening data are categorized (two categories methylation positive an negative samples) for clinical correlation studies and so far only qualitative methylation assessment has been shown to be clinically meaningful. We have shown that any variation of the melting profile of the screened sample from the characteristic unmethylated control profile reflects gains in methylation at screened loci, thus in the simplified application of MS-HRM the experimental results can be interpreted in a qualitative manner (Wojdacz *et al.*, 2010). However, MS-HRM results can also be interpreted in semi-quantitative or quantitative fashion where the methylation level of the sample is estimated on the bases of similarities of the melting profile of the unknown sample with the melting profile characteristics of a sample with know methylation level (Fig. 3.2). Nevertheless in MS-HRM as in other non-sequencing-based quantitative methylation screening techniques (e.g. MethyLight or SMART-MSP), quantification of only fully methylated epialleles can be performed (Eads *et al.*, 2000; Candiloro *et al.*, 2011). It is important to mention here that methylation level measurement may only reflect a number of fully methylated epialleles in context of contaminating fraction of the 'normal' tissue in the sample. For detailed description of limitations of the quantitative measurements of methylation in clinical samples see Wojdacz (2012a).

### Sensitivity of MS-HRM

The sensitivity of the MS-HRM in model systems the been shown to enable detection of methylation in 1–2 cells which is the state of the art detection limit expected for PCR-based method (see data in Fig. 3.2). Furthermore the detection limit of the method was not significantly affected by the use of highly degraded material such as formalin fixed paraffin embedded (FFPE) samples (Wojdacz and Dobrovic, 2007; Kristensen *et al.*, 2009). Overall, in the case of methylation studies, sample processing for the analyses such as bisulfite modification (which degrades the template) and losses of the template during purification of the bisulfite modified template have significant influence on sensitivity of the method especially if highly degraded sample material is analysed. Therefore, despite of the fact the sensitivity of the MS-HRM has been shown to be at the state of the art level, the sensitivity of this and any

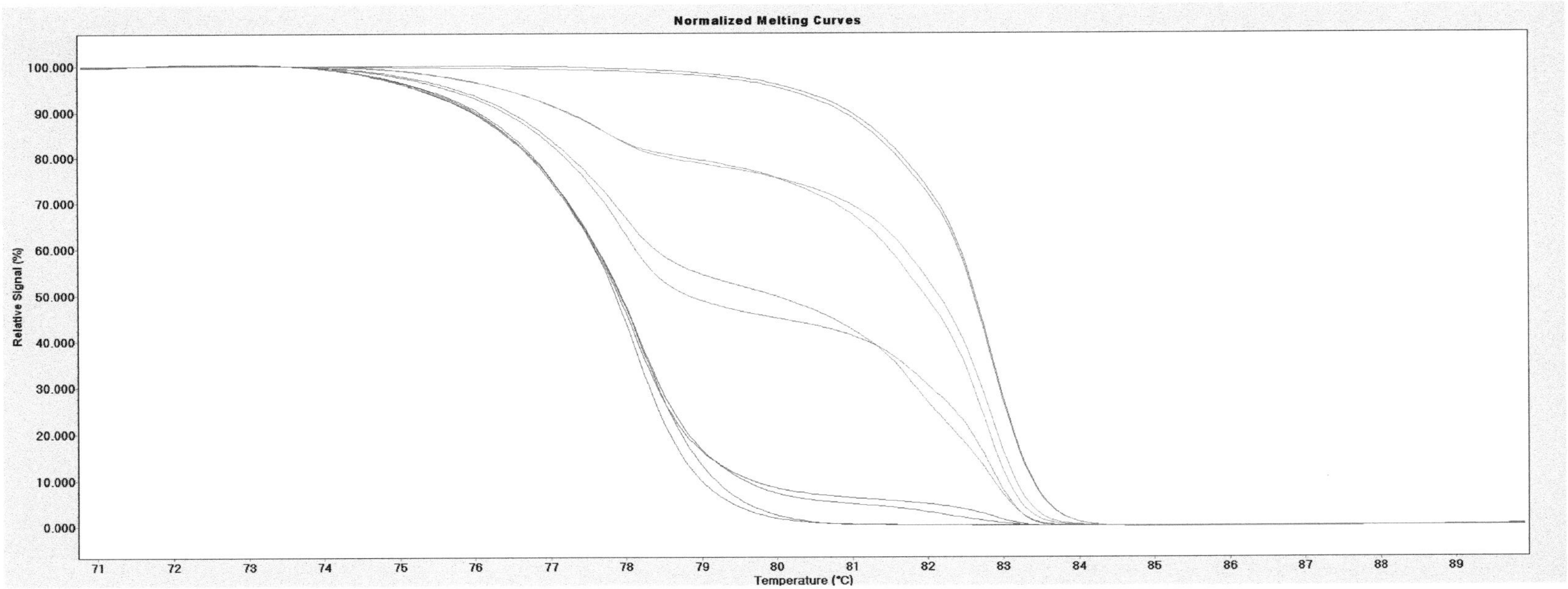


**Figure 3.2** Principles of the HRM-based quantification of methylation (the MGMT assays previously published in (Wojdacz and Dobrovic, 2007)). Different curves represent normalized HRM profiles for PCR products deriving from different mixes of methylated template diluted in unmethylated background: red, 100% methylated; yellow, 10%; green, 1%; pink, 0.1%; and blue, non-methylated. The methylation content of unknown sample can be estimated on the bases of the comparison of the HRM profile of the screened sample to the above standards. The shapes of the standard curves do not change at given experimental conditions. The 0.1% dilution point represents in this case 10 pg of methylated template, which is an equivalent of around 1.5 cell (three alleles) that exemplifies state of the art analytical sensitivity of MS-HRM protocol.

other method should be evaluated in each specific experiment with consideration of both analytical limitations of the detection method and sample material used.

## No need for data normalization

Similarly to the quantitative methods used for gene expression studies, most of the quantitative methods in methylation studies require normalization of 'the template input' for PCR amplification. MS-HRM-based quantification of methylation is different in that respect. The results of an MS-HRM experiment is a melting profile, which corresponds to the ratio of fluorescence change detected during simultaneous melting of methylated and unmethylated PCR products during the temperature gradient. Assuming that the post-PCR ratio of methylated to unmethylated PCR product reflects the initial ratio of methylated to unmethylated template in the given sample, normalization of the final result is not required and relative comparison (and estimation of methylation levels) between unknown sample and defined standards on the bases of melting profiles is possible. Two assumptions have to be made when using this model. First, PCR bias is controlled during PCR amplification and it affects the amplification of the standards (controls with known composition of methylated and unmethylated template) and unknown samples equally (thus can be disregarded from data analysis). Secondly, the melting profiles have high intra- and extra-experimental consistency under defined experimental conditions but high reproducibility of the HRM profiles using this method has been demonstrated in independent laboratories (Wojdacz and Dobrovic, 2007; Balic *et al.*, 2009; Slaats *et al.*, 2012; Migheli *et al.*, 2013).

## Ability to identify heterogeneously methylated sample

Methylation of a given locus oscillates between three stages: the locus can be methylated, devoid of methylation or heterogeneously methylated (Fig. 3.3). The physiological consequences of heterogeneous methylation are largely unknown but the fact that heterogeneous methylation is characteristic for some loci and was shown to be a intermediate stage between a non-methylated and fully methylated state of the allele, suggest a potential physiological function of this phenomenon. The identification of heterogeneous methylation presents technological challenges, as only sequencing of all epialleles in the particular sample will reflect a complexity of the heterogeneously methylated locus. The sequencing of all epialleles can currently only be performed with next-generation sequencing but the costs of this technology make it unavailable for this type of the experiment in vast majority of the laboratories.

At relatively low cost (but still significant) the patterns of heterogeneous methylation can be investigated using by cloning of PCR can obtained from bisulfite-modified template (Zhang *et al.*, 2009). However, the labour intensity of this procedure allows normally screening only limited number of clones.

Owing to quantitative nature of the measurement of the fluorescence at each nucleotide within screened sequence pyrosequencing can be applied to screening of heterogeneous methylation (Mikeska *et al.*, 2010). However, the fluorescence level on pyrosequencing diagram reflects average values for all PCR products in screened sample and therefore the difference in methylation between cpg sites within heterogeneously methylated amplicon have to be significant

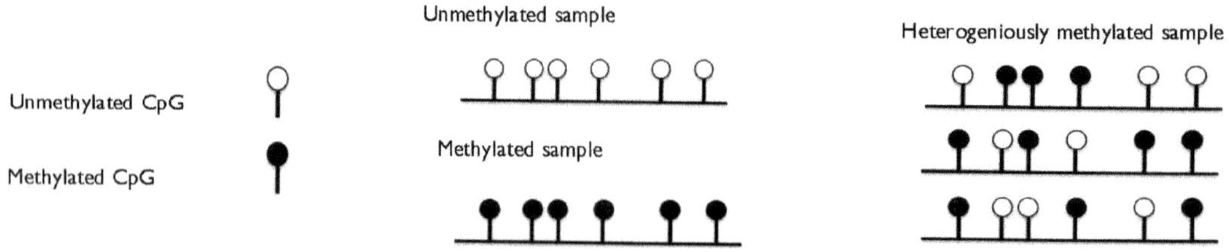


**Figure 3.3** Three potential stages of methylation of locus: methylated, non-methylated and heterogeneously methylated.

for the sample to be unambiguously classified as heterogeneously methylated.

HRM scans of the heterogeneously methylated loci display a characteristic profile (Fig. 3.4). Hence considering the limitations of the above technologies MS-HRM is the only currently available technology enabling the identification of heterogeneous methylation within the locus of interest on medium or large-scale projects. Nevertheless MS-HRM only in in digital application allows for the investigation of the pattern of methylation within screened amplicon (Candiloro *et al.*, 2008).

## Suitable for investigation of methylation status of imprinted loci

Genomic imprinting is defined as a restriction of transcription of the locus to one of the alleles paternal or maternal origin. Methylation is one of the mechanisms allowing restriction of transcription of the locus to one of the alleles. Changes of the imprinting can lead to severe developmental

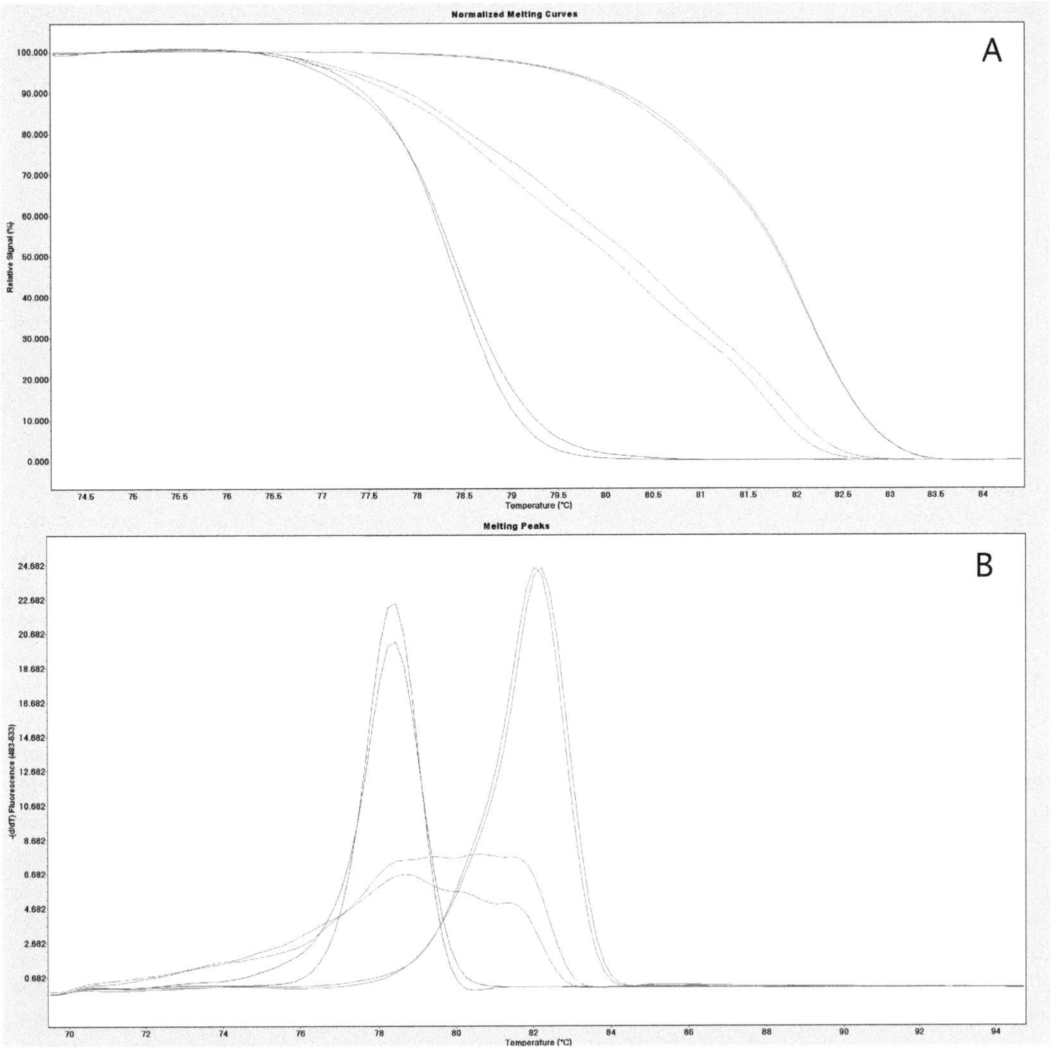


**Figure 3.4** An example of MS-HRM analyses of heterogeneously methylated loci. (A) Normalized melting curves. (B) Melting peaks. The curves characteristic for the PCR product deriving from: red, methylated sequence; blue, non-methylated; green, heterogeneously. The PCR product from heterogeneously methylated loci does not display single melting temperature but melts all across the temperature gradient.

disorders and are common during neoplastic transformation. The investigation of methylation status of imprinted loci can be challenging and technologies used in screening of the locus specific imprinting changes frequently require the use of multiple primers sets or methylation sensitive enzyme digestion followed by Southern blots (Sperandeo *et al.*, 2000; Sasamoto *et al.*, 2004). MS-HRM enables investigation of the methylation changes within a imprinted locus in a single PCR, which is significant simplification of previously used protocols (Wojdacz *et al.*, 2008a).

### Minimised risk of contamination and cost and labour efficient

The complete MS-HRM experiment can be performed in one tube with no post PCR manipulations of the PCR product. This minimizes inter-laboratory PCR contamination risks. Furthermore, only standard PCR reagents and fluorescent intercalating dye is required to perform the experiment, making the protocol cost efficient. The set up of MS-HRM experiment is no different from the set up of regular PCR amplification once bisulfite modification of the template is performed, making MS-HRM experiments no more labour intensive than standard PCR protocols. The MS-HRM data analysis is based on both real-time amplification plots and HRM profiles, therefore MS-HRM benefits from the same high degree of specificity.

### Easy to interpret result

The results of the MS-HRM experiment are visualized as normalized melting curves or derivative peaks, which are relative easy to analyse, and the data analysis can potentially be automated (Migheli *et al.*, 2013). MS-HRM-based semi or qualitative assessment of methylation is performed on the bases of the similarities of melting profiles of screened samples to standards and can be performed with high accuracy (Newman *et al.*, 2012; Migheli *et al.*, 2013). As already mentioned, data normalization, necessary when using methods based RT-PCR approaches is not needed for MS-HRM results, making processing of MS-HRM results straightforward. Furthermore the quantitative aspect of the method allows a cut off point to be defined for pathologically significant methylation and subsequent binarization of the data for analyses (Lash *et al.*, 2008).

The above features are especially important for diagnostic applications of MS-HRM. The results obtained using IVD tests have to be easy to interpret and the ambiguity in the test results interpretation must be eliminated in diagnostic laboratories. Therefore, MS-HRM is potentially highly suitable for application as an IVD methylation biomarker testing (Wojdacz *et al.*, 2010).

### Need for optimisation of empirical experimental conditions

Each MS-HRM-based locus specific assay requires careful design and empirical optimization of the PCR amplification. This can potentially be laborious and time consuming especially for new users of the technology. Nevertheless, once the MS-HRM assay is properly optimized, its performance remains relatively unaffected by user-to-user and run-to-run variations of the protocol (laboratories all over the world are currently using the assays published by our laboratory, e.g. Balic *et al.*, 2009; Migheli *et al.*, 2013).

## Final remarks

MS-HRM is one of the many methods currently used for methylation biomarker studies. This method is increasingly applied in research laboratories for clinically oriented studies (Liu *et al.*, 2010; Adachi *et al.*, 2012; Ebert *et al.*, 2012; Furst *et al.*, 2012; Mao *et al.*, 2012). The need for locus specific methylation biomarkers to be translated in clinical practice is growing, especially considering developments in the field of personalized medicine and the abundant evidence that methylation biomarkers have the potential to be widely applied in this field. MS-HRM has the potential to become widely applied in IVD testing of clinically important methylation biomarkers. However the limitations of the method indicate that most likely other methods will also needed to be used to allow wide implementation of methylation biomarker testing in clinical disease management.

## Acknowledgements

I would like to thank Andrea Tesoriero and Dr Lise Lotte Hnasen for the critical review of this manuscript. I would also like to thank The Danish Cancer Society, CIRRO – The Lundbeck Foundation Centre for Interventional Research in Radiation Oncology, The Danish Council for Strategic Research, Aarhus University, and Karolinska Institutet for the support of my work. TKW is listed as an inventor on a patent application based on aspects of MS-HRM technology.

## References

Adachi, J.I., Mishima, K., Wakiya, K., Suzuki, T., Fukuoka, K., Yanagisawa, T., Matsutani, M., Sasaki, A., and Nishikawa, R. (2012). O (6)-methylguanine-DNA methyltransferase promoter methylation in 45 primary central nervous system lymphomas: quantitative assessment of methylation and response to temozolomide treatment. J. Neurooncol. *107*, 147–153.

Balic, M., Pichler, M., Strutz, J., Heitzer, E., Ausch, C., Samonigg, H., Cote, R.J., and Dandachi, N. (2009). High quality assessment of DNA methylation in archival tissues from colorectal cancer patients using quantitative high-resolution melting analysis. J. Mol. Diagn. *11*, 102–108.

Candiloro, I.L., Mikeska, T., Hokland, P., and Dobrovic, A. (2008). Rapid analysis of heterogeneously methylated DNA using digital methylation-sensitive high resolution melting: application to the CDKN2B (p15) gene. Epigenetics Chromatin *1*, 7.

Candiloro, I.L., Mikeska, T., and Dobrovic, A. (2011). Closed-tube PCR methods for locus-specific DNA methylation analysis. Methods Mol. Biol. *791*, 55–71.

Clark, S.J., Harrison, J., Paul, C.L., and Frommer, M. (1994). High sensitivity mapping of methylated cytosines. Nucleic Acids Res. *22*, 2990–2997.

Clark, S.J., Statham, A., Stirzaker, C., Molloy, P.L., and Frommer, M. (2006). DNA methylation: bisulphite modification and analysis. Nat. Protoc. *1*, 2353–2364.

Dahl, C., and Guldberg, P. (2003). DNA methylation analysis techniques. Biogerontology *4*, 233–250.

Dobrovic, A., and Kristensen, L.S. (2009). DNA methylation, epimutations and cancer predisposition. Int. J. Biochem. Cell Biol. *41*, 34–39.

Eads, C.A., Danenberg, K.D., Kawakami, K., Saltz, L.B., Blake, C., Shibata, D., Danenberg, P.V., and Laird, P.W. (2000). MethyLight: a high-throughput assay to measure DNA methylation. Nucleic Acids Res. *28*, E32.

Ebert, M.P., Tanzer, M., Balluff, B., Burgermeister, E., Kretzschmar, A.K., Hughes, D.J., Tetzner, R., Lofton-Day, C., Rosenberg, R., Reinacher-Schick, A.C., *et al.* (2012). TFAP2E-DKK4 and chemoresistance in colorectal cancer. N. Engl. J. Med. *366*, 44–53.

Esteller, M. (2008). Epigenetics in cancer. N. Engl. J. Med. *358*, 1148–1159.

Feinberg, A.P., and Tycko, B. (2004). The history of cancer epigenetics. Nat. Rev. Cancer *4*, 143–153.

Furst, R.W., Kliem, H., Meyer, H.H., and Ulbrich, S.E. (2012). A differentially methylated single CpG-site is correlated with estrogen receptor alpha transcription. J. Steroid Biochem. Mol. Biol. *130*, 96–104.

Guldberg, P., Worm, J., and Gronbaek, K. (2002). Profiling DNA methylation by melting analysis. Methods *27*, 121–127.

Hegi, M.E., Diserens, A.C., Gorlia, T., Hamou, M.F., de Tribolet, N., Weller, M., Kros, J.M., Hainfellner, J.A., Mason, W., Mariani, L., *et al.* (2005). MGMT gene silencing and benefit from temozolomide in glioblastoma. N. Engl. J. Med. *352*, 997–1003.

Herman, J.G., Graff, J.R., Myohanen, S., Nelkin, B.D., and Baylin, S.B. (1996). Methylation-specific PCR: a novel PCR assay for methylation status of CpG islands. Proc. Natl. Acad. Sci. U.S.A. *93*, 9821–9826.

Iraola-Guzman, S., Estivill, X., and Rabionet, R. (2011). DNA methylation in neurodegenerative disorders: a missing link between genome and environment? Clin. Genet. *80*, 1–14.

Jirtle, R.L., and Skinner, M.K. (2007). Environmental epigenomics and disease susceptibility. Nat. Rev. *8*, 253–262.

Kastrup, I.B., Worm, J., Ralfkiaer, E., Hokland, P., Guldberg, P., and Gronbaek, K. (2008). Genetic and epigenetic alterations of the reduced folate carrier in untreated diffuse large B-cell lymphoma. Eur. J. Haematol. *80*, 61–66.

Kristensen, L.S., Wojdacz, T.K., Thestrup, B.B., Wiuf, C., Hager, H., and Hansen, L.L. (2009). Quality assessment of DNA derived from up to 30 years old formalin fixed paraffin embedded (FFPE) tissue for PCR-based methylation analysis using SMART-MSP and MS-HRM. BMC Cancer *9*, 453.

Kuendgen, A., and Lubbert, M. (2008). Current status of epigenetic treatment in myelodysplastic syndromes. Ann. Hematol. *87*, 601–611.

Laird, P.W. (2010). Principles and challenges of genomewide DNA methylation analysis. Nat. Rev. *11*, 191–203.

Lash, T.L., Ahern, T.P., Cronin-Fenton, D., Garne, J.P., Hamilton-Dutoit, S., Kvistgaard, M.E., Rosenberg, C.L., Silliman, R.A., and Sorensen, H.T. (2008). Modification of tamoxifen response: what have we learned? J. Clin. Oncol. *26*, 1764–1765; author reply 1765–1766.

Liu, W., Guan, M., Su, B., Ye, C., Li, J., Zhang, X., Liu, C., Li, M., Lin, Y., and Lu, Y. (2010). Quantitative assessment of AKAP12 promoter methylation in colorectal cancer using methylation-sensitive high resolution melting: correlation with Duke's stage. Cancer Biol. Ther. *9*, 862–871.

Mao, W., Rubin, J.S., Anoruo, N., Wordinger, R.J., and Clark, A.F. (2012). SFRP1 promoter methylation and expression in human trabecular meshwork cells. Exp. Eye Res. *97*, 130–136.

Migheli, F., Stoccoro, A., Coppede, F., Wan Omar, W.A., Failli, A., Consolini, R., Seccia, M., Spisni, R., Miccoli, P., Mathers, J.C., *et al.* (2013). Comparison study of MS-HRM and pyrosequencing techniques for quantification of APC and CDKN2A gene methylation. PLoS One *8*, e52501.

Mikeska, T., Candiloro, I.L., and Dobrovic, A. (2010). The implications of heterogeneous DNA methylation for the accurate quantification of methylation. Epigenomics *2*, 561–573.

Nakayama, M., Wada, M., Harada, T., Nagayama, J., Kusaba, H., Ohshima, K., Kozuru, M., Komatsu, H., Ueda, R., and Kuwano, M. (1998). Hypomethylation status of CpG sites at the promoter region and overexpression of the human MDR1 gene in acute myeloid leukaemias. Blood *92*, 4296–4307.

Newman, M., Blyth, B.J., Hussey, D.J., Jardine, D., Sykes, P.J., and Ormsby, R.J. (2012). Sensitive quantitative analysis of murine LINE1 DNA methylation using high resolution melt analysis. Epigenetics *7*, 92–105.

Ng, S.F., Lin, R.C., Laybutt, D.R., Barres, R., Owens, J.A., and Morris, M.J. (2010). Chronic high-fat diet in fathers programs beta-cell dysfunction in female rat offspring. Nature *467*, 963–966.

Palmisano, W.A., Divine, K.K., Saccomanno, G., Gilliland, F.D., Baylin, S.B., Herman, J.G., and Belinsky, S.A. (2000). Predicting lung cancer by detecting aberrant promoter methylation in sputum. Cancer Res. *60*, 5954–5958.

Sasamoto, H., Nagasaka, T., Notohara, K., Ozaki, K., Isozaki, H., Tanaka, N., and Matsubara, N. (2004). Allele-specific methylation analysis on upstream promoter region of H19 by methylation-specific PCR with confronting two-pair primers. Int. J. Oncol. *25*, 1273–1278.

Shitara, K., Muro, K., Ito, S., Sawaki, A., Tajika, M., Kawai, H., Yokota, T., Takahari, D., Shibata, T., Ura, T., *et al.* (2010). Folate intake along with genetic polymorphisms in methylenetetrahydrofolate reductase and thymidylate synthase in patients with advanced gastric cancer. Cancer Epidemiol. Biomarkers Prev. *19*, 1311–1319.

Slaats, G.G., Reinius, L.E., Alm, J., Kere, J., Scheynius, A., and Joerink, M. (2012). DNA methylation levels within the CD14 promoter region are lower in placentas of mothers living on a farm. Allergy *67*, 895–903.

Sperandeo, M.P., Ungaro, P., Vernucci, M., Pedone, P.V., Cerrato, F., Perone, L., Casola, S., Cubellis, M.V., Bruni, C.B., Andria, G., *et al.* (2000). Relaxation of insulin-like growth factor 2 imprinting and discordant methylation at KvDMR1 in two first cousins affected by Beckwith-Wiedemann and Klippel-Trenaunay-Weber syndromes. Am. J. Hum. Genet. *66*, 841–847.

Warnecke, P.M., Stirzaker, C., Melki, J.R., Millar, D.S., Paul, C.L., and Clark, S.J. (1997). Detection and measurement of PCR bias in quantitative methylation analysis of bisulphite-treated DNA. Nucleic Acids Res. *25*, 4422–4426.

Warren, J.D., Xiong, W., Bunker, A.M., Vaughn, C.P., Furtado, L.V., Roberts, W.L., Fang, J.C., Samowitz, W.S., and Heichman, K.A. (2011). Septin 9 methylated DNA is a sensitive and specific blood test for colorectal cancer. BMC Med. *9*, 133.

Wittwer, C.T., Reed, G.H., Gundry, C.N., Vandersteen, J.G., and Pryor, R.J. (2003). High-resolution genotyping by amplicon melting analysis using LCGreen. Clin. Chem. *49*, 853–860.

Wojdacz, T.K. (2012a). The limitations of locus specific methylation qualification and quantification in clinical material. Front Genet. *3*, 21.

Wojdacz, T.K. (2012b). Methylation-sensitive high-resolution melting in the context of legislative requirements for validation of analytical procedures for diagnostic applications. Expert Rev. Mol. Diagn. *12*, 39–47.

Wojdacz, T.K., and Dobrovic, A. (2007). Methylation-sensitive high resolution melting (MS-HRM): a new approach for sensitive and high-throughput assessment of methylation. Nucleic Acids Res. *35*, e41.

Wojdacz, T.K., and Hansen, L.L. (2006). Reversal of PCR bias for improved sensitivity of the DNA methylation melting curve assay. BioTechniques *41*, 274, 276, 278.

Wojdacz, T.K., Dobrovic, A., and Algar, E.M. (2008a). Rapid detection of methylation change at H19 in human imprinting disorders using methylation-sensitive high-resolution melting. Hum. Mutat. *29*, 1255–1260.

Wojdacz, T.K., Hansen, L.L., and Dobrovic, A. (2008b). A new approach to primer design for the control of PCR bias in methylation studies. BMC Res. Notes *1*, 54.

Wojdacz, T.K., Borgbo, T., and Hansen, L.L. (2009). Primer design versus PCR bias in methylation independent PCR amplifications. Epigenetics *4*, 231–234.

Wojdacz, T.K., Moller, T.H., Thestrup, B.B., Kristensen, L.S., and Hansen, L.L. (2010). Limitations and advantages of MS-HRM and bisulfite sequencing for single locus methylation studies. Expert Rev. Mol. Diagn. *10*, 575–580.

Wojdacz, T.K., Thestrup, B.B., Cold, S., Overgaard, J., and Hansen, L.L. (2011). No difference in the frequency of locus-specific methylation in the peripheral blood DNA of women diagnosed with breast cancer and age-matched controls. Future Oncol. *7*, 1451–1455.

Worm, J., Aggerholm, A., and Guldberg, P. (2001). In-tube DNA methylation profiling by fluorescence melting curve analysis. Clin. Chem. *47*, 1183–1189.

Zeller, C., Dai, W., Steele, N.L., Siddiq, A., Walley, A.J., Wilhelm-Benartzi, C.S., Rizzo, S., van der Zee, A., Plumb, J.A., and Brown, R. (2012). Candidate DNA methylation drivers of acquired cisplatin resistance in ovarian cancer identified by methylome and expression profiling. Oncogene *31*, 4567–4576.

Zhang, Y., Rohde, C., Tierling, S., Stamerjohanns, H., Reinhardt, R., Walter, J., and Jeltsch, A. (2009). DNA methylation analysis by bisulfite conversion, cloning, and sequencing of individual clones. Methods Mol. Biol. *507*, 177–187.

# Genetic and Epigenetic Biomarkers of Colorectal Cancer

# 4

Stephen A. Bustin and Jamie Murphy

## Abstract

Colorectal cancer (CRC) is one of the most prevalent cancers in the western world. Around 75% of primary CRC diagnoses are in patients with no apparent risk factor other than age, with the remaining 25% of patients having a family history of CRC that suggests a genetic contribution. Although some genetic alterations have been identified as the cause of inherited cancer risk in some families, they account for only around 6% of CRC cases. Hence additional, as yet undiscovered, genes and background genetic factors may drive the development of CRC.

CRC arises and progresses through the acquisition of genetic and epigenetic alterations that drive the transformation from normal colon epithelium to adenocarcinoma. The clinical behaviour of cancer cells depends on complex and dynamic interactions between the effects of these alterations and the individual genetic and environmental host contexts. This results in significant variability in the response of individual patients to chemotherapy treatments that are not accurately predicted by conventional prognostic stratification and adjuvant therapy selection procedures.

Consequently, there has been keen interest in the identification and functional characterization of molecular biomarkers, with the expectation that they may facilitate accurate disease diagnosis, have prognostic potential for the individual, or predict patient-specific responses to chemotherapy. Undoubtedly, they will play a role in the future diagnosis and management of CRC; however, DNA-based biomarkers are currently not widely used by physicians and most potential markers are still in the discovery phase waiting to undergo clinical validation.

## Introduction

Colorectal cancer (CRC) is one of the most common cancers in the western world (Levin *et al.*, 2008) and its incidence is rising rapidly in several Asian countries, possibly due to an increase in westernized dietary lifestyle (Yee *et al.*, 2009). However, survival statistics show significant differences between countries, with 5-year relative survival rates being 58.6% (95% CI 58.3–58.9%) and 60.0% (95% CI 59.7–60.3%) for men and women, respectively, in the USA, compared with 47.4% (95% CI 46.9–48.0%) and 45.3% (95% CI 45.0–45.6%) respectively in Europe (Coleman *et al.*, 2008). Even within a region, there is significant variation, with the male (39.8%; 95% CI 38.0–41.8%) and female (39.3%; 95% CI 37.5–41.3%) 5-year relative survival rates in Wales much lower than rates reported by its geographic neighbour, south-west England, at 50.3% (95% CI 49.0–51.5%) and 51.8% (95% CI 50.5–53.1%) respectively. Most of this variation is probably due to differences in access to diagnostic and treatment services, as survival is positively associated with gross domestic product and the amount of investment in health technology such as CT scanners (Coleman *et al.*, 2008).

It is also clear that the pathological American Joint Committee on Cancer (AJCC) stage at diagnosis has a critical impact on CRC survival rates, with a 5-year survival rate of 91% for patients with localized CRC (stage I), as compared with 70% for those with loco-regional (stage III) and 11%

for individuals diagnosed with distant metastatic disease (stage 4). Consequently, early detection of colorectal adenomas at high-risk of progression to CRC has become central to reducing CRC deaths. This has led to an increased emphasis on screening for early diagnosis of CRC and since 75% of CRC diagnoses are in patients with no apparent risk factor other than age, population screening has offered the best opportunity to reduce mortality rates to date (Pawa *et al.*, 2011). However, a general screening programme is expensive, there is no international consensus on screening procedures, and compliance within present screening programmes can be rather limited (Nielsen *et al.*, 2011). Furthermore, they are compromised by limited diagnostic accuracy and detect many adenomas that will never progress to CRC (Hoff and Dominitz, 2010). Nevertheless, many western health care systems have standardized population-based screening for asymptomatic individuals over the age of 60 at a national level, generally utilizing enzymatic or immunological faecal occult blood test systems to identify those who require endoscopic assessment of the colon and rectum (Booth, 2007).

Following a diagnosis of CRC, surgery remains the only treatment modality that has curative potential for patients with a resectable burden of disease; however, surgery may also be appropriate for individuals with incurable metastatic disease in order to palliate CRC-related symptoms. When performed with curative intent, surgical resection strategies aim to excise the CRC fully while achieving microscopically negative (R0) resection margins with >1 mm clearance of normal surrounding tissue. In practice, this is the most clinically relevant prognostic factor, since 5-year disease free survival rates following R0 resection are around 64% versus 17% after a microscopically positive margin (R1) with <1 mm clearance of normal tissue (Pawlik and Vauthey, 2008). Therefore, a great deal of attention has focused on standardization of surgical technique to facilitate R0 resection for both colonic (Hohenberger *et al.*, 2009) and rectal cancer (Heald and Ryall, 1986) patients.

However, surgical treatment strategies may be further complicated by the presence of distant metastatic disease, since at the time of diagnosis around 20% of patients have a resectable primary tumour as well as resectable liver metastases without known extra-hepatic disease (Pozzo *et al.*, 2004). For these patients combined synchronous or delayed colorectal and hepatic resectional surgery has been described (Vauthey *et al.*, 2005) after which patients may be also treated with postoperative adjuvant chemotherapy, in order to try and eradicate micrometastatic disease that is frequently present in this patient group (Nordlinger *et al.*, 2008). An additional 15–25% of patients will present with extra-hepatic metastases, and a further 20–25% will develop (extra)-hepatic metastases during post-treatment follow up (Van Cutsem *et al.*, 2010). These patients may be given (neo)-adjuvant chemotherapy to reduce primary tumour size and/or reduce the burden of metastatic disease prior to consideration of surgery (Gruenberger *et al.*, 2010). Since in 30–50% of patients with metastatic disease the liver is the only site of spread, this approach increases the likelihood of an improvement in survival rates or even cure, leading to 5-year disease-free survival rates of 25–71% (Cunningham *et al.*, 2010). Recent studies also suggest highly selected patients with extra-hepatic disease may be considered for curative metastasectomy with reported 5-year disease-free survival rates of approximately 50% (Hornbech *et al.*, 2011). However, despite these advances overall clinical outcome measures remain poor for patients with widespread disease and the management for systemic metastatic disease at diagnosis is frequently palliative.

Therefore, contemporary strategies to improve CRC survival rates have begun to focus on personalized medicine, since this approach promises accurate prediction of an individual's predisposition towards CRC or optimized detection and subsequent clinical management in the context of an individual genetic and environmental profile (Fig. 4.1). Nevertheless, its realization in clinical practice depends on the identification of early diagnostic as well as effective prognostic and predictive biomarkers for non-invasive patient assessment (La Thangue and Kerr, 2011), which the remainder of this chapter will focus upon.

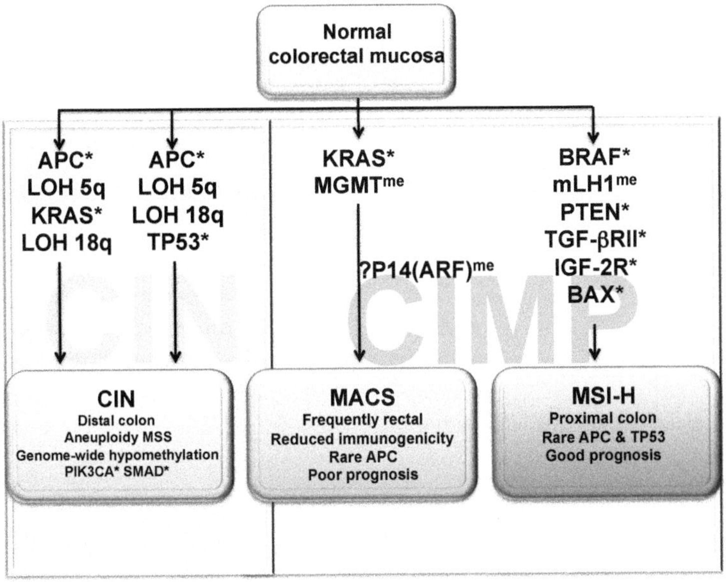


**Figure 4.1** Colorectal tumorigenesis as a multi-pathway disease. The significant heterogeneity in CRC is reflected in the different molecular features of epigenetic (indicated by me-superscript) and genetic (indicated by asterisk) alterations among these pathways. CIMP accounts for all MSI-H sporadic CRC, but only a proportion of MACS and CIN subtypes. CIN and MSI-H pathways are the most distinct, with clear differences in precursor lesions, genetics, epigenetics and outcome. The MACS pathway is more speculative, less well understood and probably more heterogeneous than shown. It has a different form of CIMP, predominant KRAS but occasional BRAF mutations, usually lacks CIN, and has a worse prognosis.

## Biomarkers

Biomarkers are defined as biological substances, characteristics, or images that provide an indication of the biological state of an organism (Group, 2001). Clinically useful biomarkers must be easy to obtain and quantify, have biological relevance and ideally represent steps in well-understood carcinogenic pathways or in host-response mechanisms. An example of a commonly used CRC biomarker is the presence of a mutation in the KRAS gene, associated with a clinically distinct prognosis, diagnosis, or response to a specific treatment (Van Schaeybroeck *et al.*, 2011). The appropriate use of a biomarker-based test requires an understanding of its sensitivity and specificity, how and in what contexts to use it, how to interpret it in various contexts, and how to properly validate it (Olson *et al.*, 2009). Cancer biomarkers have evolved from biochemical assays that measure proteins or hormones after the onset of disease to molecular assays that not only target disease-specific nucleic acids for early screening and diagnosis, but also promise more accurate classification, prognosis, risk stratification, treatment efficacy prediction and monitoring.

In CRC a wide variety of biomarkers have been reported within the tumour itself, as well in blood and faeces, with the genetic profile of the host providing additional and essential complementary information. The recent identification of CRC stem cells means that much effort is being expended in the identification of markers that facilitate development of new therapies that aim to target this fraction specifically. In addition, a range of new biomarkers, termed companion diagnostics, have been described that are designed to provide biological or clinical information that reflects the sensitivity or resistance of CRC to existing therapies and so aid clinicians in selecting the most effective therapies (Cross, 2008). A number of biomarkers have been proposed as specific predictors of chemotherapy and radiotherapy response and, in some instances, drug toxicity (Ross, 2011).

## Biology underpinning CRC biomarkers and therapy

Most CRCs are sporadic and are believed to develop slowly via a progressive accumulation of multiple genetic and epigenetic events that affect tumour suppressor genes, oncogenes and mismatch repair genes. Interestingly, reduced proliferation may be a biological feature characterizing the majority of aggressive CRCs (Anjomshoaa *et al.*, 2008), with important implications for our understanding of the biology of CRC progression and for the selection of putative novel therapy options (Pritchard and Grady, 2011). The detailed molecular analysis of CRCs has significantly changed our understanding of colorectal tumorigenesis (Roukos *et al.*, 2007). The early model of tumour progression from adenoma to carcinoma through the stepwise accumulation of genetic events to several key genes and genetic loci (Fearon and Vogelstein, 1990) has been augmented by the recognition of CRC as a morphologically and molecularly heterogeneous disease (Murphy *et al.*, 2007), rendering a purely anatomical classification obsolete (Ogino *et al.*, 2011). The cancer stem cell (CSC) model, which proposes that certain cells within the tumour mass are pluripotent and capable of self-renewal appears to describe accurately the aetiology of CRC (Sanders and Majumdar, 2011). This has important implications for cancer treatment, since most current therapies target actively proliferating cells and may not be effective against the CSCs that are responsible for recurrence. Hence the importance of a recent report that identified a gene signature, specific for adult intestinal stem cells, that predicts disease relapse in CRC patients (Merlos-Suarez *et al.*, 2011). CRCs acquire an average of 80 somatic mutations, probably within CRC stem cells, during their evolution from a benign to a metastatic state, of which an average of 14 are in candidate cancer genes (Sjoblom *et al.*, 2006; Wood *et al.*, 2007), with the most common mutations found in APC, KRAS, TP53 and BRAF. Whilst the mutated genes in different CRCs overlap to only a small extent, they cluster within a small number of intracellular pathways (Yi *et al.*, 2011), allowing their grouping into specific tumour phenotypes based on their molecular profiles. Together with DNA-methylation and chromatin-structure changes, the mutations target proteins and pathways that exert pleiotropic, context-dependent effects on critical cell phenotypes, and particular genetic and epigenetic alterations are linked to biologically and clinically distinct subsets of CRC (Fearon, 2011).

Development of CRC proceeds over a 5- to 10-year timespan, providing ample opportunity to detect adenomatous or cancerous lesions at an early stage. Since >50% of individuals will develop colorectal adenomas in their lifetime, but only 6% will develop CRC, it is essential to be able to identify those adenomas that will progress to a carcinoma (Ahlquist, 2010). To date, no molecular tests have been developed that can distinguish such precancerous lesions (Young and Bosch, 2011).

## Genomic and epigenomic instability

The molecular classification of CRC continues to evolve and currently cellular global genomic and epigenomic status is believed to delineate several pathways leading to CRC, as shown in Fig. 4.1. Each one is characterized by different clinical, pathological and biological features, although it should be remembered that these molecular classifications are dynamic and not necessarily consistent. For example, some studies report a positive correlation between the CpG island methylator phenotype (CIMP) and mutation of KRAS (Toyota *et al.*, 2000; Samowitz *et al.*, 2005a), whereas other show the opposite (van Rijnsoever *et al.*, 2002; Iacopetta *et al.*, 2006).

The majority of sporadic CRCs are microsatellite stable (MSS) and display chromosomal instability (CIN), characterized by aneuploidy, the presence of large structural or numerical alterations of the chromosomes (Lengauer *et al.*, 1997). CIN is thought to arise through defects in a number of processes, including aberrant expression or mutation of mitotic checkpoint genes, microtubule spindle defects, and telomere dysfunction. It is associated with hypomethylation (Matsuzaki *et al.*, 2005; Rodriguez *et al.*, 2006), which leads to genomic instability (Gaudet *et al.*, 2003) as well as the activation of oncogenes (Nakamura and Takenaga, 1998) and so promotes tumour development (Yamada *et al.*, 2005). Clinically, CIN is a marker of poor prognosis

(Popat and Houlston, 2005; Walther *et al.*, 2008) and appears to promote tumour progression by increasing clonal diversity (Maley *et al.*, 2006).

A second group of CRCs are characterized by microsatellite instability (MSI) and a diploid karyotype (Boland *et al.*, 1998). MSI is often referred to as a mutator phenotype (Lothe *et al.*, 1995): in sporadic CRC this phenotype is caused by epigenetic silencing of DNA mismatch repair (MMR) genes, in particular, MLH1 (Kane *et al.*, 1997), whereas patients with hereditary non-polyposis colorectal cancer (HNPCC) often have a germline mutation of one of the two major MMR genes (hMSH2 or hMLH1) (Papadopoulos and Lindblom, 1997). Sporadic MSI tumours are associated with the V600E mutation in BRAF (Nagasaka *et al.*, 2004) and mutations of tumour suppressor genes such as PTEN (Zhou *et al.*, 2002). Although MSI tumours are associated with a better prognosis (Popat *et al.*, 2005; Walther *et al.*, 2009), the predictive value of MSI is unclear, as reports of it being predictive of response to 5-fluorouracil (FU)-adjuvant therapy in AJCC stage III MSI CRCs (Elsaleh and Iacopetta, 2001; Elsaleh *et al.*, 2001) are contradicted by other studies (Ribic *et al.*, 2003; Benatti *et al.*, 2005; Jover *et al.*, 2009). This uncertainty may be caused by the use of a single marker, such as MSI, that cannot account alone for the complexity of the mechanisms underlying 5-FU cytotoxicity, with additional genome stability markers also required (Guastadisegni *et al.*, 2010).

MSI tumours are sometimes divided into two distinct phenotypes: MSI-high (MSI-H) and MSI-low (MSI-L) (Pawlik *et al.*, 2004), distinguished by expression profiling (Mori *et al.*, 2003), loss of MGMT expression (Azzoni *et al.*, 2011) and promoter methylation of p14 (ARF) (Kominami *et al.*, 2009) in MSI-L cancers. However, this distinction is not universally accepted, since definitions of MSI-H, MSI-L and MSS have varied among groups and even between different studies from the same group (Tomlinson *et al.*, 2002).

Nevertheless, genomic profiling suggests the presence of another, somewhat heterogeneous (Jones *et al.*, 2005) group, sometimes termed microsatellite and chromosome stable (MACS), characterized by MSS in the absence of chromosome instability (Georgiades *et al.*, 1999). This group has frequent deletions of 8p and 18q as well as 20q amplification (Xie *et al.*, 2007). MACS tumours also have a lower rate of loss of hMLH1 or BAX protein and significantly lower hMLH1 methylation rates than MSI-H tumours (Cai *et al.*, 2008). They often arise in the rectum and appear to be associated with characteristics such as: poor differentiation (Tang *et al.*, 2004); early onset (Chan *et al.*, 2001); an invasive phenotype that encourages early metastasis (Hawkins *et al.*, 2001); early disease recurrence; and significantly poorer survival rates (Banerjea *et al.*, 2009). The paradoxical finding that near diploid, MSS tumours display apparently early progression, with an infiltrative phenotype and widespread dissemination at the time of presentation may be due to their reduced immunogenicity, as revealed by the attenuation of the lymphocyte response and cytokine release within MACS CRC (Banerjea *et al.*, 2009). The extremely low p53 gene mutation rate (11%), but relatively high TP53 protein accumulation rate (55%) of MACS tumours suggests that stabilization of TP53 in the absence of a mutation may be an important event on a distinct pathway (Tang *et al.*, 2004).

CIN, MSI-H and MACS tumours can also be defined by their epigenetic status, which can result in either CIMP or global hypomethylation (Yamada *et al.*, 2005). CIMP is defined by frequent promoter hypermethylation and occurs in around 20–45% of CRC (Jass, 2007; Ogino and Goel, 2008), although the interpretation of studies investigating CIMP in CRC survival and the clinical utility of this designation is complicated by the different methods and differences in sample populations used to assess CIMP status (Hughes *et al.*, 2012). Classification based on hypermethylation patterns at five CpG markers result in either two subgroups, CIMP-high and CIMP-0/CIMP-low (Weisenberger *et al.*, 2006; Ogino and Goel, 2008), or three subgroups, CIMP-0, CIMP-low and CIMP-high (Barault *et al.*, 2008). Alternative groupings are based on classifications from comprehensive DNA methylation profiling and unsupervised hierarchical clustering of the methylation data, referred to as CIMP-low, CIMP-mid and CIMP-high (Ang *et al.*, 2010); and CIMP-negative, CIMP-1 and

CIMP-2 (Shen *et al.*, 2007b), respectively. CIMP is associated with specific clinical, pathological, and molecular features that include MSI-H in sporadic CRC, the BRAFV600E mutation, low frequencies of p53 mutations, proximal tumour location, mucinous histology, infiltrating lymphocytes, female sex, and later onset of disease (Toyota *et al.*, 1999; Costello *et al.*, 2000; Ogino *et al.*, 2006). CIMP tumours also display an MSI-H-independent loss of protein markers associated with differentiation, metastasis suppression, and increased CD8+ T-lymphocytes (Zlobec *et al.*, 2011). CIMP cancers develop along the serrated polyp pathway, a histopathological sequence that begins in a hyperplastic polyp, or precursor serrated aberrant crypt focus, and progresses to a form of atypical hyperplastic polyp to dysplastic serrated polyp and ultimately to serrated CIMP-high and, in most cases, also MSI-H carcinoma (Jass, 2007). This pathway is marked by an activating BRAF mutation (Kim *et al.*, 2008), with aberrant CpG-island methylation acting as the molecular engine that drives progression through the sequential steps of this signalling cascade (O'Brien, 2007). A second pathway, identified by mutations of KRAS in serrated adenoma, results in a CRC that is CIMP-low and MSS, but is delineated less completely (O'Brien *et al.*, 2006). The CIMP phenotype has been related to a poor prognosis in CRC (Van Rijnsoever *et al.*, 2003; Shen *et al.*, 2007a), with both three and 5-year disease-free survival significantly worse in rectal cancer (Jo *et al.*, 2011). A poor prognosis might result from other factors closely related to CIMP, such as the BRAF V600E mutation (Samowitz *et al.*, 2005b; Lee *et al.*, 2008), since the high rates of this mutation suggest a role for BRAF activation, although no direct cause and effect relationship has been established (Hinoue *et al.*, 2009). However, other reports suggest a better prognosis or null association (Shannon and Iacopetta, 2001; van Rijnsoever *et al.*, 2002; Ogino *et al.*, 2009b; Ahn *et al.*, 2011; Wong *et al.*, 2011). The prognostic use of CIMP may be dependent on the context of genomic instability in which it occurs, since patients with MSI-H tumours have a better prognosis than those with MSS tumours (Ward *et al.*, 2003; Ogino *et al.*, 2007; Barault *et al.*, 2008; Kim *et al.*, 2009; Dahlin *et al.*, 2010), although again this is not found in every study (Ferracin *et al.*, 2008; Kakar *et al.*, 2008). Nevertheless, patients with CIMP-positive CRCs do not appear to benefit from 5-FU-based adjuvant chemotherapy (Jover *et al.*, 2011).

When CRCs are categorized according to their MSI and CIMP status, five (Jass, 2007) or six (Ogino and Goel, 2008) groups are identified, which have diverse clinical and pathological features (Fig. 4.2). The molecular pathways that make up these groups are determined at an early evolutionary stage and are fully established within precancerous lesions.

## Genetic markers associated with CRC tumorigenesis (Fig. 4.1)

### 18q allelic imbalance

Around 85% of microsatellite stable CRC are characterized by allelic imbalance (AI) and chromosome 18q allelic loss was one of the first molecular markers associated with poor prognosis in CRC patients with AJCC stage II (Jen *et al.*, 1994) and stage III disease (Jernvall *et al.*, 1999). Furthermore, allelic loss at D18S61 and D18S851 has been associated with a poor response to chemotherapy (Barratt *et al.*, 2002) and at D18S851 with liver metastasis (Tanaka *et al.*, 2006). Interestingly, tumours expressing low levels of SMAD4 protein or mRNA are associated with liver metastasis (Tanaka *et al.*, 2006) and with significantly shorted 5-year disease-free survival (Alhopuro *et al.*, 2005b). Nevertheless, in line with many molecular analyses, additional work has resulted in a far more complex picture that defies easy interpretation. Many reports show no difference in 5-year survival rate between patients with and without 18q AI (Arango *et al.*, 2005; Bertagnolli *et al.*, 2011) and no association of loss of D18S851 and lymph node-positivity (Ghadimi *et al.*, 2003). Remarkably, there is also a report that loss of heterozygosity (LOH) at D18S61 predicts a more favourable outcome, especially in AJCC stage II CRC (Pilozzi *et al.*, 2011).

There are numerous possible explanations for these discordant findings that strongly accentuate the need for caution when using a single marker for prognosis management and emphasize that accurate individual prognostication will have to take

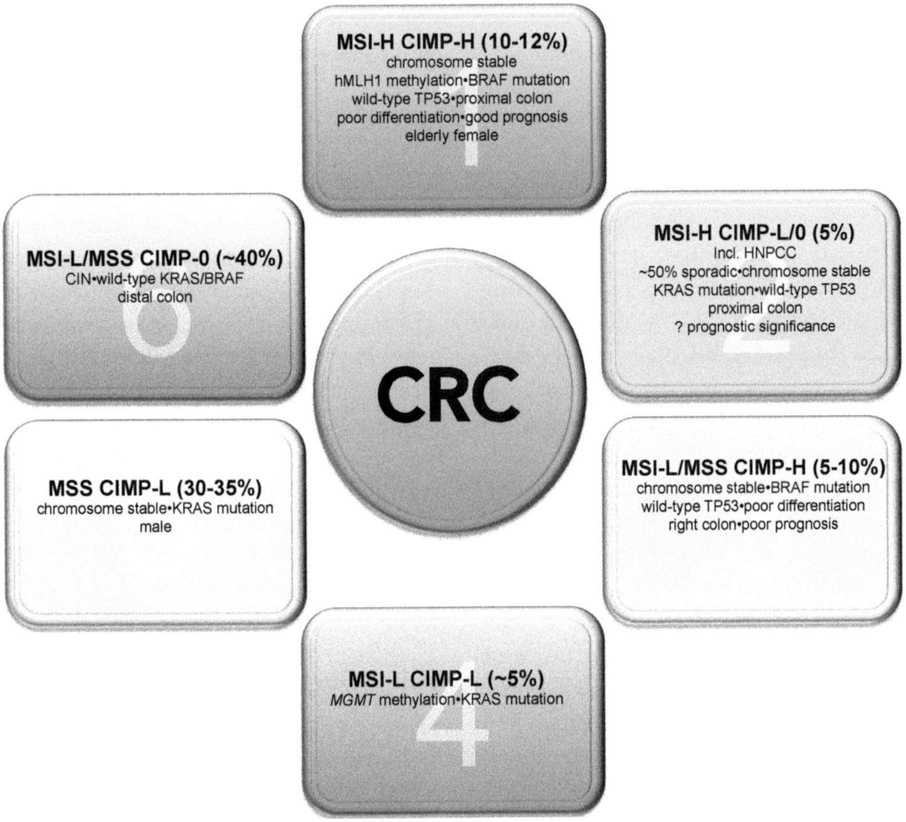


**Figure 4.2** Molecular classification of CRC according to MSI and CIMP status. Groups 1–6 are modified from descriptions published previously (Jass, 2007; Ogino and Goel, 2008). The differences between MSI-L and MSS and CIMP-L and CIMP-0, respectively are subtle. Since groups 4, 5, and 6 share similar clinical, pathological and molecular features, they may be combined into a single subtype, MSI-L/MSS CIMP-L/0 (Ogino and Goel, 2008).

into account several markers. There is extensive intratumoral genetic heterogeneity, particularly for 18q LOH (Losi *et al.*, 2005), in advanced CRCs, resulting in several topographically distinct genotypic subclones within the tumour (Baisse *et al.*, 2001). Hence analysis of one or two biopsies may not result in an accurate assessment of the metastatic potential of the whole tumour. Other explanations include differences in the scoring of AI, use of different genetic markers that result in the analysis of AI in different regions on chromosome 18q, stage-specific effects of 18q, loss of 18q LOH significance in multivariate models as well as questions about what actually is being measured by 18q AI (Tejpar *et al.*, 2010).

CRC development is driven by several key pathways, in particular the WNT, transforming growth factor β (TGF_β), epidermal growth factor receptor (EGFR) signalling and phosphatidylinositol 3-kinase (PI3K) pathways, as well as by mutations in several tumour-suppressor genes (Fearon, 2011).

## WNT

Although up to 70% of sporadic CRCs acquire mutations in the adenomatous polyposis coli (APC) gene at the earliest stages of neoplasia and activating mutations in the β-catenin (CTNNB1) gene have been observed, APC or CTNNB1 mutations are currently of no clinical use for treatment selection, prognosis or early cancer detection (Pritchard and Grady, 2011). Similarly, whilst SMAD4 expression levels may predict response to 5-FU (Boulay *et al.*, 2002; Alhopuro *et al.*, 2005a), none of the genetic markers associated with the TGF-β pathway have a definite clinical role.

### KRAS and BRAF

In contrast, the EGFR signalling mediators KRAS and BRAF are mutated in 40% and 15% of CRCs, respectively. Mutations in KRAS codon 12 or 13 are relatively early events and result in constitutive signalling (McCormick, 1995) as well as driving epigenetic deregulation of the transcriptome (Schafer and Sers, 2011). BRAF is a direct downstream effector of KRAS and the most common mutation is a single base change resulting in the substitution of glutamic acid for valine at codon 600 (Siena *et al.*, 2009). KRAS and BRAF mutations are mutually exclusive (Chan *et al.*, 2003; Fransen *et al.*, 2004) and are more frequent in MSI-H and CIMP tumours. BRAF mutant CRCs may constitute a discrete disease subset characterized by a distinct pattern of metastatic spread (Tran *et al.*, 2011). There is also evidence that up-regulation of the EGFR ligands epiregulin and amphiregulin are associated with an anti-EGFR drug response (Jacobs *et al.*, 2009).

### PI3K

Around 40% of CRCs harbour mutations in PI3K pathway genes, with the most frequent ones found in the p110a catalytic subunit of PIK3CA (Samuels, 2004) and the tumour suppressor PTEN (Danielsen *et al.*, 2008). Mutations in PIK3CA (Kato *et al.*, 2007) and absence of PTEN expression (Li *et al.*, 2009) have been associated with a shorter survival, possibly in association with KRAS wild-type status (Ogino *et al.*, 2009c), although phosphorylated AKT expression has been associated with good prognosis (Baba *et al.*, 2011) and overall survival of patients receiving adjuvant chemotherapy and/or radiotherapy is significantly longer in CRC with amplification of the PIK3CA gene (Jehan *et al.*, 2009). Mutations in PIK3CA are associated with CIMP-H (Whitehall *et al.*, 2011), do not to predict liver metastasis development (Bruin *et al.*, 2011) and are not consistent between primary tumour and associated metastasis (Laszlo, 2010). Hence, despite some promise, currently, there is not sufficient evidence from clinical studies to support the use of PI3K pathway mutations as predictive or prognostic biomarkers (Pritchard and Grady, 2011).

### MGMT

$O^6$-methylguanine-DNA methyltransferase (MGMT) specifies a DNA repair protein and is often methylated during CRC development (Hibi *et al.*, 2009). The resulting loss of mutagenesis policing is associated with an increase in point mutations of KRAS (de Vogel *et al.*, 2009) and has been implicated in the establishment of a field defect i.e. the replacement of the normal epithelial cell population with a histologically non-dysplastic one that harbours genetic or epigenetic alterations (de Vogel *et al.*, 2009; Svrcek *et al.*, 2010). However, despite its role in establishing aberrations in CRC, neither MGMT promoter methylation nor MGMT loss serves as a prognostic biomarker in CRC (Shima *et al.*, 2011).

## Proteomic biomarkers

Whilst major genetic changes and mutations have been well characterized in CRC, much less is known at the protein and proteome level. Conventional enzyme-linked immunosorbent assays that use a panel of protein biomarkers to detect inflammatory bowel disease in exfoliated cell samples collected from the surface of the rectal mucosa (Anderson *et al.*, 2011) may also be useful for cancer screening in this relatively high-risk population. However, technical issues, mainly related with sensitivity and reproducibility, have limited such proteomic approaches. Nevertheless, recent advances in proteomic techniques and mass spectrometry systems have resulted in the characterization of CRC proteomes (Kang *et al.*, 2011; O'Dwyer *et al.*, 2011) and rekindled the hunt for proteomic biomarkers in CRC (Ang *et al.*, 2011). Hence a promising approach is to identify proteins secreted by the cancer (Klein-Scory *et al.*, 2010) or cell surface protein biomarkers with extracellular domains that could be targets for current or emerging technologies, especially novel molecular imaging modalities (de Wit *et al.*, 2011). Since mutated proteins drive tumorigenesis, their accurate detection and quantification is likely to prove useful for diagnostic applications. A recent report demonstrates that it is possible to use mass spectrometry to quantify the number and fraction of mutant RAS proteins present in CRC cell lines and tissue (Wang *et al.*, 2011). The relevance of detecting the presence or absence of a single protein is less clear. Although detection of single proteins such as CCR7 (Gunther *et al.*, 2005), S100A4 (Hemandas *et al.*, 2006), CK20 (Wong *et al.*, 2009), HIWI (Zeng *et al.*, 2011), REG I_ and HIP/PAP (Zheng *et al.*, 2011) or E-cadherin and β-catenin (Toth *et al.*, 2011) is claimed to be a reliable prognostic marker, the lessons from gene expression-based detection methods (Lurje *et al.*, 2007) suggest that reliance on a single marker is inadvisable. There are two interesting studies reporting the use of multiple protein biomarkers that are potentially useful for early diagnosis of CRC. One study used mass spectrometry to identify a serum protein fingerprint in CRC patients that can distinguish CRC from healthy controls with high sensitivity (92.85%) and specificity (91.25%) (Liu *et al.*, 2011). The other report identified a five-peptide classifier obtained from a CRC phage expression library that was biopanned using serum pools of CRC patients and healthy controls and was able to discriminate between early CRC patients and healthy controls, with sensitivities of 90.0–92.7% and specificities of 91.7–93.3% (Chang *et al.*, 2011). Clearly, there is huge promise afforded by a lengthening list of potential protein biomarkers and relevant detection techniques. However, it is also obvious that more research efforts and technical advances are needed, especially when investigating the range of low-abundance proteins, which are likely to play strategic roles in CRC diagnostics and progression (Barderas *et al.*, 2010).

## Biomarker location

### Faeces

Approximately 1.5 × $10^6$ colonic epithelial cells can be isolated per gram of stool (Iyengar *et al.*, 1991). Hence faecal analysis constitutes a potent and non-invasive method for the detection, monitoring and management of CRC (Ahlquist, 2002), with the potential to detect cancer-specific mutations significantly earlier than with endoscopic identification (Ogreid and Hamre, 2007). However, molecular analysis of exfoliated epithelial cells and free nucleic acids shed in the faeces poses a significant technical challenge, since faecal DNA is made up from an immense number and variety of bacteria and other cells including normal colorectal epithelial cells as well as lymphocytes. Furthermore, faeces contain nucleases as well as intrinsic substances that inhibit many molecular assays, especially PCR-based tests. The ideal biomarker for the detection of CRC and premalignant lesions would be consistently positive in the presence of and negative in the absence of cancerous lesions, stable despite faecal toxicities, easily recoverable from the stool, and reproducibly assayed (Osborn and Ahlquist, 2005). Detection of mutated DNA holds the most promise, as DNA is fairly stable (Nechvatal *et al.*, 2008; Koga *et al.*, 2009) and mutations are either present or absent. Nonetheless, high testing volume, frequently low tumour content, and the spectrum of rare mutations in many target genes can make mutation detection challenging (Arcila *et al.*, 2011) and it is important to choose the method of detection carefully (Minamoto *et al.*, 2000). Several studies have investigated the relevance of detecting single markers in human stool following the first successful report of KRAS identification for CRC patients in 1992 (Sidransky *et al.*, 1992). However, as discussed earlier, the genetic heterogeneity of CRC implies that a panel of DNA markers is more likely to be informative than a single marker (Kahlenberg *et al.*, 2003). This is demonstrated by a recent chip-based temperature gradient capillary electrophoresis technique, which identified KRAS mutations the stools of 57% of CRC patients and in 7% of controls (Zhang *et al.*, 2011), results similar to those obtained using a different set of techniques, where mutations were found in 41% of cancers and 5% of controls (Chien *et al.*, 2007). Other studies also do not support the use of KRAS alone as a molecular marker for CRC screening (Atkin and Martin, 2001; Haug *et al.*, 2007), suggesting that the most likely use for KRAS is as a component of multitargeted assays.

Several publications have used different methods to investigate the potential utility of multiple DNA markers from faeces for the detection of CRC. An early report that utilized KRAS, TP53,

APC and Bat-26 claimed sensitivities of 91% and 83% for cancer and adenomas, respectively and a target marker-dependent specificity of up to 100% (Ahlquist *et al.*, 2000). A second report that analysed the same markers found that the assay identified 71% of patients with CRC, but did not report the assay's specificity (Dong *et al.*, 2001). A third comparison detected mutations in 55% of cancer patients, compared to 11% of healthy volunteers (Onouchi *et al.*, 2008). Although both specificity and sensitivity currently make these and similar (Rengucci *et al.*, 2001; Calistri *et al.*, 2003) assays clinically impractical, they point towards the potential of this approach, assuming that the right combination of markers can be found.

Cancer-specific epigenetic events such as altered DNA methylation patterns provide another potential source for informative biomarkers for colorectal adenomas and CRC (Kim *et al.*, 2010). Again it is unlikely that a single epigenetic marker will provide the sensitivity and specificity required for clinically useful screening (Lenhard *et al.*, 2005). As such, a recent study investigated promoter hypermethylation of the CNRIP1, FBN1, INA, MAL, SNCA, and SPG20 genes in human faecal cells and found that this panel of epigenetic markers was characteristically altered in colorectal adenomas (35–91%) and cancers (65–94%), whereas normal mucosa samples were rarely (0–5%) methylated. The combined sensitivity of at least two positives among the six markers was 94% for CRCs and 93% for adenoma samples, with a specificity of 98% (Lind *et al.*, 2011a). Hypermethylation of SPG20 promoter is also detectable in 89% of CRC, 78% of adenomas but only 1% of normal colonic mucosa. Importantly, this biomarker is detectable in corresponding faecal samples (Lind *et al.*, 2011b).

## Blood

Haematogenous spread of CRC cells from a primary tumour is a crucial step in the metastasis cascade, and circulating tumour cells (CTCs) have long been considered an indicator of tumour aggressiveness (Sleijfer *et al.*, 2007). Hence they could provide a potential source of cells for real-time monitoring of CRC patients through the course of their disease, enable the detection and molecular characterization of early dissemination of cancers, and allow monitoring of treatment response or resistance. Consequently, a lot of attention has focused on the development of assays for their reliable detection and quantification. The rarity of CTCs in peripheral blood (estimated as one tumour cell per $10^9$ normal blood cells in patients with known metastatic cancer (Maheswaran and Haber, 2010)) means that their detection requires a combination of reliable enrichment steps as well as robust, sensitive and specific detection techniques (Sun *et al.*, 2011). There are several highly specific cytometric analysis methods available for the identification of CTCs, some of which can rapidly analyse large volumes of sample (fibreoptic array scanning technology: FAST), detect viable cells (EPISPOT), allow genetic analysis (fluorescence *in situ* hybridization: FISH) or are semi-automated (CellSearch) (Alunni-Fabbroni and Sandri, 2010). In general, their main drawback is lack of sensitivity or that detection is limited to specific subsets of CTCs (e.g. EpCam$^{+ve}$). However, the choice of antibody is crucial, and there is no consensus about which antibody or collection of antibodies is optimal (Antolovic *et al.*, 2010; Konigsberg *et al.*, 2010). This is highlighted by the use of EpCAM as a marker of CTCs, since one important characteristic of this biomarker is that its expression levels are variable (van der Gun *et al.*, 2010) and its down-regulation indicates increased metastatic potential in CRCs (Gosens *et al.*, 2007), making it a less than useful selective marker. Furthermore, the absence of universal markers to discriminate colorectal CTCs and the heterogeneity of the tumours themselves make assay implementation into routine clinical practice highly challenging.

These difficulties make the detection of circulating cell-free DNA in plasma or serum a potentially attractive target as an alternative or additional source of DNA to screen for CRC-specific genetic alterations (Fleischhacker and Schmidt, 2007). There is evidence that serum cell-free DNA levels are increased early (Danese *et al.*, 2010) in the CRC pathway (Schwarzenbach *et al.*, 2008), that plasma levels decrease progressively in the follow-up period in tumour-free patients (Frattini *et al.*, 2006), that they are elevated in plasma from patients with metastatic disease (Guadalajara *et al.*, 2008) and that all types of alterations described in CRC can be detected (Lecomte *et al.*, 2010; Spindler *et al.*, 2012). However, there are questions about the reliability of using cell-free DNA as a biomarker (Diehl *et al.*, 2005) and the increased fragmentation characteristic of large or metastatic tumours makes it essential to develop technologies that can detect the number of mutant DNA molecules and the specific mutation in the same sample (Mouliere *et al.*, 2011). Furthermore, there is significant DNA extraction method-dependent variability (Fong *et al.*, 2009) and the results of serum and plasma DNA levels do not necessarily agree. Technical limitations related to the clotting cascade explain some of this variability (Thijssen *et al.*, 2002), as it has been shown that the amount of DNA in serum can be up to 24 times higher than in plasma due to the release of DNA from destroyed leucocytes during the clotting process (Chan *et al.*, 2005).

## Primary tumour

Since the first application of CRC microarray analysis (Hegde *et al.*, 2001; Takemasa *et al.*, 2001), numerous studies have used mutation (Di Martino *et al.*, 2011), methylation (Mori *et al.*, 2011), antibody (Zhou *et al.*, 2011), mRNA (Nannini *et al.*, 2009), miRNA (Luo *et al.*, 2011) and tissue (Spisak *et al.*, 2009) microarrays to profile CRC. These studies have had a profound impact on our understanding of the biology of CRC and have provided insights into their biological heterogeneity, facilitated identification of novel oncogenes and tumour suppressors and defined pathways that interact to drive the growth of individual cancers (Cowin *et al.*, 2010).

Apart from being used to identify mutations in CRC-associated genes such as APC (Cowie *et al.*, 2004) and KRAS, microarrays have been utilized in the search for prognostic markers. A meta-analysis of data obtained from 31 comparative genomic hybridization (CGH) studies associates loss of 4p with the transition from AJCC stage I to stage II-IV and correlates deletion of 8p and gains of 7p and 17q with the transition from primary tumour to liver metastasis (Diep *et al.*, 2006). More recent microarray-based CGH has identified two recurrently altered genomic regions as independent indicators of poor prognosis (Kim *et al.*, 2006b). A region on chromosome 3q26 containing PIK3CA is amplified in 38% of cancers, and gains are associated with high levels of PIK3CA protein expression (Jehan *et al.*, 2009). Survival of patients receiving adjuvant chemotherapy and/or radiotherapy is significantly longer in PIK3CA-amplified cancers than in patients with cancers without amplification independent of stage, grade, histology subtype, gender, and age categories. Another study identified DNA copy number loss at 18q12.2, harbouring the BRUNOL4 gene that encodes the Bruno-like 4 splicing factor, as an independent prognostic indicator (Poulogiannis *et al.*, 2010).

It is clear that targeted analysis of individual biological systems has yielded results that are ambiguous, can be contradictory and do not readily translate into clinical practice. For example, the matrix metalloproteinase (MMP)/tissue inhibitor of metalloproteinase (TIMP) system has a major, if unclear, role in tumour invasion and metastasis. The main groups of human MMPs include the collagenases (MMPs 1, 8 and 13), stromelysins (MMPs 3, 10 and 11), gelatinases (MMPs 2 and 9), membrane-type MMPs (MMPs 14–17, 24 and 25) and matrilysins (MMPs 7 and 26). MMPs 19, 20, 21, 23 and 26–28 are grouped separately and TIMPs are the main physiological regulators of the MMPs. Genetically, there may be an association between MMP single-nucleotide polymorphisms (SNPs) and CRC risk, prognosis and therapy response. However, there is no agreement on which SNPs are relevant, with one report suggesting that a specific SNP in MMP9 may be associated with tumorigenesis but not metastatic progression (Park *et al.*, 2011), whereas another report associates SNP-independent MMP9 levels, but not (a different) MMP9 SNP with worse survival rates (Langers *et al.*, 2008). Both reports also differ in the MMP2 SNPs they claim to be associated with poorer survival, with a third report demonstrating that Chinese Han individuals, with yet another MMP9 SNP, are more susceptible to CRC (Fang *et al.*, 2010).

## Oncology

5-FU with or without leucovorin (LV) has been the foundation for the adjuvant treatment of CRC for nearly five decades. Unfortunately, patients undergoing this regimen have an overall response rate of only 10% and a median survival of up to 12 months (Prenen *et al.*, 2009). The introduction of the topoisomerase-1 inhibitor irinotecan (Saltz *et al.*, 2000), the platinum derivative oxaliplatin (OHP) (de Gramont *et al.*, 2000) and the oral fluoropyrimidine capecitabine (Van Cutsem *et al.*, 2000) has increased the repertoire of drugs for the treatment of advanced CRC and the neoadjuvant treatment of rectal cancer (Hirsch and Zafar, 2011). Currently, combinations of OHP with various schedules of 5-FU and LV ('FOLFOX4') are first-line treatments for unresectable CRC and achieve much higher response rates than 5-FU, LV and irinotecan combined (Goldberg *et al.*, 2004).

An early meta-analysis found that KRAS G12V mutations were prognostic for AJCC stage III disease (Andreyev *et al.*, 1998), with another study indicating the same for stage II disease (Belly *et al.*, 2001). However, recent data from two major trials demonstrate that KRAS mutational status was not associated with any significant influence with regard to disease-free or overall survival for patients treated with 5-FU (Ogino *et al.*, 2009a; Roth *et al.*, 2010). In the adjuvant setting, there appears to be no link between KRAS status and response to standard chemotherapy (Ocvirk *et al.*, 2010), although one group reports that patients with KRAS wild-type tumours benefit from 5-FU/LV therapy (Ahnen *et al.*, 1998). A recent study suggests that the glucocorticoid-induced protein-coding gene (DEXI), which is methylated in around 49% of CRC, is associated with poor response and outcome in irinotecan-based chemotherapy (Miyaki *et al.*, 2012).

One reason for the uncontrolled proliferation of cancer cells is the increased expression of growth factor receptors, for example the EGFR, which is overexpressed in around 75% of CRCs (Ciardiello and Tortora, 2008), although it is worth stating that EGFR expression levels vary with the detection method used (Nannini *et al.*, 2011). Its association with CRC development has made the EGFR an attractive target for therapy (Yarom and Jonker, 2011) and two monoclonal antibodies against the EGFR, cetuximab and panitumumab, have been developed (Van Cutsem and Geboes, 2007). Vascular endothelial growth factor (VEGF), a factor associated with the development of aberrant blood vessels required for tumour progression and a poor prognosis (Hyodo *et al.*, 1998), is targeted by another antibody, bevacizumab (Mulder *et al.*, 2011). The combination of monoclonal antibodies with conventional polychemotherapy has proven efficacy and has increased the median disease-free survival of AJCC stage IV CRC patients to approximately 24 months (Prenen *et al.*, 2009). Most recently, the phase III VELOUR study showed significant improvements in overall survival, progression-free survival, and response rates with the use of the anti-VEGF angiogenesis inhibitor aflibercept plus FOLFIRI in patients with metastatic colorectal cancer following failure of an oxaliplatin regimen (Gaya and Tse, 2012) and the FDA has approved its use with FOLFIRI in these patients.

Nevertheless, whilst a meta-analysis of data from seven randomized controlled trials of chemotherapy with or without cetuximab in patients with AJCC stage III and IV CRC found that cetuximab increases the likelihood of a response to treatment and makes progression-free survival more likely, improvements in overall survival rates were of borderline significance (Liu *et al.*, 2010). Furthermore, several phase III trials investigating the efficacy of adding irinotecan or bevacizumab to fluorouracil-based regimens in the adjuvant setting have failed to demonstrate a substantial improvement in outcomes for patients with AJCC stage III and IV CRC (Peeters *et al.*, 2010; Fischer von Weikersthal *et al.*, 2011; Maughan *et al.*, 2011; Molinari *et al.*, 2011; Papadimitriou *et al.*, 2011; Price *et al.*, 2011). Certainly, the clinical management of AJCC stage III and IV CRC has progressed to a complex range of pharmaceutical and interventional therapies that presents clinicians with many questions and challenges with regards to the best choice treatment combination, or the optimal selection of patients eligible for the various treatment options (Field and Lipton, 2007).

It is probable that chemotherapy response is linked to linked to variations in underlying genetic characteristics of cancer and host, although our knowledge of this subject remains poor (Walther *et al.*, 2009). Attempts to identify predictive factors for efficacy of antiangiogenic therapies have been disappointing (Dienstmann *et al.*, 2011) and no factor has been identified to predict the efficacy of bevacizumab (Asghar *et al.*, 2010; Cacheux *et al.*, 2011). However, a link has been proposed with mutated KRAS, present in 35–40% of patients with advanced CRC (Normanno *et al.*, 2009), believed to activate the EGFR signalling pathway independently of ligand stimulation of the receptor that is essential for maintenance of the transformed and invasive phenotype of human colon cancer cells (Pollock *et al.*, 2005). Conveniently, KRAS mutations are distributed homogeneously throughout tumour tissue (Ekelund *et al.*, 2011), implying that a single biopsy should be sufficient to determine the KRAS status of a cancer.

Mutated KRAS has been reported to be a factor of non-response, or even of deleterious response to the use both anti-EGFR monoclonal antibodies, cetuximab and panitumumab (Lievre *et al.*, 2006; Khambata-Ford *et al.*, 2007; Amado *et al.*, 2008; De Roock *et al.*, 2008; Karapetis *et al.*, 2008; Bokemeyer *et al.*, 2009, 2011; Van Cutsem *et al.*, 2011; Kohne *et al.*, 2012; Stintzing *et al.*, 2012; Wang *et al.*, 2012). However, different mutant KRAS proteins affect patient survival or downstream signalling in different ways (Ihle *et al.*, 2012) and patients with G13D-mutated tumours may respond differently (De Roock *et al.*, 2010b). A meta-analysis of 11 studies confirmed the benefit with regard to both progression-free survival and overall survival among patients with wild type KRAS (Dahabreh *et al.*, 2011), although another meta-analysis concluded that the addition of anti-EGFR treatment improves progression-free, but not median overall survival in patients with wild type KRAS status (Lin *et al.*, 2011a). This prompted the question as to what constitutes an appropriate primary endpoint for the evaluation of biological agents (Van Loon and Venook, 2011). A report suggesting that panitumumab may be effective for patients with refractory metastatic CRC who have experienced treatment failure with standard therapy, including cetuximab-based regimens (Saif *et al.*, 2010), has also not been reproduced (Wadlow *et al.*, 2011).

Unfortunately, there is a huge amount of variation and lack of published information in the study methods, with patient numbers, duration of follow-up, absolute 5-year disease-free survival rates and percentage of patients undergoing resection of liver/lung metastases after treatment not detailed in the meta-analyses. Another problem is chemotherapy-associated hepatotoxicity induced by combination therapy: a randomized trial investigating the effects of oxaliplatin- or irinotecan-based chemotherapy and either bevacizumab alone or bevacizumab plus panitumumab found that panitumumab shortened progression-free survival and increased toxicity (Hecht *et al.*, 2009). Other investigators assessing the effect of chemotherapy plus bevacizumab, or this combination plus cetuximab found that the addition of cetuximab decreased median progression-free survival and increased the likelihood of serious adverse events, arguing against combining anti-VEGF and anti-EGFR monoclonal antibodies in metastatic CRC (Tol *et al.*, 2009a). The situation in rectal cancer is equally unclear, as one study found KRAS mutation status to be a predictive marker of pathologic response to neoadjuvant cetuximab-based chemoradiation in patients with locally advanced T4 rectal cancers (Grimminger *et al.*, 2011), whereas another concluded that the presence of KRAS mutations are not associated with impaired response to cetuximab-based chemoradiotherapy and 3-year disease-free survival (Erben *et al.*, 2011). The US Food and Drug Administration (FDA) has recently (May 2012) approved a rapid (<5 hours) genetic test developed by Qiagen for wild-type KRAS to identify patients whose tumours will respond to cetuximab. Cetuximab was also given a new indication for first-line treatment in combination with the chemotherapy regimen FOLFIRI (irinotecan, 5-fluorouracil, leucovorin) for patients with metastatic colorectal cancer who have EGFR+ve, wild-type KRAS tumours, based on results from the phase III CRYSTAL trial (Raoul *et al.*, 2009). Again, it is important to stress that although a KRAS mutation excludes treatment with cetuximab for late-stage colorectal cancer patients, wild-type KRAS does not guarantee response to anti-EGFR treatment.

In any case, most patients (80%) with KRAS wild-type tumours still do not respond, suggesting that additional genetic or epigenetic factors are critical determinants of therapy response.

Indeed, alterations in other effectors downstream of the EGFR, such as BRAF, and deregulation of the PIK3CA/PTEN pathway have been independently found to be associated with resistance to therapy. As discussed earlier, BRAF acts as a downstream effector of KRAS signalling, and BRAF mutations, which are mutually exclusive with those of KRAS (Fransen *et al.*, 2004), are important in colorectal tumorigenesis (Rajagopalan *et al.*, 2002). The most frequently reported BRAF mutation, V600E, is tightly associated with sporadic MSI caused by hMLH1 promoter hypermethylation (French *et al.*, 2008), induces a distinct expression profile compared with wild-type BRAF tumours (Kim *et al.*, 2006a), and results in the inappropriate activation of the MEK–ERK pathway (Ikenoue *et al.*, 2003). Studies have shown (i) an association between BRAF mutation and poor prognosis in patients with AJCC stage II–III (French *et al.*, 2008), stage I–IV (Ogino *et al.*, 2009b) and stage IV (Van Cutsem *et al.*, 2011; Yokota *et al.*, 2011) CRC and (ii) that this abrogates the better prognosis associated with MSI (French *et al.*, 2008; Ogino *et al.*, 2009b). These results contrast with another study of AJCC stage I–IV patients, which reported no influence of the V600E status on the 5-year survival of MSI tumours (Samowitz *et al.*, 2005b). The studies also differ in that two report an association of the V600E mutation with poor survival in MSS CRC (Samowitz *et al.*, 2005b; Shaukat *et al.*, 2010), whereas the other found no difference in this subset of patients (French *et al.*, 2008). These discrepancies may be due to stage differences in the study populations, different racial groups and only the latter being in a randomized prospective clinical trial setting. Consequently, it remains unclear whether patients with BRAF-mutated tumours experience a survival benefit from treatment with anti-EGFR antibodies (Yokota, 2011). However, one report concludes that BRAF mutations are a negative prognostic marker as well as predictive of a lack of response to bevacizumab and cetuximab in patients with AJCC stage IV CRC (Tol *et al.*, 2009b) and, whilst not prognostic of relapse-free survival in stage II and stage III CRC patients, is prognostic for overall survival, particularly in patients with MSS tumours (Roth *et al.*, 2010). A recent study suggests that a combination of BRAF, NRAS, and PIK3CA exon 20 mutations (De Roock *et al.*, 2010a), or loss of PTEN combined with mutations of KRAS, BRAF and PIK3CA (Sartore-Bianchi *et al.*, 2009) is significantly associated with a low response rate (Bohanes *et al.*, 2011).

Taking all the evidence together, the prognostic or predictive relevance of these mutations remains unclear. First, a systematic review of the literature suggests that the evidence regarding an association of BRAF mutations with poorer treatment response are of 'low quality', that the evidence for NRAS and PIK3CA exon 20 mutations is based on a small number of identified mutations and the data on PTEN and AKT are limited by variable methods for assessing protein expression (Lin *et al.*, 2011b). Second, a retrospective evaluation of the efficacy of a cetuximab-containing treatment in AJCC stage IV CRC found that, while KRAS status significantly correlated with a worse outcome in patients treated with cetuximab, no definitive inference could be drawn about the role of BRAF mutations and PTEN loss of expression (Inno *et al.*, 2011). Third, it appears that in order to assess whether the mutation status of BRAF and other markers may be a valid prognostic marker in the adjuvant setting, their association with different molecular subgroups may have to be considered (Tejpar *et al.*, 2010) and there is an urgent need for greater comparability and transparency of published studies. Furthermore, all of these studies need to be validated by large prospective randomized clinical trials, since the molecular alterations could be negative prognostic biomarkers independent of the targeted therapies, rather than being predictive biomarkers. In addition, PTEN expression is not a good candidate biomarker as assessment of its expression is a subjective, graded analysis with relatively low concordance between primary and metastatic tumours. Nevertheless, an assessment of the cost-effectiveness of predictive testing for KRAS and BRAF mutations, prior to cetuximab treatment of chemorefractory metastatic CRC patients, concludes that it is economically favourable to identify patients with KRAS and BRAF wild-type status (Blank *et al.*, 2011; Konigsberg *et al.*, 2012).

Similar controversy surrounds the role of thymidylate synthase (TYMS), whose activity is inhibited by 5-FU, leading to cell cycle arrest and apoptosis (Pinedo and Peters, 1988). Although early *in vitro* evidence links TYMS expression and 5-FU sensitivity, conflicting *in vivo* data make the role of this gene or its gene product as a prognostic or predictive marker controversial in the adjuvant setting (Tejpar *et al.*, 2010), despite the recent suggestion that increased expression of TYMS improves outcome stratification for patients with CRC liver metastases treated with resection and modern chemotherapy regimens (Maithel *et al.*, 2011).

Because of these limitations, it is important that clear guidelines for the clinical use of molecular biomarkers are established. The National Comprehensive Cancer Network (NCCN) has published the following clinical practice guidelines for colon (http://www.nccn.org/professionals/physician_gls/pdf/colon.pdf) and rectal (http://www.nccn.org/professionals/physician_gls/pdf/rectal.pdf) cancers, which currently (January 2012) state that:

- In patients with AJCC stage II colon cancer, MMR protein or MSI-H testing should be considered to guide decisions on whether to administer adjuvant single-agent fluoropyrimidine chemotherapy, since patients with MSI-H tumours do not benefit from adjuvant single-agent fluoropyrimidine therapy.
- KRAS mutational analysis for patients with metastatic colorectal cancers is recommended as a predictive marker for non-response to EGFR- targeted therapy with cetuximab or panitumumab. Patients whose tumours harbour certain mutations in the KRAS gene do not respond to cetuximab or panitumumab and should not be treated with either of these agents.
- BRAF mutational analysis is also included as an option for patients with KRAS wild-type metastatic colorectal cancers as a strong negative prognostic factor and for possibly predicting a lower likelihood of benefit from EGFR-targeted therapy after progression on first-line therapy.

These guidelines are frequently updated; making them a valuable tool that enables clinicians to interpret a rapidly changing evidence base, identify the biomarkers that are appropriate for integration into practice and change clinical practice to reflect this information.

## Practical limitations

The excitement about biomarkers originating from research laboratories must be balanced by a critical review of the available evidence before establishing their use in clinical decision-making (Bustin, 2008). As has become clear, many of the studies reporting the discovery of biomarkers useful for CRC management are retrospective, involve small series of patients and are unable to predict disease progression accurately with clinically adequate resolution and reproducibility (Lurje *et al.*, 2007). Sample size is important, as small studies can give inflated, over-promising results as a result of selection bias (Ioannidis *et al.*, 2003). While small study populations are likely to be more homogeneous and thus molecular classifiers may be more efficient, they are frequently underpowered and thus unable to discriminate informative molecular signatures and may incorrectly reach negative conclusions. Study endpoints are an additional critical aspect of oncologic clinical trials that require careful attention when assessing the relevance of biomarkers and the efficacy of treatment (Wu *et al.*, 2011). As a consequence, although a combinatorial approach to molecular prognostics, similar to that established for breast cancer patients, may have significance and be used in future for CRC patient management (Kahlenberg *et al.*, 2003), currently research has failed to yield consistent sets of externally validated markers (Lai *et al.*, 2003; Govindarajan and Paty, 2011) that clinicians can use for clinical decision making (De Roock *et al.*, 2009).

Indeed, despite robust data for their clinical validity, the examples of MMR, KRAS, and BRAF also highlight the risks of early biomarker adoption in the setting of complex molecular pathways and incomplete evidence. As discussed above, the clinical implications of MMR deficiency and BRAF V600E mutation, which coexist in most sporadic MMR-deficient colorectal tumours, currently remain poorly understood and are likely to be mediated by an interplay of factors, including

CpG island methylation status (Ogino *et al.*, 2009b; Ahn *et al.*, 2011). Hence, although it is reasonable to use CIN and MSI-H status in clinical trials to stratify patients, the relative contributions of CIN, MSI-H and CIMP to the outcome of patients with CRC need to be further investigated to better understand the effects and interactions of each variable (Walther *et al.*, 2009).

There is another reason to advocate caution with the use of commercial tests. It is obvious that the evidence underlying the clinical relevance of molecular biomarkers is dynamic and judgements change with the constant publication of corroborating or contradictory data sets. For example, two small single-arm non-randomized studies suggested that the BRAF V600E mutation was a predictive marker for non-response to EGFR-targeted therapies (Di Nicolantonio *et al.*, 2008; Laurent-Puig *et al.*, 2009). More recently, randomized studies have called this interpretation into question (Roock *et al.*, 2010; Tol *et al.*, 2010). These ambiguities are reflected in the NCCN Guidelines, which were changed from advising that patients with tumours harbouring a BRAF V600E mutation seemed unlikely to benefit from anti EGFR therapy, to stating that the data reporting any benefit of anti-EGFR therapy in patients with this mutation were uncertain. It is of course not in the interest of commercial suppliers, nor is it probably practically possible for them, to continuously update their products and seek regulatory body approval based on the latest published information. The resulting serious potential for inappropriate clinical implementation of biomarker data is self-evident.

## Biological variability

The biological variability that is characteristic of human beings introduces an important confounding issue into the identification of universal molecular biomarkers. Colorectal tissue consists of individual cells from a variety of lineages that reside in a changeable natural environment and function as coordinated and dynamic partners. Each cell is comprised of a highly stochastic internal environment with constantly changing concentrations of metabolites, RNA, regulatory molecules and proteins (Raj *et al.*, 2006). In addition, SNPs, epigenetic differences and structural variation such as copy number polymorphisms, inversions, deletions and duplications generate unpredictable and unique genomic backgrounds. This results in multifaceted and variable behaviour patterns regulated by genetic and phenotypic variation as well as being subject to the intrinsic stochastic kinetic noise of biochemical reactions (Raser and O'Shea, 2005).

Crucially, host genetics plays a critical and increasingly recognized role in determining tumour behaviour as well as patient response to adjuvant therapies. In practice, this means that malignancy is influenced not just by environmental stimuli or multiple genetic and epigenetic events that arise within the malignant epithelium, but also by the genetic background of the host (Hunter, 2006; Hunter and Crawford, 2006). In addition, there is an important contribution from the tumour microenvironment and metastasis can be modified by stromal events.

Consequently, new molecular therapeutics are likely to find optimal activity in particular, well-defined subgroups of patients. Furthermore, whereas there are numerous tumour-type specific changes on the path to malignancy and metastasis, it appears that there may be fewer, and common changes in the cancer microenvironment, raising the hope that therapeutic targeting of these events could be generally applicable (Joyce, 2005). Interestingly, if polymorphisms rather than oncogenic events induce the metastasis-predictive gene signatures, then it may not be necessary to obtain tumour biopsies for prognosis. Instead, normal colonic biopsies or even more easily accessible tissues, e.g. blood, could be useful for identifying patients at risk before they develop tumours. These issues are well worth examining and it is essential to obtain epidemiological evidence linking human polymorphisms with metastasis. In any event, the molecular networks involved in regulating the metastatic cascade are slowly being untangled and molecular medicine, by helping with more accurate prognostic stratification of CRC patients, will play a leading role in the move towards individualized, genome-based, combinatorial cancer therapy. Together with an increasing array of additional tools, such as tissue-specific genetic knockouts, siRNA-mediated knockdown and the development of protein, tissue and interaction arrays, these techniques serve to increase our understanding of the metastatic process and the role played by the tumour microenvironment.

## Conclusions and outlook

CRC arises from complex, variable and patient-specific interactions between genetic, epigenetic and environmental factors. The development, improvement and increased sophistication of a wide range of molecular tools has accelerated the transition to large-scale systemic approaches for the study of this disease, making it possible to map genetic and epigenetic changes and analyse the expression patterns associated with its molecular pathogenesis. As a consequence, the last 20 years have witnessed a remarkable increase in knowledge of the genome, transcriptome, proteome and metabolome of CRCs. The identification of critical molecular pathways associated with CRC tumorigenesis and metastasis has demonstrated the feasibility of combining large-scale molecular analysis of (epi)-genetic and gene expression profiles with classic morphological and clinical methods of staging and grading cancer for better outcome prediction. This is driving the pursuit of a change of emphasis for tumour classification from morphological to molecular markers and has led to the identification of numerous potential prognostic biomarkers.

Nevertheless, despite the promise of markers such as chromosomal alteration or genomic and epigenomic instability, e.g. MSI-H cancers having a better and CIN tumours having a worse prognosis, they have not yet been adopted into routine clinical decision-making (Pritchard and Grady, 2011). Furthermore, although numerous individual markers and several expression profiles have been reported as being independent predictors of disease outcome, none have been universally validated. Studies often generate vast amounts of information without any clear evidence that this is any more relevant than available strategies for practical patient management. Hence, despite the introduction of targeted drugs for the treatment of advanced CRC and improved overall survival for non-resectable disease, cure rates remain low. Currently KRAS is the only sufficiently validated predictive molecular marker for anti-EGFR directed therapy, with the predictive value of BRAF mutations unclear.

There are numerous studies under way aimed at incorporating putative predictive molecular markers into the clinical decision making process, and it seems safe to suggest that the use of molecular markers in routine clinical practice will increase as more markers are identified and validated. Furthermore, technical progress continues apace, with 2012 witnessing the introduction of new sequencings instruments that enable a whole human genome to be sequenced in hours at a run cost of $1000. Together with the identification of new, selective inhibitors of the signalling pathways critical to CRC, e.g. BRAF inhibitor PLX-4032 or PI3K pathway inhibitor XL147, and the development of innovative technologies for the simultaneous detection of biomarkers and therapeutic drugs (Yang *et al.*, 2012), this will lead to the enhanced potential for targeted disease management in the individual patient.

### References

Ahlquist, D.A. (2002). Stool-based DNA tests for colorectal cancer: clinical potential and early results. Rev. Gastroenterol. Disord. 2 (Suppl. 1), S20–S26.

Ahlquist, D.A. (2010). Molecular detection of colorectal neoplasia. Gastroenterology *138*, 2127–2139.

Ahlquist, D.A., Skoletsky, J.E., Boynton, K.A., Harrington, J.J., Mahoney, D.W., Pierceall, W.E., Thibodeau, S.N., and Shuber, A.P. (2000). Colorectal cancer screening by detection of altered human DNA in stool: feasibility of a multitarget assay panel. Gastroenterology *119*, 1219–1227.

Ahn, J.B., Chung, W.B., Maeda, O., Shin, S.J., Kim, H.S., Chung, H.C., Kim, N.K., and Issa, J.P. (2011). DNA methylation predicts recurrence from resected stage III proximal colon cancer. Cancer *117*, 1847–1854.

Ahnen, D.J., Feigl, P., Quan, G., Fenoglio-Preiser, C., Lovato, L.C., Bunn, P.A.J., Stemmerman, G., Wells, J.D., Macdonald, J.S., and Meyskens, F.L.J. (1998). Ki-ras mutation and p53 overexpression predict the clinical behavior of colorectal cancer: a Southwest Oncology Group study. Cancer Res. *58*, 1149–1158.

Alhopuro, P., Alazzouzi, H., Sammalkorpi, H., Davalos, V., Salovaara, R., Hemminki, A., Jarvinen, H., Mecklin, J.P., Schwartz, S.J., Aaltonen, L.A., *et al.* (2005a). SMAD4 levels and response to 5-fluorouracil in colorectal cancer. Clin. Cancer Res. *11*, 6311–6316.

Alunni-Fabbroni, M., and Sandri, M.T. (2010). Circulating tumour cells in clinical practice: methods of detection and possible characterization. Methods *50*, 289–297.

Amado, R.G., Wolf, M., Peeters, M., Van Cutsem, E., Siena, S., Freeman, D.J., Juan, T., Sikorski, R., Suggs, S.,

Radinsky, R., *et al.* (2008). Wild-type KRAS is required for panitumumab efficacy in patients with metastatic colorectal cancer. J. Clin. Oncol. *26*, 1626–1634.

Anderson, N., Suliman, I., Bandaletova, T., Obichere, A., Lywood, R., and Loktionov, A. (2011). Protein biomarkers in exfoliated cells collected from the human rectal mucosa: implications for colorectal disease detection and monitoring. Int. J. Colorectal Dis. *26*, 1287–1297.

Andreyev, H.J., Norman, A.R., Cunningham, D., Oates, J.R., and Clarke, P.A. (1998). Kirsten ras mutations in patients with colorectal cancer: the multicenter 'RASCAL' study. J. Natl. Cancer Inst. *90*, 675–684.

Ang, C.S., Phung, J., and Nice, E.C. (2011). The discovery and validation of colorectal cancer biomarkers. Biomed. Chromatogr. *25*, 82–99.

Ang, P.W., Loh, M., Liem, N., Lim, P.L., Grieu, F., Vaithilingam, A., Platell, C., Yong, W.P., Iacopetta, B., and Soong, R. (2010). Comprehensive profiling of DNA methylation in colorectal cancer reveals subgroups with distinct clinicopathological and molecular features. BMC Cancer *10*, 227.

Anjomshoaa, A., Lin, Y.H., Black, M.A., McCall, J.L., Humar, B., Song, S., Fukuzawa, R., Yoon, H.S., Holzmann, B., Friederichs, J., *et al.* (2008). Reduced expression of a gene proliferation signature is associated with enhanced malignancy in colon cancer. Br. J. Cancer *99*, 966–973.

Antolovic, D., Galindo, L., Carstens, A., Rahbari, N., Buchler, M.W., Weitz, J., and Koch, M. (2010). Heterogeneous detection of circulating tumor cells in patients with colorectal cancer by immunomagnetic enrichment using different EpCAM-specific antibodies. BMC Biotechnol. *10*, 35.

Arango, D., Laiho, P., Kokko, A., Alhopuro, P., Sammalkorpi, H., Salovaara, R., Nicorici, D., Hautaniemi, S., Alazzouzi, H., Mecklin, J.P., *et al.* (2005). Gene-expression profiling predicts recurrence in Dukes' C colorectal cancer. Gastroenterology *129*, 874–884.

Arcila, M., Lau, C., Nafa, K., and Ladanyi, M. (2011). Detection of KRAS and BRAF mutations in colorectal carcinoma roles for high-sensitivity locked nucleic acid-PCR sequencing and broad-spectrum mass spectrometry genotyping. J. Mol. Diagn. *13*, 64–73.

Asghar, U., Hawkes, E., and Cunningham, D. (2010). Predictive and prognostic biomarkers for targeted therapy in metastatic colorectal cancer. Clin. Colorectal Cancer *9*, 274–281.

Atkin, W., and Martin, J.P. (2001). Stool DNA-based colorectal cancer detection: finding the needle in the haystack. J. Natl. Cancer Inst. *93*, 798–799.

Azzoni, C., Bottarelli, L., Cecchini, S., Silini, E.M., Bordi, C., and Sarli, L. (2011). Sporadic colorectal carcinomas with low-level microsatellite instability: a distinct subgroup with specific clinicopathological and molecular features. Int. J. Colorectal Dis. *26*, 445–453.

Baba, Y., Nosho, K., Shima, K., Hayashi, M., Meyerhardt, J.A., Chan, A.T., Giovannucci, E., Fuchs, C.S., and Ogino, S. (2011). Phosphorylated AKT expression is associated with PIK3CA mutation, low stage, and favorable outcome in 717 colorectal cancers. Cancer *117*, 1399–1408.

Baisse, B., Bouzourene, H., Saraga, E.P., Bosman, F.T., and Benhattar, J. (2001). Intratumor genetic heterogeneity in advanced human colorectal adenocarcinoma. Int. J. Cancer *93*, 346–352.

Banerjea, A., Hands, R.E., Powar, M.P., Bustin, S.A., and Dorudi, S. (2009). Microsatellite and chromosomal stable colorectal cancers demonstrate poor immunogenicity and early disease recurrence. Colorectal Dis. *11*, 601–608.

Barault, L., Charon-Barra, C., Jooste, V., de la Vega, M.F., Martin, L., Roignot, P., Rat, P., Bouvier, A.M., Laurent-Puig, P., Faivre, J., Chapusot, C., and Piard, F. (2008). Hypermethylator phenotype in sporadic colon cancer: study on a population-based series of 582 cases. Cancer Res. *68*, 8541–8546.

Barderas, R., Babel, I., and Casal, J.I. (2010). Colorectal cancer proteomics, molecular characterization and biomarker discovery. Proteomics Clin. Appl. *4*, 159–178.

Barratt, P.L., Seymour, M.T., Stenning, S.P., Georgiades, I., Walker, C., Birbeck, K., and Quirke, P. (2002). DNA markers predicting benefit from adjuvant fluorouracil in patients with colon cancer: a molecular study. Lancet *360*, 1381–1391.

Belly, R.T., Rosenblatt, J.D., Steinmann, M., Toner, J., Sun, J., Shehadi, J., Peacock, J.L., Raubertas, R.F., Jani, N., and Ryan, C.K. (2001). Detection of mutated K12-ras in histologically negative lymph nodes as an indicator of poor prognosis in stage II colorectal cancer. Clin. Colorectal Cancer *1*, 110–116.

Benatti, P., Gafa, R., Barana, D., Marino, M., Scarselli, A., Pedroni, M., Maestri, I., Guerzoni, L., Roncucci, L., Menigatti, M., Roncari, B., Maffei, S., Rossi, G., Ponti, G., Santini, A., Losi, L., Di Gregorio, C., Oliani, C., Ponz de Leon, M., and Lanza, G. (2005). Microsatellite instability and colorectal cancer prognosis. Clin. Cancer Res. *11*, 8332–8340.

Bertagnolli, M.M., Redston, M., Compton, C.C., Niedzwiecki, D., Mayer, R.J., Goldberg, R.M., Colacchio, T.A., Saltz, L.B., and Warren, R.S. (2011). Microsatellite instability and loss of heterozygosity at chromosomal location 18q: prospective evaluation of biomarkers for stages II and III colon cancer--a study of CALGB 9581 and 89803. J. Clin. Oncol. *29*, 3153–3162.

Biomarkers Definitions Working Group (2001). Biomarkers and surrogate endpoints: preferred definitions and conceptual framework. Clin. Pharmacol. Ther. *69*, 89–95.

Blank, P.R., Moch, H., Szucs, T.D., and Schwenkglenks, M. (2011). KRAS and BRAF mutation analysis in metastatic colorectal cancer: a cost-effectiveness analysis from a Swiss perspective. Clin. Cancer Res. *17*, 6338–6346.

Bohanes, P., LaBonte, M.J., Winder, T., and Lenz, H.J. (2011). Predictive molecular classifiers in colorectal cancer. Semin. Oncol. *38*, 576–587.

Bokemeyer, C., Bondarenko, I., Makhson, A., Hartmann, J.T., Aparicio, J., de Braud, F., Donea, S., Ludwig, H., Schuch, G., Stroh, C., *et al.* (2009). Fluorouracil, leucovorin, and oxaliplatin with and without cetuximab in the first-line treatment of metastatic colorectal cancer. J. Clin. Oncol. *27*, 663–671.

Bokemeyer, C., Bondarenko, I., Hartmann, J.T., de Braud, F., Schuch, G., Zubel, A., Celik, I., Schlichting, M., and Koralewski, P. (2011). Efficacy according to biomarker status of cetuximab plus FOLFOX-4 as first-line treatment for metastatic colorectal cancer: the OPUS study. Ann. Oncol. *22*, 1535–1546.

Boland, C.R., Thibodeau, S.N., Hamilton, S.R., Sidransky, D., Eshleman, J.R., Burt, R.W., Meltzer, S.J., Rodriguez-Bigas, M.A., Fodde, R., Ranzani, G.N., *et al.* (1998). A National Cancer Institute Workshop on Microsatellite Instability for cancer detection and familial predisposition: development of international criteria for the determination of microsatellite instability in colorectal cancer. Cancer Res. *58*, 5248–5257.

Booth, R.A. (2007). Minimally invasive biomarkers for detection and staging of colorectal cancer. Cancer Lett. *249*, 87–96.

Boulay, J.L., Mild, G., Lowy, A., Reuter, J., Lagrange, M., Terracciano, L., Laffer, U., Herrmann, R., and Rochlitz, C. (2002). SMAD4 is a predictive marker for 5-fluorouracil-based chemotherapy in patients with colorectal cancer. Br. J. Cancer *87*, 630–634.

Bruin, S.C., He, Y., Mikolajewska-Hanclich, I., Liefers, G.J., Klijn, C., Vincent, A., Verwaal, V.J., de Groot, K.A., Morreau, H., van Velthuysen, M.L., *et al.* (2011). Molecular alterations associated with liver metastases development in colorectal cancer patients. Br. J. Cancer *105*, 281–287.

Bustin, S. (2008). Molecular medicine, gene-expression profiling and molecular diagnostics: putting the cart before the horse. Biomark. Med. *2*, 201–207.

Cacheux, W., Tourneau, C.L., Baranger, B., Mignot, L., and Mariani, P. (2011). Targeted biotherapy in metastatic colorectal carcinoma: Current practice. J. Visc. Surg. *148*, 12–18.

Cai, G., Xu, Y., Lu, H., Shi, Y., Lian, P., Peng, J., Du, X., Zhou, X., Guan, Z., Shi, D., and Cai, S. (2008). Clinicopathologic and molecular features of sporadic microsatellite- and chromosomal-stable colorectal cancers. Int. J. Colorectal Dis. *23*, 365–373.

Calistri, D., Rengucci, C., Bocchini, R., Saragoni, L., Zoli, W., and Amadori, D. (2003). Fecal multiple molecular tests to detect colorectal cancer in stool. Clin. Gastroenterol. Hepatol. *1*, 377–383.

Chan, K.C., Yeung, S.W., Lui, W.B., Rainer, T.H., and Lo, Y.M. (2005). Effects of preanalytical factors on the molecular size of cell-free DNA in blood. Clin. Chem. *51*, 781–784.

Chan, T.L., Curtis, L.C., Leung, S.Y., Farrington, S.M., Ho, J.W., Chan, A.S., Lam, P.W., Tse, C.W., Dunlop, M.G., Wyllie, A.H., and *et al.* (2001). Early-onset colorectal cancer with stable microsatellite DNA and near-diploid chromosomes. Oncogene *20*, 4871–4876.

Chan, T.L., Zhao, W., Leung, S.Y., and Yuen, S.T. (2003). BRAF and KRAS mutations in colorectal hyperplastic polyps and serrated adenomas. Cancer Res. *63*, 4878–4881.

Chang, W., Wu, L., Cao, F., Liu, Y., Ma, L., Wang, M., Zhao, D., Li, P., Zhang, Q., Tan, X., *et al.* (2011). Development of autoantibody signatures as biomarkers for early detection of colorectal carcinoma. Clin. Cancer Res. *17*, 5715–5724.

Chien, C.C., Chen, S.H., Liu, C.C., Lee, C.L., Yang, R.N., Yang, S.H., and Huang, C.J. (2007). Correlation of K-ras codon 12 mutations in human feces and ages of patients with colorectal cancer (CRC). Transl. Res. *149*, 96–102.

Ciardiello, F., and Tortora, G. (2008). EGFR antagonists in cancer treatment. N. Engl. J. Med. *358*, 1160–1174.

Coleman, M.P., Quaresma, M., Berrino, F., Lutz, J.M., De Angelis, R., Capocaccia, R., Baili, P., Rachet, B., Gatta, G., Hakulinen, T., *et al.* (2008). Cancer survival in five continents: a worldwide population-based study (CONCORD). Lancet Oncol. *9*, 730–756.

Costello, J.F., Fruhwald, M.C., Smiraglia, D.J., Rush, L.J., Robertson, G.P., Gao, X., Wright, F.A., Feramisco, J.D., Peltomaki, P., Lang, J.C., *et al.* (2000). Aberrant CpG-island methylation has non-random and tumour-type-specific patterns. Nat. Genet. *24*, 132–138.

Cowie, S., Drmanac, S., Swanson, D., Delgrosso, K., Huang, S., du Sart, D., Drmanac, R., Surrey, S., and Fortina, P. (2004). Identification of APC gene mutations in colorectal cancer using universal microarray-based combinatorial sequencing-by-hybridization. Hum. Mutat. *24*, 261–271.

Cowin, P.A., Anglesio, M., Etemadmoghadam, D., and Bowtell, D.D. (2010). Profiling the cancer genome. Annu. Rev. Genomics Hum. Genet. *11*, 133–159.

Cross, J. (2008). DxS Ltd. Pharmacogenomics *9*, 463–467.

Cunningham, D., Atkin, W., Lenz, H.J., Lynch, H.T., Minsky, B., Nordlinger, B., and Starling, N. (2010). Colorectal cancer. Lancet *375*, 1030–1047.

Dahabreh, I.J., Terasawa, T., Castaldi, P.J., and Trikalinos, T.A. (2011). Systematic review: Anti-epidermal growth factor receptor treatment effect modification by KRAS mutations in advanced colorectal cancer. Ann. Intern. Med. *154*, 37–49.

Dahlin, A.M., Palmqvist, R., Henriksson, M.L., Jacobsson, M., Eklof, V., Rutegard, J., Oberg, A., and Van Guelpen, B.R. (2010). The role of the CpG island methylator phenotype in colorectal cancer prognosis depends on microsatellite instability screening status. Clin. Cancer Res. *16*, 1845–1855.

Danese, E., Montagnana, M., Minicozzi, A.M., De Matteis, G., Scudo, G., Salvagno, G.L., Cordiano, C., Lippi, G., and Guidi, G.C. (2010). Real-time polymerase chain reaction quantification of free DNA in serum of patients with polyps and colorectal cancers. Clin. Chem. Lab Med *48*, 1665–1668.

Danielsen, S.A., Lind, G.E., Bjornslett, M., Meling, G.I., Rognum, T.O., Heim, S., and Lothe, R.A. (2008). Novel mutations of the suppressor gene PTEN in colorectal carcinomas stratified by microsatellite instability- and TP53 mutation- status. Hum. Mutat. *29*, E252–62.

De Roock, W., Piessevaux, H., De Schutter, J., Janssens, M., De Hertogh, G., Personeni, N., Biesmans, B., Van Laethem, J.L., Peeters, M., Humblet, Y., *et al.* (2008). KRAS wild-type state predicts survival and is associated to early radiological response in metastatic colorectal cancer treated with cetuximab. Ann. Oncol. *19*, 508–515.

De Roock, W., Biesmans, B., De Schutter, J., and Tejpar, S. (2009). Clinical biomarkers in oncology: focus on colorectal cancer. Mol. Diagn. Ther. *13*, 103–114.

De Roock, W., Claes, B., Bernasconi, D., De Schutter, J., Biesmans, B., Fountzilas, G., Kalogeras, K.T., Kotoula, V., Papamichael, D., Laurent-Puig, P., *et al.* (2010a). Effects of KRAS, BRAF, NRAS, and PIK3CA mutations on the efficacy of cetuximab plus chemotherapy in chemotherapy-refractory metastatic colorectal cancer: a retrospective consortium analysis. Lancet Oncol. *11*, 753–762.

De Roock, W., Jonker, D.J., Di Nicolantonio, F., Sartore-Bianchi, A., Tu, D., Siena, S., Lamba, S., Arena, S., Frattini, M., Piessevaux, H., *et al.* (2010b). Association of KRAS p.G13D mutation with outcome in patients with chemotherapy-refractory metastatic colorectal cancer treated with cetuximab. JAMA *304*, 1812–1820.

Di Martino, M.T., Arbitrio, M., Leone, E., Guzzi, P.H., Rotundo, M.S., Ciliberto, D., Tomaino, V., Fabiani, F., Talarico, D., Sperlongano, P., *et al.* (2011). Single nucleotide polymorphisms of ABCC5 and ABCG1 transporter genes correlate to irinotecan-associated gastrointestinal toxicity in colorectal cancer patients: a DMET microarray profiling study. Cancer Biol. Ther. *12*, 780–787.

Di Nicolantonio, F., Martini, M., Molinari, F., Sartore-Bianchi, A., Arena, S., Saletti, P., De Dosso, S., Mazzucchelli, L., Frattini, M., Siena, S., *et al.* (2008). Wild-type BRAF is required for response to panitumumab or cetuximab in metastatic colorectal cancer. J. Clin. Oncol. *26*, 5705–5712.

Diehl, F., Li, M., Dressman, D., He, Y., Shen, D., Szabo, S., Diaz, L.A. Jr., Goodman, S.N., David, K.A., Juhl, H., *et al.* (2005). Detection and quantification of mutations in the plasma of patients with colorectal tumors. Proc. Natl. Acad. Sci. U.S.A. *102*, 16368–16373.

Dienstmann, R., Vilar, E., and Tabernero, J. (2011). Molecular predictors of response to chemotherapy in colorectal cancer. Cancer J. *17*, 114–126.

Diep, C.B., Kleivi, K., Ribeiro, F.R., Teixeira, M.R., Lindgjaerde, O.C., and Lothe, R.A. (2006). The order of genetic events associated with colorectal cancer progression inferred from meta-analysis of copy number changes. Genes Chromosomes Cancer *45*, 31–41.

Dong, S.M., Traverso, G., Johnson, C., Geng, L., Favis, R., Boynton, K., Hibi, K., Goodman, S.N., D'Allessio, M., Paty, P., *et al.* (2001). Detecting colorectal cancer in stool with the use of multiple genetic targets. J. Natl. Cancer Inst. *93*, 858–865.

Ekelund, S., Papadogiannakis, N., Olivecrona, H., and Lindforss, U. (2011). Tissue sampling for mutation analysis in colorectal cancer: K-ras is homogeneously distributed throughout the tumor tissue. Oncol. Rep. *25*, 253–258.

Elsaleh, H., and Iacopetta, B. (2001). Microsatellite instability is a predictive marker for survival benefit from adjuvant chemotherapy in a population-based series of stage III colorectal carcinoma. Clin. Colorectal Cancer *1*, 104–109.

Elsaleh, H., Powell, B., McCaul, K., Grieu, F., Grant, R., Joseph, D., and Iacopetta, B. (2001). P53 alteration and microsatellite instability have predictive value for survival benefit from chemotherapy in stage III colorectal carcinoma. Clin. Cancer Res. *7*, 1343–1349.

Erben, P., Strobel, P., Horisberger, K., Popa, J., Bohn, B., Hanfstein, B., Kahler, G., Kienle, P., Post, S., Wenz, F., *et al.* (2011). KRAS and BRAF mutations and PTEN expression do not predict efficacy of cetuximab-based chemoradiotherapy in locally advanced rectal cancer. Int. J. Radiat. Oncol. Biol. Phys. *81*, 1032–1038.

Fang, W.L., Liang, W.B., He, H., Zhu, Y., Li, S.L., Gao, L.B., and Zhang, L. (2010). Association of matrix metalloproteinases 1, 7, and 9 gene polymorphisms with genetic susceptibility to colorectal carcinoma in a Han Chinese population. DNA Cell. Biol. *29*, 657–661.

Fearon, E.R. (2011). Molecular genetics of colorectal cancer. Annu. Rev. Pathol. *6*, 479–507.

Fearon, E.R., and Vogelstein, B. (1990). A genetic model for colorectal tumorigenesis. Cell *61*, 759–767.

Ferracin, M., Gafa, R., Miotto, E., Veronese, A., Pultrone, C., Sabbioni, S., Lanza, G., and Negrini, M. (2008). The methylator phenotype in microsatellite stable colorectal cancers is characterized by a distinct gene expression profile. J. Pathol. *214*, 594–602.

Field, K., and Lipton, L. (2007). Metastatic colorectal cancer-past, progress and future. World J. Gastroenterol. *13*, 3806–3815.

Fischer von Weikersthal, L., Schalhorn, A., Stauch, M., Quietzsch, D., Maubach, P.A., Lambertz, H., Oruzio, D., Schlag, R., Weigang-Kohler, K., Vehling-Kaiser, U., *et al.* (2011). Phase III trial of irinotecan plus infusional 5-fluorouracil/folinic acid versus irinotecan plus oxaliplatin as first-line treatment of advanced colorectal cancer. Eur. J. Cancer *47*, 206–214.

Fleischhacker, M., and Schmidt, B. (2007). Circulating nucleic acids (CNAs) and cancer--a survey. Biochim. Biophys. Acta *1775*, 181–232.

Fong, S.L., Zhang, J.T., Lim, C.K., Eu, K.W., and Liu, Y. (2009). Comparison of 7 methods for extracting cell-free DNA from serum samples of colorectal cancer patients. Clin. Chem. *55*, 587–589.

Fransen, K., Klintenas, M., Osterstrom, A., Dimberg, J., Monstein, H.J., and Soderkvist, P. (2004). Mutation analysis of the BRAF, ARAF and RAF-1 genes in human colorectal adenocarcinomas. Carcinogenesis *25*, 527–533.

Frattini, M., Gallino, G., Signoroni, S., Balestra, D., Battaglia, L., Sozzi, G., Leo, E., Pilotti, S., and Pierotti, M.A. (2006). Quantitative analysis of plasma DNA in colorectal cancer patients: a novel prognostic tool. Ann. N. Y. Acad. Sci. *1075*, 185–190.

French, A.J., Sargent, D.J., Burgart, L.J., Foster, N.R., Kabat, B.F., Goldberg, R., Shepherd, L., Windschitl, H.E., and Thibodeau, S.N. (2008). Prognostic significance of defective mismatch repair and BRAF V600E in patients with colon cancer. Clin. Cancer Res. *14*, 3408–3415.

Gaudet, F., Hodgson, J.G., Eden, A., Jackson-Grusby, L., Dausman, J., Gray, J.W., Leonhardt, H., and Jaenisch, R. (2003). Induction of tumors in mice by genomic hypomethylation. Science *300*, 489–492.

Gaya, A., and Tse, V. (2012). A preclinical and clinical review of aflibercept for the management of cancer. Cancer Treat. Rev. *38*, 484–493.

Georgiades, I.B., Curtis, L.J., Morris, R.M., Bird, C.C., and Wyllie, A.H. (1999). Heterogeneity studies identify a subset of sporadic colorectal cancers without evidence for chromosomal or microsatellite instability. Oncogene *18*, 7933–7940.

Ghadimi, B.M., Grade, M., Liersch, T., Langer, C., Siemer, A., Fuzesi, L., and Becker, H. (2003). Gain of chromosome 8q23–24 is a predictive marker for lymph node positivity in colorectal cancer. Clin. Cancer Res. *9*, 1808–1814.

Goldberg, R.M., Sargent, D.J., Morton, R.F., Fuchs, C.S., Ramanathan, R.K., Williamson, S.K., Findlay, B.P., Pitot, H.C., and Alberts, S.R. (2004). A randomized controlled trial of fluorouracil plus leucovorin, irinotecan, and oxaliplatin combinations in patients with previously untreated metastatic colorectal cancer. J. Clin. Oncol. *22*, 23–30.

Gosens, M.J., van Kempen, L.C., van de Velde, C.J., van Krieken, J.H., and Nagtegaal, I.D. (2007). Loss of membranous Ep-CAM in budding colorectal carcinoma cells. Mod. Pathol. *20*, 221–232.

Govindarajan, A., and Paty, P.B. (2011). Predictive markers of colorectal cancer liver metastases. Future Oncol. *7*, 299–307.

de Gramont, A., Figer, A., Seymour, M., Homerin, M., Hmissi, A., Cassidy, J., Boni, C., Cortes-Funes, H., Cervantes, A., Freyer, G., *et al.* (2000). Leucovorin and fluorouracil with or without oxaliplatin as first-line treatment in advanced colorectal cancer. J. Clin. Oncol. *18*, 2938–2947.

Grimminger, P.P., Danenberg, P., Dellas, K., Arnold, D., Rodel, C., Machiels, J.P., Haustermans, K., Debucquoy, A., Velenik, V., Sempoux, C., *et al.* (2011). Biomarkers for cetuximab-based neoadjuvant radiochemotherapy in locally advanced rectal cancer. Clin. Cancer Res. *17*, 3469–3477.

Gruenberger, B., Schueller, J., Heubrandtner, U., Wrba, F., Tamandl, D., Kaczirek, K., Roka, R., Freimann-Pircher, S., and Gruenberger, T. (2010). Cetuximab, gemcitabine, and oxaliplatin in patients with unresectable advanced or metastatic biliary tract cancer: a phase 2 study. Lancet Oncol. *11*, 1142–1148.

Guadalajara, H., Dominguez-Berzosa, C., Garcia-Arranz, M., Herreros, M.D., Pascual, I., Sanz-Baro, R., Garcia-Olmo, D.C., and Garcia-Olmo, D. (2008). The concentration of deoxyribonucleic acid in plasma from 73 patients with colorectal cancer and apparent clinical correlations. Cancer Detect. Prev. *32*, 39–44.

Guastadisegni, C., Colafranceschi, M., Ottini, L., and Dogliotti, E. (2010). Microsatellite instability as a marker of prognosis and response to therapy: a meta-analysis of colorectal cancer survival data. Eur. J. Cancer *46*, 2788–2798.

van der Gun, B.T., Melchers, L.J., Ruiters, M.H., de Leij, L.F., McLaughlin, P.M., and Rots, M.G. (2010). EpCAM in carcinogenesis: the good, the bad or the ugly. Carcinogenesis *31*, 1913–1921.

Gunther, K., Leier, J., Henning, G., Dimmler, A., Weissbach, R., Hohenberger, W., and Forster, R. (2005). Prediction of lymph node metastasis in colorectal carcinoma by expressionof chemokine receptor CCR7. Int. J. Cancer *116*, 726–733.

Haug, U., Hillebrand, T., Bendzko, P., Low, M., Rothenbacher, D., Stegmaier, C., and Brenner, H. (2007). Mutant-enriched PCR and allele-specific hybridization reaction to detect K-ras mutations in stool DNA: high prevalence in a large sample of older adults. Clin. Chem. *53*, 787–790.

Hawkins, N.J., Tomlinson, I., Meagher, A., and Ward, R.L. (2001). Microsatellite-stable diploid carcinoma: a biologically distinct and aggressive subset of sporadic colorectal cancer. Br. J. Cancer *84*, 232–236.

Heald, R.J., and Ryall, R.D. (1986). Recurrence and survival after total mesorectal excision for rectal cancer. Lancet *1*, 1479–1482.

Hecht, J.R., Mitchell, E., Chidiac, T., Scroggin, C., Hagenstad, C., Spigel, D., Marshall, J., Cohn, A., McCollum, D., Stella, P., *et al.* (2009). A randomized phase IIIB trial of chemotherapy, bevacizumab, and panitumumab compared with chemotherapy and bevacizumab alone for metastatic colorectal cancer. J. Clin. Oncol. *27*, 672–680.

Hegde, P., Qi, R., Gaspard, R., Abernathy, K., Dharap, S., Earle-Hughes, J., Gay, C., Nwokekeh, N.U., Chen, T., Saeed, A.I., *et al.* (2001). Identification of tumor markers in models of human colorectal cancer using a 19,200-element complementary DNA microarray. Cancer Res. *61*, 7792–7797.

Hemandas, A.K., Salto-Tellez, M., Maricar, S.H., Leong, A.F., and Leow, C.K. (2006). Metastasis-associated protein S100A4--a potential prognostic marker for colorectal cancer. J. Surg. Oncol. 93, 498–503.

Hibi, K., Goto, T., Mizukami, H., Kitamura, Y., Sakata, M., Saito, M., Ishibashi, K., Kigawa, G., Nemoto, H., and Sanada, Y. (2009). MGMT gene is aberrantly methylated from the early stages of colorectal cancers. Hepatogastroenterology *56*, 1642–1644.

Hinoue, T., Weisenberger, D.J., Pan, F., Campan, M., Kim, M., Young, J., Whitehall, V.L., Leggett, B.A., and Laird, P.W. (2009). Analysis of the association between CIMP and BRAF in colorectal cancer by DNA methylation profiling. PLoS One *4*, e8357.

Hirsch, B.R., and Zafar, S.Y. (2011). Capecitabine in the management of colorectal cancer. Cancer Manag. Res. *3*, 79–89.

Hoff, G., and Dominitz, J.A. (2010). Contrasting US and European approaches to colorectal cancer screening: which is best? Gut *59*, 407–414.

Hohenberger, W., Weber, K., Matzel, K., Papadopoulos, T., and Merkel, S. (2009). Standardized surgery for colonic cancer: complete mesocolic excision and central ligation--technical notes and outcome. Colorectal Dis. *11*, 354–64; discussion 364–5.

Hornbech, K., Ravn, J., and Steinbruchel, D.A. (2011). Outcome after pulmonary metastasectomy: analysis of 5 years consecutive surgical resections 2002–2006. J. Thorac. Oncol. *6*, 1733–1740.

Hughes, L.A., Khalid-de Bakker, C.A., Smits, K.M., van den Brandt, P.A., Jonkers, D., Ahuja, N., Herman, J.G., Weijenberg, M.P., and van Engeland, M. (2012). The CpG island methylator phenotype in colorectal cancer: Progress and problems. Biochim. Biophys. Acta *1825*, 77–85.

Hunter, K. (2006). Host genetics influence tumour metastasis. Nat. Rev. Cancer *6*, 141–146.

Hunter, K.W., and Crawford, N.P. (2006). Germ line polymorphism in metastatic progression. Cancer Res. *66*, 1251–1254.

Hyodo, I., Doi, T., Endo, H., Hosokawa, Y., Nishikawa, Y., Tanimizu, M., Jinno, K., and Kotani, Y. (1998). Clinical significance of plasma vascular endothelial growth factor in gastrointestinal cancer. Eur. J. Cancer *34*, 2041–2045.

Iacopetta, B., Grieu, F., Li, W., Ruszkiewicz, A., Caruso, M., Moore, J., Watanabe, G., and Kawakami, K. (2006). APC gene methylation is inversely correlated with features of the CpG island methylator phenotype in colorectal cancer. Int. J. Cancer *119*, 2272–2278.

Ihle, N.T., Byers, L.A., Kim, E.S., Saintigny, P., Lee, J.J., Blumenschein, G.R., Tsao, A., Liu, S., Larsen, J.E., Wang, J., *et al.* (2012). Effect of KRAS oncogene substitutions on protein behavior: implications for signaling and clinical outcome. J. Natl. Cancer Inst. *104*, 228–239.

Ikenoue, T., Hikiba, Y., Kanai, F., Tanaka, Y., Imamura, J., Imamura, T., Ohta, M., Ijichi, H., Tateishi, K., Kawakami, T., *et al.* (2003). Functional analysis of mutations within the kinase activation segment of B-Raf in human colorectal tumors. Cancer Res. *63*, 8132–8137.

Inno, A., Salvatore, M.D., Cenci, T., Martini, M., Orlandi, A., Strippoli, A., Ferrara, A.M., Bagala, C., Cassano, A., Larocca, L.M., *et al.* (2011). Is there a role for IGF1R and c-MET pathways in resistance to cetuximab in metastatic colorectal cancer? Clin. Colorectal Cancer 10, 325–332.

Ioannidis, J.P., Trikalinos, T.A., Ntzani, E.E., and Contopoulos-Ioannidis, D.G. (2003). Genetic associations in large versus small studies: an empirical assessment. Lancet *361*, 567–571.

Iyengar, V., Albaugh, G.P., Lohani, A., and Nair, P.P. (1991). Human stools as a source of viable colonic epithelial cells. FASEB J. *5*, 2856–2859.

Jacobs, B., De Roock, W., Piessevaux, H., Van Oirbeek, R., Biesmans, B., De Schutter, J., Fieuws, S., Vandesompele, J., Peeters, M., Van Laethem, J.L., *et al.* (2009). Amphiregulin and epiregulin mRNA expression in primary tumors predicts outcome in metastatic colorectal cancer treated with cetuximab. J. Clin. Oncol. *27*, 5068–5074.

Jass, J.R. (2007). Classification of colorectal cancer based on correlation of clinical, morphological and molecular features. Histopathology *50*, 113–130.

Jehan, Z., Bavi, P., Sultana, M., Abubaker, J., Bu, R., Hussain, A., Alsbeih, G., Al-Sanea, N., Abduljabbar, A., Ashari, L.H., *et al.* (2009). Frequent PIK3CA gene amplification and its clinical significance in colorectal cancer. J. Pathol. *219*, 337–346.

Jen, J., Kim, H., Piantadosi, S., Liu, Z.F., Levitt, R.C., Sistonen, P., Kinzler, K.W., Vogelstein, B., and Hamilton, S.R. (1994). Allelic loss of chromosome 18q and prognosis in colorectal cancer. N. Engl. J. Med. *331*, 213–221.

Jernvall, P., Makinen, M.J., Karttunen, T.J., Makela, J., and Vihko, P. (1999). Loss of heterozygosity at 18q21 is indicative of recurrence and therefore poor prognosis in a subset of colorectal cancers. Br. J. Cancer *79*, 903–908.

Jo, P., Jung, K., Grade, M., Conradi, L.C., Wolff, H.A., Kitz, J., Becker, H., Ruschoff, J., Hartmann, A., Beissbarth, T., *et al.* (2011). CpG island methylator phenotype infers a poor disease-free survival in locally advanced rectal cancer. Surgery *151*, 564–570.

Jones, A.M., Douglas, E.J., Halford, S.E., Fiegler, H., Gorman, P.A., Roylance, R.R., Carter, N.P., and Tomlinson, I.P. (2005). Array-CGH analysis of microsatellite-stable, near-diploid bowel cancers and comparison with other types of colorectal carcinoma. Oncogene *24*, 118–129.

Jover, R., Zapater, P., Castells, A., Llor, X., Andreu, M., Cubiella, J., Balaguer, F., Sempere, L., Xicola, R.M., Bujanda, L., *et al.* (2009). The efficacy of adjuvant chemotherapy with 5-fluorouracil in colorectal cancer depends on the mismatch repair status. Eur. J. Cancer *45*, 365–373.

Jover, R., Nguyen, T.P., Perez-Carbonell, L., Zapater, P., Paya, A., Alenda, C., Rojas, E., Cubiella, J., Balaguer, F., Morillas, J.D., *et al.* (2011). 5-Fluorouracil adjuvant chemotherapy does not increase survival in patients with CpG island methylator phenotype colorectal cancer. Gastroenterology *140*, 1174–1181.

Joyce, J.A. (2005). Therapeutic targeting of the tumor microenvironment. Cancer Cell *7*, 513–520.

Kahlenberg, M.S., Sullivan, J.M., Witmer, D.D., and Petrelli, N.J. (2003). Molecular prognostics in colorectal cancer. Surg. Oncol. *12*, 173–186.

Kakar, S., Deng, G., Sahai, V., Matsuzaki, K., Tanaka, H., Miura, S., and Kim, Y.S. (2008). Clinicopathologic characteristics, CpG island methylator phenotype, and BRAF mutations in microsatellite-stable colorectal cancers without chromosomal instability. Arch. Pathol. Lab. Med. *132*, 958–964.

Kane, M.F., Loda, M., Gaida, G.M., Lipman, J., Mishra, R., Goldman, H., Jessup, J.M., and Kolodner, R. (1997). Methylation of the hMLH1 promoter correlates with lack of expression of hMLH1 in sporadic colon tumors and mismatch repair-defective human tumor cell lines. Cancer Res. *57*, 808–811.

Kang, U.B., Yeom, J., Kim, H.J., Kim, H., and Lee, C. (2011). Expression profiling of more than 3500 proteins of MSS-type colorectal cancer by stable isotope labeling and mass spectrometry. J. Proteomics *75*, 3050–3062.

Karapetis, C.S., Khambata-Ford, S., Jonker, D.J., O'Callaghan, C.J., Tu, D., Tebbutt, N.C., Simes, R.J., Chalchal, H., Shapiro, J.D., Robitaille, S., *et al.* (2008). K-ras mutations and benefit from cetuximab in advanced colorectal cancer. N. Engl. J. Med. *359*, 1757–1765.

Kato, S., Iida, S., Higuchi, T., Ishikawa, T., Takagi, Y., Yasuno, M., Enomoto, M., Uetake, H., and Sugihara, K. (2007). PIK3CA mutation is predictive of poor survival in patients with colorectal cancer. Int. J. Cancer *121*, 1771–1778.

Khambata-Ford, S., Garrett, C.R., Meropol, N.J., Basik, M., Harbison, C.T., Wu, S., Wong, T.W., Huang, X., Takimoto, C.H., Godwin, A.K., *et al.* (2007). Expression of epiregulin and amphiregulin and K-ras mutation status predict disease control in metastatic colorectal cancer patients treated with cetuximab. J. Clin. Oncol. *25*, 3230–3237.

Kim, I.J., Kang, H.C., Jang, S.G., Kim, K., Ahn, S.A., Yoon, H.J., Yoon, S.N., and Park, J.G. (2006a). Oligonucleotide microarray analysis of distinct gene expression patterns in colorectal cancer tissues harboring BRAF and K-ras mutations. Carcinogenesis *27*, 392–404.

Kim, J.H., Shin, S.H., Kwon, H.J., Cho, N.Y., and Kang, G.H. (2009). Prognostic implications of CpG island hypermethylator phenotype in colorectal cancers. Virchows Arch. *455*, 485–494.

Kim, M.S., Lee, J., and Sidransky, D. (2010). DNA methylation markers in colorectal cancer. Cancer Metastasis Rev. *29*, 181–206.

Kim, M.Y., Yim, S.H., Kwon, M.S., Kim, T.M., Shin, S.H., Kang, H.M., Lee, C., and Chung, Y.J. (2006b). Recurrent genomic alterations with impact on survival in colorectal cancer identified by genome-wide array comparative genomic hybridization. Gastroenterology *131*, 1913–1924.

Kim, Y.H., Kakar, S., Cun, L., Deng, G., and Kim, Y.S. (2008). Distinct CpG island methylation profiles and BRAF mutation status in serrated and adenomatous colorectal polyps. Int. J. Cancer *123*, 2587–2593.

Klein-Scory, S., Kubler, S., Diehl, H., Eilert-Micus, C., Reinacher-Schick, A., Stuhler, K., Warscheid, B., Meyer, H.E., Schmiegel, W., and Schwarte-Waldhoff, I. (2010). Immunoscreening of the extracellular proteome of colorectal cancer cells. BMC Cancer *10*, 70.

Koga, Y., Yasunaga, M., Moriya, Y., Akasu, T., Fujita, S., Yamamoto, S., Baba, H., and Matsumura, Y. (2009). Detection of the DNA point mutation of colorectal cancer cells isolated from feces stored under different conditions. Jpn. J. Clin. Oncol. *39*, 62–69.

Kohne, C.H., Hofheinz, R., Mineur, L., Letocha, H., Greil, R., Thaler, J., Fernebro, E., Gamelin, E., Decosta, L., and Karthaus, M. (2012). First-line panitumumab plus irinotecan/5-fluorouracil/leucovorin treatment in patients with metastatic colorectal cancer. J. Cancer Res. Clin. Oncol. *138*, 65–72.

Kominami, K., Nagasaka, T., Cullings, H.M., Hoshizima, N., Sasamoto, H., Young, J., Leggett, B.A., Tanaka, N., and Matsubara, N. (2009). Methylation in p14(ARF) is frequently observed in colorectal cancer with low-level microsatellite instability. J. Int. Med. Res. *37*, 1038–1045.

Konigsberg, R., Gneist, M., Jahn-Kuch, D., Pfeiler, G., Hager, G., Hudec, M., Dittrich, C., and Zeillinger, R. (2010). Circulating tumor cells in metastatic colorectal cancer: efficacy and feasibility of different enrichment methods. Cancer Lett. *293*, 117–123.

Konigsberg, R., Hulla, W., Klimpfinger, M., Reiner-Concin, A., Steininger, T., Buchler, W., Terkola, R., and Dittrich, C. (2012). Clinical and economic aspects of KRAS mutational status as predictor for epidermal growth factor receptor inhibitor therapy in metastatic colorectal cancer patients. Oncology *81*, 359–364.

La Thangue, N.B., and Kerr, D.J. (2011). Predictive biomarkers: a paradigm shift towards personalized cancer medicine. Nat. Rev. Clin. Oncol. *8*, 587–596.

Lai, D., King, T.M., Moye, L.A., and Wei, Q. (2003). Sample size for biomarker studies: more subjects or more measurements per subject? Ann. Epidemiol. *13*, 204–208.

Langers, A.M., Sier, C.F., Hawinkels, L.J., Kubben, F.J., van Duijn, W., van der Reijden, J.J., Lamers, C.B., Hommes, D.W., and Verspaget, H.W. (2008). MMP-2 genophenotype is prognostic for colorectal cancer survival, whereas MMP-9 is not. Br. J. Cancer *98*, 1820–1823.

Laszlo, L. (2010). Predictive and prognostic factors in the complex treatment of patients with colorectal cancer. Magy Onkol. *54*, 383–394.

Laurent-Puig, P., Cayre, A., Manceau, G., Buc, E., Bachet, J.B., Lecomte, T., Rougier, P., Lievre, A., Landi, B., Boige, V., *et al.* (2009). Analysis of PTEN, BRAF, and EGFR status in determining benefit from cetuximab therapy in wild-type KRAS metastatic colon cancer. J. Clin. Oncol. *27*, 5924–5930.

Lecomte, T., Ceze, N., Dorval, E., and Laurent-Puig, P. (2010). Circulating free tumor DNA and colorectal cancer. Gastroenterol. Clin. Biol. *34*, 662–681.

Lee, S., Cho, N.Y., Choi, M., Yoo, E.J., Kim, J.H., and Kang, G.H. (2008). Clinicopathological features of CpG island methylator phenotype-positive colorectal cancer and its adverse prognosis in relation to KRAS/BRAF mutation. Pathol. Int. *58*, 104–113.

Lengauer, C., Kinzler, K.W., and Vogelstein, B. (1997). Genetic instability in colorectal cancers. Nature *386*, 623–627.

Lenhard, K., Bommer, G.T., Asutay, S., Schauer, R., Brabletz, T., Goke, B., Lamerz, R., and Kolligs, F.T. (2005). Analysis of promoter methylation in stool: a novel method for the detection of colorectal cancer. Clin. Gastroenterol. Hepatol. 3, 142–149.

Levin, B., Lieberman, D.A., McFarland, B., Andrews, K.S., Brooks, D., Bond, J., Dash, C., Giardiello, F.M., Glick, S., Johnson, D., *et al.* (2008). Screening and surveillance for the early detection of colorectal cancer and adenomatous polyps, 2008: a joint guideline from the American Cancer Society, the US Multi-Society Task Force on Colorectal Cancer, and the American College of Radiology. Gastroenterology *134*, 1570–1595.

Li, X.H., Zheng, H.C., Takahashi, H., Masuda, S., Yang, X.H., and Takano, Y. (2009). PTEN expression and mutation in colorectal carcinomas. Oncol. Rep. *22*, 757–764.

Lievre, A., Bachet, J.B., Le Corre, D., Boige, V., Landi, B., Emile, J.F., Cote, J.F., Tomasic, G., Penna, C., Ducreux, M., *et al.* (2006). KRAS mutation status is predictive of response to cetuximab therapy in colorectal cancer. Cancer Res. *66*, 3992–3995.

Lin, A.Y., Buckley, N.S., Lu, A.T., Kouzminova, N.B., and Salpeter, S.R. (2011a). Effect of KRAS mutational status in advanced colorectal cancer on the outcomes of anti-epidermal growth factor receptor monoclonal antibody therapy: a systematic review and meta-analysis. Clin. Colorectal Cancer *10*, 63–69.

Lin, J.S., Webber, E.M., Senger, C.A., Holmes, R.S., and Whitlock, E.P. (2011b). Systematic review of pharmacogenetic testing for predicting clinical benefit to anti-EGFR therapy in metastatic colorectal cancer. Am. J. Cancer Res. *1*, 650–662.

Lind, G.E., Danielsen, S.A., Ahlquist, T., Merok, M.A., Andresen, K., Skotheim, R.I., Hektoen, M., Rognum, T.O., Meling, G.I., Hoff, G., *et al.* (2011a). Identification of an epigenetic biomarker panel with high sensitivity and specificity for colorectal cancer and adenomas. Mol. Cancer *10*, 85.

Lind, G.E., Raiborg, C., Danielsen, S.A., Rognum, T.O., Thiis-Evensen, E., Hoff, G., Nesbakken, A., Stenmark, H., and Lothe, R.A. (2011b). SPG20, a novel biomarker for early detection of colorectal cancer, encodes a regulator of cytokinesis. Oncogene *30*, 3967–3978.

Liu, C., Pan, C., Shen, J., Wang, H., and Yong, L. (2011). MALDI-TOF MS combined with magnetic beads for detecting serum protein biomarkers and establishment of boosting decision tree model for diagnosis of colorectal cancer. Int. J. Med. Sci. *8*, 39–47.

Liu, L., Cao, Y., Tan, A., Liao, C., and Gao, F. (2010). Cetuximab-based therapy versus non-cetuximab therapy for advanced cancer: a meta-analysis of 17 randomized controlled trials. Cancer Chemother. Pharmacol. *65*, 849–861.

Losi, L., Baisse, B., Bouzourene, H., and Benhattar, J. (2005). Evolution of intratumoral genetic heterogeneity during colorectal cancer progression. Carcinogenesis *26*, 916–922.

Lothe, R.A., Andersen, S.N., Hofstad, B., Meling, G.I., Peltomaki, P., Heim, S., Brogger, A., Vatn, M., Rognum, T.O., and Borresen, A.L. (1995). Deletion of 1p loci and microsatellite instability in colorectal polyps. Genes Chromosomes Cancer *14*, 182–188.

Luo, X., Burwinkel, B., Tao, S., and Brenner, H. (2011). MicroRNA signatures: novel biomarker for colorectal cancer? Cancer Epidemiol. Biomarkers Prev. *20*, 1272–1286.

Lurje, G., Zhang, W., and Lenz, H.J. (2007). Molecular prognostic markers in locally advanced colon cancer. Clin. Colorectal Cancer *6*, 683–690.

McCormick, F. (1995). Ras-related proteins in signal transduction and growth control. [Review] [8 refs]. Mol. Reprod. Dev. *42*, 500–506.

Maheswaran, S., and Haber, D.A. (2010). Circulating tumor cells: a window into cancer biology and metastasis. Curr. Opin. Genet. Dev. *20*, 96–99.

Maithel, S.K., Gonen, M., Ito, H., Dematteo, R.P., Allen, P.J., Fong, Y., Blumgart, L.H., Jarnagin, W.R., and D'Angelica, M.I. (2011). Improving the clinical risk score: an analysis of molecular biomarkers in the era of modern chemotherapy for resectable hepatic colorectal cancer metastases. Surgery *151*, 162–170.

Maley, C.C., Galipeau, P.C., Finley, J.C., Wongsurawat, V.J., Li, X., Sanchez, C.A., Paulson, T.G., Blount, P.L., Risques, R.A., Rabinovitch, P.S., *et al.* (2006). Genetic clonal diversity predicts progression to esophageal adenocarcinoma. Nat. Genet. *38*, 468–473.

Matsuzaki, K., Deng, G., Tanaka, H., Kakar, S., Miura, S., and Kim, Y.S. (2005). The relationship between global methylation level, loss of heterozygosity, and microsatellite instability in sporadic colorectal cancer. Clin. Cancer Res. *11*, 8564–8569.

Maughan, T.S., Adams, R.A., Smith, C.G., Meade, A.M., Seymour, M.T., Wilson, R.H., Idziaszczyk, S., Harris, R., Fisher, D., Kenny, S.L., *et al.* (2011). Addition of cetuximab to oxaliplatin-based first-line combination chemotherapy for treatment of advanced colorectal cancer: results of the randomised phase 3 MRC COIN trial. Lancet *377*, 2103–2114.

Merlos-Suarez, A., Barriga, F.M., Jung, P., Iglesias, M., Cespedes, M.V., Rossell, D., Sevillano, M., Hernando-Momblona, X., da Silva-Diz, V., Munoz, P., Clevers, H., Sancho, E., Mangues, R., and Batlle, E. (2011). The intestinal stem cell signature identifies colorectal cancer stem cells and predicts disease relapse. Cell Stem Cell *8*, 511–524.

Minamoto, T., Mai, M., and Ronai, Z. (2000). K-ras mutation: early detection in molecular diagnosis and risk assessment of colorectal, pancreas, and lung cancers--a review. Cancer Detect. Prev. *24*, 1–12.

Miyaki, Y., Suzuki, K., Koizumi, K., Kato, T., Saito, M., Kamiyama, H., Maeda, T., Shibata, K., Shiya, N., and Konishi, F. (2012). Identification of a potent epigenetic biomarker for resistance to camptothecin and poor outcome to irinotecan-based chemotherapy in colon cancer. Int. J. Oncol. *40*, 217–226.

Molinari, F., Felicioni, L., Buscarino, M., De Dosso, S., Buttitta, F., Malatesta, S., Movilia, A., Luoni, M., Boldorini, R., Alabiso, O., *et al.* (2011). Increased detection sensitivity for KRAS mutations enhances the prediction of anti-EGFR monoclonal antibody resistance in metastatic colorectal cancer. Clin. Cancer Res. *17*, 4901–4914.

Mori, Y., Selaru, F.M., Sato, F., Yin, J., Simms, L.A., Xu, Y., Olaru, A., Deacu, E., Wang, S., Taylor, J.M., *et al.* (2003). The impact of microsatellite instability on the molecular phenotype of colorectal tumors. Cancer Res. *63*, 4577–4582.

Mori, Y., Olaru, A.V., Cheng, Y., Agarwal, R., Yang, J., Luvsanjav, D., Yu, W., Selaru, F.M., Hutfless, S., Lazarev, M., *et al.* (2011). Novel candidate colorectal cancer biomarkers identified by methylation microarray-based scanning. Endocr. Relat. Cancer *18*, 465–478.

Mouliere, F., Robert, B., Arnau Peyrotte, E., Del Rio, M., Ychou, M., Molina, F., Gongora, C., and Thierry, A.R. (2011). High fragmentation characterizes tumour-derived circulating DNA. PLoS One *6*, e23418.

Mulder, K., Scarfe, A., Chua, N., and Spratlin, J. (2011). The role of bevacizumab in colorectal cancer: understanding its benefits and limitations. Expert Opin. Biol. Ther. *11*, 405–413.

Murphy, J., Dorudi, S., and Bustin, S.A. (2007). Molecular staging of colorectal cancer: new paradigm or waste of time? Expert Opin. Med. Diagn. *1*, 31–45.

Nagasaka, T., Sasamoto, H., Notohara, K., Cullings, H.M., Takeda, M., Kimura, K., Kambara, T., MacPhee, D.G., Young, J., Leggett, B.A., *et al.* (2004). Colorectal cancer with mutation in BRAF, KRAS, and wild-type with respect to both oncogenes showing different patterns of DNA methylation. J. Clin. Oncol. *22*, 4584–4594.

Nakamura, N., and Takenaga, K. (1998). Hypomethylation of the metastasis-associated S100A4 gene correlates with gene activation in human colon adenocarcinoma cell lines. Clin. Exp. Metastasis *16*, 471–479.

Nannini, M., Pantaleo, M.A., Maleddu, A., Astolfi, A., Formica, S., and Biasco, G. (2009). Gene expression profiling in colorectal cancer using microarray technologies: results and perspectives. Cancer Treat. Rev. 35, 201–209.

Nannini, M., Pantaleo, M.A., Paterini, P., Piazzi, G., Ceccarelli, C., La Rovere, S., Maleddu, A., and Biasco, G. (2011). Molecular detection of epidermal growth factor receptor in colorectal cancer: does it still make sense? Colorectal Dis. *13*, 542–548.

Nechvatal, J.M., Ram, J.L., Basson, M.D., Namprachan, P., Niec, S.R., Badsha, K.Z., Matherly, L.H., Majumdar, A.P., and Kato, I. (2008). Fecal collection, ambient preservation, and DNA extraction for PCR amplification of bacterial and human markers from human feces. J. Microbiol. Methods *72*, 124–132.

Nielsen, H.J., Jakobsen, K.V., Christensen, I.J., and Brunner, N. (2011). Screening for colorectal cancer: possible improvements by risk assessment evaluation? Scand. J. Gastroenterol. *46*, 1283–1294.

Nordlinger, B., Sorbye, H., Glimelius, B., Poston, G.J., Schlag, P.M., Rougier, P., Bechstein, W.O., Primrose, J.N., Walpole, E.T., Finch-Jones, M., *et al.* (2008). Perioperative chemotherapy with FOLFOX4 and surgery versus surgery alone for resectable liver metastases from colorectal cancer (EORTC Intergroup trial 40983): a randomised controlled trial. Lancet *371*, 1007–1016.

Normanno, N., Tejpar, S., Morgillo, F., De Luca, A., Van Cutsem, E., and Ciardiello, F. (2009). Implications for KRAS status and EGFR-targeted therapies in metastatic CRC. Nat. Rev. Clin. Oncol. *6*, 519–527.

O'Brien, M.J. (2007). Hyperplastic and serrated polyps of the colorectum. Gastroenterol. Clin. North Am. *36*, 947–968.

O'Brien, M.J., Yang, S., Mack, C., Xu, H., Huang, C.S., Mulcahy, E., Amorosino, M., and Farraye, F.A. (2006). Comparison of microsatellite instability, CpG island methylation phenotype, BRAF and KRAS status in serrated polyps and traditional adenomas indicates separate pathways to distinct colorectal carcinoma end points. Am. J. Surg. Pathol. *30*, 1491–1501.

Ocvirk, J., Brodowicz, T., Wrba, F., Ciuleanu, T.E., Kurteva, G., Beslija, S., Koza, I., Papai, Z., Messinger, D., Yilmaz, U., *et al.* (2010). Cetuximab plus FOLFOX6 or FOLFIRI in metastatic colorectal cancer: CECOG trial. World J. Gastroenterol. *16*, 3133–3143.

O'Dwyer, D., Ralton, L.D., O'Shea, A., and Murray, G.I. (2011). The proteomics of colorectal cancer: identification of a protein signature associated with prognosis. PLoS One *6*, e27718.

Ogino, S., and Goel, A. (2008). Molecular classification and correlates in colorectal cancer. J. Mol. Diagn. *10*, 13–27.

Ogino, S., Odze, R.D., Kawasaki, T., Brahmandam, M., Kirkner, G.J., Laird, P.W., Loda, M., and Fuchs, C.S. (2006). Correlation of pathologic features with CpG island methylator phenotype (CIMP) by quantitative DNA methylation analysis in colorectal carcinoma. Am. J. Surg. Pathol. *30*, 1175–1183.

Ogino, S., Meyerhardt, J.A., Kawasaki, T., Clark, J.W., Ryan, D.P., Kulke, M.H., Enzinger, P.C., Wolpin, B.M., Loda, M., and Fuchs, C.S. (2007). CpG island methylation, response to combination chemotherapy, and patient survival in advanced microsatellite stable colorectal carcinoma. Virchows Arch. *450*, 529–537.

Ogino, S., Meyerhardt, J.A., Irahara, N., Niedzwiecki, D., Hollis, D., Saltz, L.B., Mayer, R.J., Schaefer, P., Whittom, R., Hantel, A., *et al.* (2009a). KRAS mutation in stage III colon cancer and clinical outcome following intergroup trial CALGB 89803. Clin. Cancer Res. *15*, 7322–7329.

Ogino, S., Nosho, K., Kirkner, G.J., Kawasaki, T., Meyerhardt, J.A., Loda, M., Giovannucci, E.L., and Fuchs, C.S. (2009b). CpG island methylator phenotype, microsatellite instability, BRAF mutation and clinical outcome in colon cancer. Gut *58*, 90–96.

Ogino, S., Nosho, K., Kirkner, G.J., Shima, K., Irahara, N., Kure, S., Chan, A.T., Engelman, J.A., Kraft, P., Cantley, L.C., *et al.* (2009c). PIK3CA mutation is associated with poor prognosis among patients with curatively resected colon cancer. J. Clin. Oncol. *27*, 1477–1484.

Ogino, S., Chan, A.T., Fuchs, C.S., and Giovannucci, E. (2011). Molecular pathological epidemiology of colorectal neoplasia: an emerging transdisciplinary and interdisciplinary field. Gut *60*, 397–411.

Ogreid, D., and Hamre, E. (2007). Stool DNA analysis detects premorphological colorectal neoplasia: a case report. Eur. J. Gastroenterol. Hepatol. *19*, 725–727.

Olson, S., Robinson, S., and Griffin, R. (2009). Accelerating the Development of Biomarkers for Drug Safety: Workshop Summary (The National Academies Press, Washington DC).

Onouchi, S., Matsushita, H., Moriya, Y., Akasu, T., Fujita, S., Yamamoto, S., Hasegawa, H., Kitagawa, Y., and Matsumura, Y. (2008). New method for colorectal cancer diagnosis based on SSCP analysis of DNA from exfoliated colonocytes in naturally evacuated feces. Anticancer Res. *28*, 145–150.

Osborn, N.K., and Ahlquist, D.A. (2005). Stool screening for colorectal cancer: molecular approaches. Gastroenterology *128*, 192–206.

Papadimitriou, C.A., Papakostas, P., Karina, M., Malettou, L., Dimopoulos, M.A., Pentheroudakis, G., Samantas, E., Bamias, A., Miliaras, D., Basdanis, G., *et al.* (2011). A randomized phase III trial of adjuvant chemotherapy with irinotecan, leucovorin and fluorouracil versus leucovorin and fluorouracil for stage II and III colon cancer: a Hellenic Cooperative Oncology Group study. BMC Med. *9*, 10.

Papadopoulos, N., and Lindblom, A. (1997). Molecular basis of HNPCC: mutations of MMR genes. Hum. Mutat. *10*, 89–99.

Park, K.S., Kim, S.J., Kim, K.H., and Kim, J.C. (2011). Clinical characteristics of TIMP2, MMP2, and MMP9 gene polymorphisms in colorectal cancer. J. Gastroenterol. Hepatol. *26*, 391–397.

Pawa, N., Arulampalam, T., and Norton, J.D. (2011). Screening for colorectal cancer: established and emerging modalities. Nat. Rev. Gastroenterol. Hepatol. *8*, 711–722.

Pawlik, T.M., and Vauthey, J.N. (2008). Surgical margins during hepatic surgery for colorectal liver metastases: complete resection not millimeters defines outcome. Ann. Surg. Oncol. *15*, 677–679.

Pawlik, T.M., Raut, C.P., and Rodriguez-Bigas, M.A. (2004). Colorectal carcinogenesis: MSI-H versus MSI-L. Dis. Markers. *20*, 199–206.

Peeters, M., Price, T.J., Cervantes, A., Sobrero, A.F., Ducreux, M., Hotko, Y., Andre, T., Chan, E., Lordick, F., Punt, C.J., *et al.* (2010). Randomized phase III study of panitumumab with fluorouracil, leucovorin, and irinotecan (FOLFIRI) compared with FOLFIRI alone as second-line treatment in patients with metastatic colorectal cancer. J. Clin. Oncol. *28*, 4706–4713.

Pilozzi, E., Ferri, M., Onelli, M.R., Mercantini, P., Corigliano, N., Duranti, E., Dionisi, L., Felicioni, F., Virgilio, E., Ziparo, V., *et al.* (2011). Prognostic significance of 18q LOH in sporadic colorectal carcinoma. Am. Surg. *77*, 38–43.

Pinedo, H.M., and Peters, G.F. (1988). Fluorouracil: biochemistry and pharmacology. J. Clin. Oncol. *6*, 1653–1664.

Pollock, C.B., Shirasawa, S., Sasazuki, T., Kolch, W., and Dhillon, A.S. (2005). Oncogenic K-RAS is required to maintain changes in cytoskeletal organization, adhesion, and motility in colon cancer cells. Cancer Res. *65*, 1244–1250.

Popat, S., and Houlston, R.S. (2005). A systematic review and meta-analysis of the relationship between chromosome 18q genotype, DCC status and colorectal cancer prognosis. Eur. J. Cancer *41*, 2060–2070.

Popat, S., Hubner, R., and Houlston, R.S. (2005). Systematic review of microsatellite instability and colorectal cancer prognosis. J. Clin. Oncol. *23*, 609–618.

Poulogiannis, G., Ichimura, K., Hamoudi, R.A., Luo, F., Leung, S.Y., Yuen, S.T., Harrison, D.J., Wyllie, A.H., and Arends, M.J. (2010). Prognostic relevance of DNA copy number changes in colorectal cancer. J. Pathol. *220*, 338–347.

Pozzo, C., Basso, M., Cassano, A., Quirino, M., Schinzari, G., Trigila, N., Vellone, M., Giuliante, F., Nuzzo, G., and Barone, C. (2004). Neoadjuvant treatment of unresectable liver disease with irinotecan and 5-fluorouracil plus folinic acid in colorectal cancer patients. Ann. Oncol. *15*, 933–939.

Prenen, H., Tejpar, S., and Van Cutsem, E. (2009). Impact of molecular markers on treatment selection in advanced colorectal cancer. Eur. J. Cancer 45 (Suppl. 1), 70–78.

Price, T.J., Hardingham, J.E., Lee, C.K., Weickhardt, A., Townsend, A.R., Wrin, J.W., Chua, A., Shivasami, A., Cummins, M.M., Murone, C., *et al.* (2011). Impact of KRAS and BRAF gene mutation status on outcomes from the phase III AGITG MAX trial of capecitabine alone or in combination with bevacizumab and mitomycin in advanced colorectal cancer. J. Clin. Oncol. *29*, 2675–2682.

Pritchard, C.C., and Grady, W.M. (2011). Colorectal cancer molecular biology moves into clinical practice. Gut *60*, 116–129.

Raj, A., Peskin, C.S., Tranchina, D., Vargas, D.Y., and Tyagi, S. (2006). Stochastic mRNA synthesis in mammalian cells. PLoS Biol. *4*, e309.

Rajagopalan, H., Bardelli, A., Lengauer, C., Kinzler, K.W., Vogelstein, B., and Velculescu, V.E. (2002). Tumorigenesis: RAF/RAS oncogenes and mismatch-repair status. Nature *418*, 934.

Raoul, J.L., Van Laethem, J.L., Peeters, M., Brezault, C., Husseini, F., Cals, L., Nippgen, J., Loos, A.H., and Rougier, P. (2009). Cetuximab in combination with irinotecan/5-fluorouracil/folinic acid (FOLFIRI) in the initial treatment of metastatic colorectal cancer: a multicentre two-part phase I/II study. BMC Cancer *9*, 112.

Raser, J.M., and O'Shea, E.K. (2005). Noise in gene expression: origins, consequences, and control. Science *309*, 2010–2013.

Rengucci, C., Maiolo, P., Saragoni, L., Zoli, W., Amadori, D., and Calistri, D. (2001). Multiple detection of genetic alterations in tumors and stool. Clin. Cancer Res. *7*, 590–593.

Ribic, C.M., Sargent, D.J., Moore, M.J., Thibodeau, S.N., French, A.J., Goldberg, R.M., Hamilton, S.R., Laurent-Puig, P., Gryfe, R., Shepherd, L.E., *et al.* (2003). Tumor microsatellite-instability status as a predictor of benefit from fluorouracil-based adjuvant chemotherapy for colon cancer. N. Engl. J. Med. *349*, 247–257.

van Rijnsoever, M., Grieu, F., Elsaleh, H., Joseph, D., and Iacopetta, B. (2002). Characterisation of colorectal cancers showing hypermethylation at multiple CpG islands. Gut *51*, 797–802.

Rodriguez, J., Frigola, J., Vendrell, E., Risques, R.A., Fraga, M.F., Morales, C., Moreno, V., Esteller, M., Capella, G., Ribas, M., *et al.* (2006). Chromosomal instability correlates with genome-wide DNA demethylation in human primary colorectal cancers. Cancer Res. *66*, 8462–9468.

Roock, W.D., Vriendt, V.D., Normanno, N., Ciardiello, F., and Tejpar, S. (2010). KRAS, BRAF, PIK3CA, and PTEN mutations: implications for targeted therapies in metastatic colorectal cancer. Lancet. Oncol. *12*, 594–603.

Ross, J.S. (2011). Biomarker-based selection of therapy for colorectal cancer. Biomark. Med. *5*, 319–332.

Roth, A.D., Tejpar, S., Delorenzi, M., Yan, P., Fiocca, R., Klingbiel, D., Dietrich, D., Biesmans, B., Bodoky, G., Barone, C., *et al.* (2010). Prognostic role of KRAS and BRAF in stage II and III resected colon cancer: results of the translational study on the PETACC-3, EORTC 40993, SAKK 60–00 trial. J. Clin. Oncol. *28*, 466–474.

Roukos, D.H., Murray, S., and Briasoulis, E. (2007). Molecular genetic tools shape a roadmap towards a more accurate prognostic prediction and personalized management of cancer. Cancer Biol. Ther. *6*, 308–312.

Saif, M.W., Kaley, K., Chu, E., and Copur, M.S. (2010). Safety and efficacy of panitumumab therapy after progression with cetuximab: experience at two institutions. Clin. Colorectal Cancer *9*, 315–318.

Saltz, L.B., Cox, J.V., Blanke, C., Rosen, L.S., Fehrenbacher, L., Moore, M.J., Maroun, J.A., Ackland, S.P., Locker, P.K., Pirotta, N., *et al.* (2000). Irinotecan plus fluorouracil and leucovorin for metastatic colorectal cancer. Irinotecan Study Group. N. Engl. J. Med. *343*, 905–914.

Samowitz, W.S., Albertsen, H., Herrick, J., Levin, T.R., Sweeney, C., Murtaugh, M.A., Wolff, R.K., and Slattery, M.L. (2005a). Evaluation of a large, population-based sample supports a CpG island methylator phenotype in colon cancer. Gastroenterology *129*, 837–845.

Samowitz, W.S., Sweeney, C., Herrick, J., Albertsen, H., Levin, T.R., Murtaugh, M.A., Wolff, R.K., and Slattery, M.L. (2005b). Poor survival associated with the BRAF V600E mutation in microsatellite-stable colon cancers. Cancer Res. *65*, 6063–6069.

Samuels, M.A. (2004). Cytoprotection in head and neck cancer: issues in oral care. J. Support Oncol. *2*, 9–12.

Sanders, M.A., and Majumdar, A.P. (2011). Colon cancer stem cells: implications in carcinogenesis. Front. Biosci. *16*, 1651–1662.

Sartore-Bianchi, A., Di Nicolantonio, F., Nichelatti, M., Molinari, F., De Dosso, S., Saletti, P., Martini, M., Cipani, T., Marrapese, G., Mazzucchelli, L., *et al.* (2009). Multi-determinants analysis of molecular alterations for predicting clinical benefit to EGFR-targeted monoclonal antibodies in colorectal cancer. PLoS One *4*, e7287.

Schafer, R., and Sers, C. (2011). RAS oncogene-mediated deregulation of the transcriptome: from molecular signature to function. Adv. Enzyme Regul. *51*, 126–136.

Schwarzenbach, H., Stoehlmacher, J., Pantel, K., and Goekkurt, E. (2008). Detection and monitoring of cell-free DNA in blood of patients with colorectal cancer. Ann. N. Y. Acad. Sci. *1137*, 190–196.

Shannon, B.A., and Iacopetta, B.J. (2001). Methylation of the hMLH1, p16, and MDR1 genes in colorectal carcinoma: associations with clinicopathological features. Cancer Lett. *167*, 91–97.

Shaukat, A., Arain, M., Thaygarajan, B., Bond, J.H., and Sawhney, M. (2010). Is BRAF mutation associated with interval colorectal cancers? Dig. Dis. Sci. *55*, 2352–2356.

Shen, L., Catalano, P.J., Benson, A.B.R., O'Dwyer, P., Hamilton, S.R., and Issa, J.P. (2007a). Association between DNA methylation and shortened survival in patients with advanced colorectal cancer treated with 5-fluorouracil based chemotherapy. Clin. Cancer Res. *13*, 6093–6098.

Shen, L., Toyota, M., Kondo, Y., Lin, E., Zhang, L., Guo, Y., Hernandez, N.S., Chen, X., Ahmed, S., Konishi, K., *et al.* (2007b). Integrated genetic and epigenetic analysis identifies three different subclasses of colon cancer. Proc. Natl. Acad. Sci. U.S.A. *104*, 18654–18659.

Shima, K., Morikawa, T., Baba, Y., Nosho, K., Suzuki, M., Yamauchi, M., Hayashi, M., Giovannucci, E., Fuchs, C.S., and Ogino, S. (2011). MGMT promoter methylation, loss of expression and prognosis in 855 colorectal cancers. Cancer Causes Control *22*, 301–309.

Sidransky, D., Tokino, T., Hamilton, S.R., Kinzler, K.W., Levin, B., Frost, P., and Vogelstein, B. (1992). Identification of ras oncogene mutations in the stool of patients with curable colorectal tumors. Science *256*, 102–105.

Siena, S., Sartore-Bianchi, A., Di Nicolantonio, F., Balfour, J., and Bardelli, A. (2009). Biomarkers predicting clinical outcome of epidermal growth factor receptor-targeted therapy in metastatic colorectal cancer. J. Natl. Cancer Inst. *101*, 1308–1324.

Sjoblom, T., Jones, S., Wood, L.D., Parsons, D.W., Lin, J., Barber, T.D., Mandelker, D., Leary, R.J., Ptak, J., Silliman, N., *et al.* (2006). The consensus coding sequences of human breast and colorectal cancers. Science *314*, 268–274.

Sleijfer, S., Gratama, J.W., Sieuwerts, A.M., Kraan, J., Martens, J.W., and Foekens, J.A. (2007). Circulating tumour cell detection on its way to routine diagnostic implementation? Eur. J. Cancer *43*, 2645–2650.

Spindler, K.L., Pallisgaard, N., Vogelius, I., and Jakobsen, A. (2012). Quantitative cell free DNA, KRAS and BRAF mutations in plasma from patients with metastatic colorectal cancer during treatment with cetuximab and irinotecan. Clin. Cancer Res. *18*, 1177–1185.

Spisak, S., Galamb, B., Wichmann, B., Sipos, F., Galamb, O., Solymosi, N., Nemes, B., Tulassay, Z., and Molnar, B. (2009). Tissue microarray (TMA) validated progression markers in colorectal cancer using antibody microarrays. Orv. Hetil. *150*, 1607–1613.

Stintzing, S., Fischer von Weikersthal, L., Decker, T., Vehling-Kaiser, U., Jager, E., Heintges, T., Stoll, C., Giessen, C., Modest, D.P., Neumann, J., *et al.* (2012). FOLFIRI plus cetuximab versus FOLFIRI plus bevacizumab as first-line treatment for patients with metastatic colorectal cancer-subgroup analysis of patients with KRAS: mutated tumours in the randomised German AIO study KRK-0306. Ann Oncol *23*, 1693–1699.

Sun, Y.F., Yang, X.R., Zhou, J., Qiu, S.J., Fan, J., and Xu, Y. (2011). Circulating tumor cells: advances in detection methods, biological issues, and clinical relevance. J. Cancer Res. Clin. Oncol. *137*, 1151–1173.

Svrcek, M., Buhard, O., Colas, C., Coulet, F., Dumont, S., Massaoudi, I., Lamri, A., Hamelin, R., Cosnes, J., Oliveira, C., *et al.* (2010). Methylation tolerance due to an O6-methylguanine DNA methyltransferase (MGMT) field defect in the colonic mucosa: an initiating step in the development of mismatch repair-deficient colorectal cancers. Gut *59*, 1516–1526.

Takemasa, I., Higuchi, H., Yamamoto, H., Sekimoto, M., Tomita, N., Nakamori, S., Matoba, R., Monden, M., and Matsubara, K. (2001). Construction of preferential cDNA microarray specialized for human colorectal carcinoma: molecular sketch of colorectal cancer. Biochem. Biophys. Res. Commun. *285*, 1244–1249.

Tanaka, T., Watanabe, T., Kazama, Y., Tanaka, J., Kanazawa, T., Kazama, S., and Nagawa, H. (2006). Chromosome 18q deletion and Smad4 protein inactivation correlate with liver metastasis: A study matched for T- and N-classification. Br. J. Cancer *95*, 1562–1567.

Tang, R., Changchien, C.R., Wu, M.C., Fan, C.W., Liu, K.W., Chen, J.S., Chien, H.T., and Hsieh, L.L. (2004). Colorectal cancer without high microsatellite instability and chromosomal instability--an alternative genetic pathway to human colorectal cancer. Carcinogenesis *25*, 841–846.

Tejpar, S., Bertagnolli, M., Bosman, F., Lenz, H.J., Garraway, L., Waldman, F., Warren, R., Bild, A., Collins-Brennan, D., Hahn, H., *et al.* (2010). Prognostic and predictive biomarkers in resected colon cancer: current status and future perspectives for integrating genomics into biomarker discovery. Oncologist *15*, 390–404.

Thijssen, M.A., Swinkels, D.W., Ruers, T.J., and de Kok, J.B. (2002). Difference between free circulating plasma and serum DNA in patients with colorectal liver metastases. Anticancer Res. *22*, 421–425.

Tol, J., Koopman, M., Cats, A., Rodenburg, C.J., Creemers, G.J., Schrama, J.G., Erdkamp, F.L., Vos, A.H., van Groeningen, C.J., Sinnige, H.A., *et al.* (2009a). Chemotherapy, bevacizumab, and cetuximab in metastatic colorectal cancer. N. Engl. J. Med. *360*, 563–572.

Tol, J., Nagtegaal, I.D., and Punt, C.J. (2009b). BRAF mutation in metastatic colorectal cancer. N. Engl. J. Med. *361*, 98–99.

Tol, J., Dijkstra, J.R., Klomp, M., Teerenstra, S., Dommerholt, M., Vink-Borger, M.E., van Cleef, P.H., van Krieken, J.H., Punt, C.J., and Nagtegaal, I.D. (2010). Markers for EGFR pathway activation as predictor of outcome in metastatic colorectal cancer patients treated with or without cetuximab. Eur. J. Cancer *46*, 1997–2009.

Tomlinson, I., Halford, S., Aaltonen, L., Hawkins, N., and Ward, R. (2002). Does MSI-low exist? J. Pathol. *197*, 6–13.

Toth, L., Andras, C., Molnar, C., Tanyi, M., Csiki, Z., Molnar, P., and Szanto, J. (2011). Investigation of beta-catenin and E-cadherin expression in Dukes B2 stage colorectal cancer with tissue microarray method. Is it a marker of metastatic potential in rectal cancer? Pathol Oncol Res *18*, 429–437.

Toyota, M., Ahuja, N., Ohe-Toyota, M., Herman, J.G., Baylin, S.B., and Issa, J.P. (1999). CpG island methylator phenotype in colorectal cancer. Proc. Natl. Acad. Sci. U.S.A. *96*, 8681–8686.

Toyota, M., Ohe-Toyota, M., Ahuja, N., and Issa, J.P. (2000). Distinct genetic profiles in colorectal tumors with or without the CpG island methylator phenotype. Proc. Natl. Acad. Sci. U.S.A. *97*, 710–715.

Tran, B., Kopetz, S., Tie, J., Gibbs, P., Jiang, Z.Q., Lieu, C.H., Agarwal, A., Maru, D.M., Sieber, O., and Desai, J. (2011). Impact of BRAF mutation and microsatellite instability on the pattern of metastatic spread and prognosis in metastatic colorectal cancer. Cancer *117*, 4623–4632.

Van Cutsem, E., Findlay, M., Osterwalder, B., Kocha, W., Dalley, D., Pazdur, R., Cassidy, J., Dirix, L., Twelves, C., Allman, D., *et al.* (2000). Capecitabine, an oral fluoropyrimidine carbamate with substantial activity in advanced colorectal cancer: results of a randomized phase II study. J. Clin. Oncol. *18*, 1337–1345.

Van Cutsem, E., and Geboes, K. (2007). The multidisciplinary management of gastrointestinal cancer. The integration of cytotoxics and biologicals in the treatment of metastatic colorectal cancer. Best Pract. Res. Clin. Gastroenterol. *21*, 1089–1108.

Van Cutsem, E., Nordlinger, B., and Cervantes, A. (2010). Advanced colorectal cancer: ESMO Clinical Practice Guidelines for treatment. Ann. Oncol. *21* (Suppl. 5), v93–97.

Van Cutsem, E., Kohne, C.H., Lang, I., Folprecht, G., Nowacki, M.P., Cascinu, S., Shchepotin, I., Maurel, J., Cunningham, D., Tejpar, S., *et al.* F. (2011). Cetuximab plus irinotecan, fluorouracil, and leucovorin as first-line treatment for metastatic colorectal cancer: updated analysis of overall survival according to tumor KRAS and BRAF mutation status. J. Clin. Oncol. *29*, 2011–2019.

Van Loon, K., and Venook, A.P. (2011). Adjuvant treatment of colon cancer: what is next? Curr. Opin. Oncol. *23*, 403–409.

Van Rijnsoever, M., Elsaleh, H., Joseph, D., McCaul, K., and Iacopetta, B. (2003). CpG island methylator phenotype is an independent predictor of survival benefit from 5-fluorouracil in stage III colorectal cancer. Clin. Cancer Res. *9*, 2898–2903.

Van Schaeybroeck, S., Allen, W.L., Turkington, R.C., and Johnston, P.G. (2011). Implementing prognostic and predictive biomarkers in CRC clinical trials. Nat. Rev. Clin. Oncol. *8*, 222–232.

Vauthey, J.N., Zorzi, D., and Pawlik, T.M. (2005). Making unresectable hepatic colorectal metastases resectable--does it work? Semin. Oncol. *32*, S118–22.

de Vogel, S., Weijenberg, M.P., Herman, J.G., Wouters, K.A., de Goeij, A.F., van den Brandt, P.A., de Bruine, A.P., and van Engeland, M. (2009). MGMT and MLH1 promoter methylation versus APC, KRAS and BRAF gene mutations in colorectal cancer: indications for distinct pathways and sequence of events. Ann. Oncol. *20*, 1216–1222.

Wadlow, R.C., Hezel, A.F., Abrams, T.A., Blaszkowsky, L.S., Fuchs, C.S., Kulke, M.H., Kwak, E.L., Meyerhardt, J.A., Ryan, D.P., Szymonifka, J., *et al.* (2011). Panitumumab in patients with KRAS wild-type colorectal cancer after progression on cetuximab. Oncologist *17*, 14.

Walther, A., Houlston, R., and Tomlinson, I. (2008). Association between chromosomal instability and prognosis in colorectal cancer: a meta-analysis. Gut *57*, 941–950.

Walther, A., Johnstone, E., Swanton, C., Midgley, R., Tomlinson, I., and Kerr, D. (2009). Genetic prognostic and predictive markers in colorectal cancer. Nat. Rev. Cancer *9*, 489–499.

Wang, L., Chen, X., Li, W., and Sheng, Z. (2012). Antiepidermal growth factor receptor monoclonal antibody improves survival outcomes in the treatment of patients with metastatic colorectal cancer. Anticancer Drugs *23*, 155–160.

Wang, Q., Chaerkady, R., Wu, J., Hwang, H.J., Papadopoulos, N., Kopelovich, L., Maitra, A., Matthaei, H., Eshleman, J.R., Hruban, R.H., *et al.* (2011). Mutant proteins as cancer-specific biomarkers. Proc. Natl. Acad. Sci. U.S.A. *108*, 2444–2449.

Ward, R.L., Cheong, K., Ku, S.L., Meagher, A., O'Connor, T., and Hawkins, N.J. (2003). Adverse prognostic effect of methylation in colorectal cancer is reversed by microsatellite instability. J. Clin. Oncol. *21*, 3729–3736.

Weisenberger, D.J., Siegmund, K.D., Campan, M., Young, J., Long, T.I., Faasse, M.A., Kang, G.H., Widschwendter, M., Weener, D., Buchanan, D., *et al.* (2006). CpG island methylator phenotype underlies sporadic microsatellite instability and is tightly associated with BRAF mutation in colorectal cancer. Nat. Genet. *38*, 787–793.

Whitehall, V.L., Rickman, C., Bond, C.E., Ramsnes, I., Greco, S.A., Umapathy, A., McKeone, D., Faleiro, R.J., Buttenshaw, R.L., Worthley, D.L., *et al.* (2011). Oncogenic PIK3CA mutations in colorectal cancers and polyps. Int. J. Cancer *131*, 813–820.

de Wit, M., Jimenez, C.R., Carvalho, B., Belien, J.A., Delis-van Diemen, P.M., Mongera, S., Piersma, S.R., Vikas, M., Navani, S., Ponten, F., *et al.* (2011). Cell surface proteomics identifies glucose transporter type 1 and prion protein as candidate biomarkers for colorectal adenoma-to-carcinoma progression. Gut *61*, 855–864.

Wong, J.J., Hawkins, N.J., Ward, R.L., and Hitchins, M.P. (2011). Methylation of the 3p22 region encompassing MLH1 is representative of the CpG island methylator phenotype in colorectal cancer. Mod. Pathol. *24*, 396–411.

Wong, S.C., Chan, C.M., Ma, B.B., Hui, E.P., Ng, S.S., Lai, P.B., Cheung, M.T., Lo, E.S., Chan, A.K., Lam, M.Y., *et al.* (2009). Clinical significance of cytokeratin 20-positive circulating tumor cells detected by a refined immunomagnetic enrichment assay in colorectal cancer patients. Clin. Cancer Res. *15*, 1005–1012.

Wood, L.D., Parsons, D.W., Jones, S., Lin, J., Sjoblom, T., Leary, R.J., Shen, D., Boca, S.M., Barber, T., Ptak, J., *et al.* (2007). The genomic landscapes of human breast and colorectal cancers. Science *318*, 1108–1113.

Wu, W., Shi, Q., and Sargent, D.J. (2011). Statistical considerations for the next generation of clinical trials. Semin. Oncol. *38*, 598–604.

Xie, D., Wu, H.X., Liu, Y.D., Zeng, S.D., and Lin, F. (2007). The significance and characteristics of chromosomal abnormalities in patients with microsatellite and chromosome stable colorectal carcinoma. Zhonghua Yi Xue Za Zhi *87*, 11–15.

Yamada, Y., Jackson-Grusby, L., Linhart, H., Meissner, A., Eden, A., Lin, H., and Jaenisch, R. (2005). Opposing effects of DNA hypomethylation on intestinal and liver carcinogenesis. Proc. Natl. Acad. Sci. U.S.A. *102*, 13580–13585.

Yang, C., Xu, C., Wang, X., and Hu, X. (2012). Quantum-dot-based biosensor for simultaneous detection of biomarker and therapeutic drug: first steps toward an assay for quantitative pharmacology. Analyst *137*, 1205–1209.

Yarom, N., and Jonker, D.J. (2011). The role of the epidermal growth factor receptor in the mechanism and treatment of colorectal cancer. Discov. Med. *11*, 95–105.

Yee, Y.K., Tan, V.P., Chan, P., Hung, I.F., Pang, R., and Wong, B.C. (2009). Epidemiology of colorectal cancer in Asia. J. Gastroenterol. Hepatol. *24*, 1810–1816.

Yi, J.M., Dhir, M., Van Neste, L., Downing, S.R., Jeschke, J., Glockner, S.C., de Freitas Calmon, M., Hooker, C.M., Funes, J.M., Boshoff, C., *et al.* (2011). Genomic and epigenomic integration identifies a prognostic signature in colon cancer. Clin. Cancer Res. *17*, 1535–1545.

Yokota, T. (2011). Are KRAS/BRAF mutations potent prognostic and/or predictive biomarkers in colorectal cancers? Anticancer Agents Med Chem *12*, 163–171.

Yokota, T., Ura, T., Shibata, N., Takahari, D., Shitara, K., Nomura, M., Kondo, C., Mizota, A., Utsunomiya, S., Muro, K., *et al.* (2011). BRAF mutation is a powerful prognostic factor in advanced and recurrent colorectal cancer. Br. J. Cancer *104*, 856–862.

Young, G.P., and Bosch, L.J. (2011). Fecal tests: from blood to molecular markers. Curr. Colorectal Cancer Rep. 7, 62–70.

Zeng, Y., Qu, L.K., Meng, L., Liu, C.Y., Dong, B., Xing, X.F., Wu, J., and Shou, C.C. (2011). HIWI expression profile in cancer cells and its prognostic value for patients with colorectal cancer. Chin. Med. J. (Engl) *124*, 2144–2149.

Zhang, H., Wang, X., Ma, Q., Zhou, Z., and Fang, J. (2011). Rapid detection of low-abundance K-ras mutation in stools of colorectal cancer patients using chip-based temperature gradient capillary electrophoresis. Lab. Invest. *91*, 788–798.

Zheng, H.C., Sugawara, A., Okamoto, H., Takasawa, S., Takahashi, H., Masuda, S., and Takano, Y. (2011). Expression profile of the REG gene family in colorectal carcinoma. J. Histochem. Cytochem. *59*, 106–115.

Zhou, J., Belov, L., Solomon, M.J., Chan, C., Clarke, S.J., and Christopherson, R.I. (2011). Colorectal cancer cell surface protein profiling using an antibody microarray and fluorescence multiplexing. J. Vis. Exp. *55*, 3322.

Zhou, X.P., Loukola, A., Salovaara, R., Nystrom-Lahti, M., Peltomaki, P., de la Chapelle, A., Aaltonen, L.A., and Eng, C. (2002). PTEN mutational spectra, expression levels, and subcellular localization in microsatellite stable and unstable colorectal cancers. Am. J. Pathol. *161*, 439–447.

Zlobec, I., Bihl, M., Foerster, A., Rufle, A., and Lugli, A. (2011). Comprehensive analysis of CpG island methylator phenotype (CIMP)-high, -low, and -negative colorectal cancers based on protein marker expression and molecular features. J. Pathol. *225*, 336–343.

# Part II

# Molecular Diagnostics of Infectious Diseases: Past, Present and Future

# Molecular Diagnosis in Medical Microbiology: The Horizon Draws Near

5

Gemma L. Vanstone, Rebecca L. Gorton, Bambos M. Charalambous and Timothy D. McHugh

**Abstract**

Diagnostic microbiology is at the threshold of a new era; the advent of molecular tools, increased automation and use of informatics to allow interrogation of complex data sets has presented the opportunity to rethink our approach to diagnosis of bacterial and fungal infections. In this review we consider the advances that are in practice today in clinical laboratories. Molecular tools have huge potential in improving patient management and public health. We must be alert to the danger that the technology blinds us to the underlying biology, and it is essential that we consider the new technologies with the same critical rigour that we would apply to conventional assays and in the context of the clinical setting. The horizon is close and full of opportunity.

## Introduction

Diagnostic microbiology is at the threshold of a new era. After decades of incremental improvements in culture methodology and analysis of phenotype by biochemical tests, the advent of molecular tools has presented the opportunity to rethink our approach to diagnosis of bacterial and fungal infections (virology has consistently been ahead of the curve in this area). The ability of conventional microbiology to meet the needs of modern clinical practice has been improved by increased automation (Bourbeau and Ledebrei, 2013) and informatics allowing interpretation of complex phenotypic screens (Clark *et al.*, 2013). However, these approaches are limited by biology; organisms have a predefined growth rate which can only be manipulated within strict parameters. Furthermore the relationship between bacterial load and pathogenesis varies greatly between pathogens, for example the infective dose of *Shigella* infection is ten organisms whereas $10^6$ is the dose required in food poisoning with salmonella; for *Streptococcus pneumoniae* we consider a load of less than $10^5$ organisms to represent carriage rather than infection (Schmid-Hempel and Frank, 2007). The detection and analysis of the genome, transcriptome or proteome can not only enhance the level of detection of a pathogen in any particular clinical scenario, but can enhance the information available to contribute to patient management. This might include antimicrobial drug sensitivity, strain or species sub-typing or detection of virulence genotypes or phenotypes. In this review we focus on specific examples to illustrate the role that molecular methods can have, their advantages and limitations and, as our title suggests, the opportunities on the near horizon.

Of course, the most significant step in the molecular diagnosis of microbial infections was the development of the polymerase chain reaction (PCR) in 1985 (Saiki *et al.*, 1985), although it was the 1990s which saw the real explosion in use of PCR in diagnostic settings. In retrospect, although many published manuscripts reported the use of PCR in routine diagnostic settings (Espy *et al.*, 2006) much of this work was exploratory as the community learnt how to implement the assays and more importantly interpret the results. Examples from our own practice include the use of a previously published PCR to diagnose cerebral toxoplasmosis, in which we reported a molecular cross reaction

:Hugh *et al.*, 1995). This led us t rules around the validation hed or developed in house, ve testing against organisms …ıferential diagnosis regardless … ιaxonomic association with the target pathogen, we adopted a 'pathology not phylogeny' mantra in seeking controls for our assays. Another example is the adoption of PCR to improve the diagnostic yield in lower respiratory tract infection, first achieved using conventional PCR (Creer *et al.*, 2006) but extended with real-time PCR to address bacterial load as a pathogenic driver in chronic obstructive pulmonary disease (COPD; (Garcha *et al.*, 2012)). The limitation of conventional PCR is the need to visualize the amplimers on a gel; however, real-time PCR has enabled automation and consequently the move from low-throughput specialist tests to high-throughput tests both for screening (Wassenberg *et al.*, 2012) and routine diagnosis (Lawn *et al.*, 2013; Ling and McHugh, 2013). The patent restrictions and disputes around PCR led to innovation and several commercial systems have thrived as alternative nucleic acid amplification techniques (NAATs), e.g. strand displacement amplification (McHugh *et al.*, 2004).

A natural progression from the ability to amplify specific DNA fragments is the suggestion that these may be interrogated for genomic information and amplification of the 16S or 18S encoding genes has been adopted widely to inform the classification of a pathogen (Williams *et al.*, 2007; Jenkins *et al.*, 2012). DNA sequence analysis can also provide an indication of the drug resistance potential of an isolate although in many cases the genotype does not provide the full explanation for the phenotype (Dantas and Sommer, 2012). Again technology has broadened the possibilities here as the accessibility of whole-genome sequencing improves on an almost daily basis the questions we were able to ask of a single gene or even a selection of genes, for example in multilocus sequence typing, are expanded so we can interrogate the whole genome. Further to this, the technologies allow us to address the panbiome associated with a particular clinical presentation.

Just as conventional microbiology does not give the whole picture of the pathogenesis of an infectious event, nor does the genome. It is for this reason that transcriptomic and proteomic approaches are of increasing importance. Measurement of pathogen specific mRNA as an indicator of the presence of viable organisms and their metabolic state is a well established strategy; however, the labile nature and low abundance of mRNA has made it difficult to incorporate into diagnostic assays. Thus the mRNA approach has largely been confined to research studies. This may be changed with the increased use of RNASeq, which combines the transcriptomic approach with next generation sequencing technology (Wang *et al.*, 2009). Use of other RNA moieties has been proposed and one such example if the use of 16S RNA as a target for monitoring therapy in tuberculosis (Honeyborne *et al.*, 2011).

One of the challenges of microbiology is the timely identification of organisms, conventional techniques can routinely take days to achieve a result and genotypic methods, such as 16S gene sequencing, have proven technically exacting. In recent years the development of soft desorption/ionization and advanced software has enabled the evolution of MALDI-TOF MS bench top systems suitable for use in the routine diagnostic laboratory (Wang *et al.*, 2013). The impact of MALDI-TOF MS is 2-fold: through speedier identification, less than 5 minutes, and reduced costs, approximately 3 pence per test (El-Bouri *et al.*, 2012; Wang *et al.*, 2013). Complex microbiomes, such as found in the respiratory tract, can be characterized rapidly and inexpensively. Speedier identification can also direct antimicrobial therapy 24–48 hours prior to susceptibility data being available (Burckhardt and Zimmermann, 2011; Romero-Gomez *et al.*, 2012).

A major advantage of MALDI-TOF MS is the extensive capacity of reference databases. The Bruker Biotyper 3.0 (Bruker Daltonics) and Saramis™ (AnagnosTec) databases currently include reference spectra for over 4000 and 2000 species respectively which have enhanced the identification of microorganisms, such as *Burkholderia cepacia*, previously misidentified by biochemical assays (Marko *et al.*, 2012).

## Enhancing diagnosis

There are a wide range of clinical settings where molecular techniques can enhance our ability to detect and diagnose bacterial infections. These include infections where the bacterial load is clinically significant but below the level of conventional detection, serious infections where a rapid result can change clinical management and therefore make a direct benefit to patient care and for the detection of slow growing fastidious organisms that are difficult (or sometimes impossible!) to grow in a routine diagnostic laboratory.

### Pathogen load below that of conventional detection

16S rDNA sequencing has played a pivotal role in the accurate diagnosis of infections where low bacterial load or recent antibiotic therapy can give misleading culture negative results (Huggett *et al.*, 2008). Identification of bacteria using 16S rDNA sequencing utilizes inherent properties of the gene that codes for the 16S rRNA, which has both highly conserved and hypervariable regions. Primers are designed that bind to regions of the gene within the highly conserved areas, ensuring that DNA will be amplified from almost all species of bacteria. The region of the gene between the conserved primers contains hypervariable regions, that when sequenced provide species-specific sequences that can be compared to known sequences already entered into vast online databases.

The diagnosis of prosthetic joint infection (PJI) is an area where 16S rDNA sequencing has proven useful for detecting the aetiological cause of infection. Microbiological cultures of joint fluids can often result in both false-positive results due to contaminants that can enter the culturing process and false-negative results due to either low bacterial load, or recent antibiotic therapy. 16S rDNA sequencing has been shown to have a higher specificity and positive predictive value compared to culture (Marin *et al.*, 2012), indicating that for the diagnosis and successful treatment of serious PJI, molecular techniques should be considered when culture results do not appear to match clinical observations.

Like all techniques, there are of course limitations of 16S rDNA sequencing for the diagnosis of infection. The use of conserved 16S primers for the PCR together with Sanger-sequencing technology (which is still the most common method used in front-line diagnostic microbiology laboratories) does not allow for the identification of bacterial species from mixed infections, and differentiation between live and dead organisms cannot be made. There are some quality control issues that need to be addressed: for example, there is little consensus on the exact parameters used to interpret results between laboratories that perform diagnostic 16S sequencing. In addition, the quality of the results obtained can rely heavily on the specimen type and the database used; commonly used databases such as EMBL and GenBank are often open-access where reference sequences are uploaded with out peer-review. These databases are attractive because they are large, with millions of reference sequences to compare results against; however, it is not always possible to be sure of the authenticity of the reference sequences and studies have shown that a significant proportion of reference entries contain anomalies within the exact sequence data (Ashelford *et al.*, 2005). With increasing use of molecular identification methods in diagnostic microbiology laboratories, improved standardization between institutions are an important consideration.

Pyrosequencing is an alternative approach to sequencing of nucleic acid that addresses some of the limitations of Sanger-sequencing technology. Based on the detection of pyrophosphate release upon nucleotide incorporation into a growing nucleotide chain, rather than chain termination with dideoxynucleotides, it results in short-read sequences that can be combined together if needed. Compared to Sanger-sequencing, it allows for a much higher throughput, is easier to perform, is cheaper and quicker and can be used to sequence individual DNA sequences from mixed populations. Currently it is much more commonly used in the field of clinical virology than clinical bacteriology; however, its ease of use and ever reducing cost will make it possible for even small diagnostic laboratories to diagnose infections rapidly and there are already many of examples of this technique being applied to clinical bacteriology.

An emerging field within molecular microbiology that utilizes the advantages of pyrosequencing is that of metagenomics, which combines the data achieved from high-throughput, simultaneous sequencing of thousands of random fragments of DNA from complex polymicrobial samples with sophisticated software that is able to piece together all of the fragments to give detailed sequence information on the entire microbial community present. Sequence information can be achieved directly from mixed populations of organisms without the need to culture them in the laboratory at all. The field of metagenomics was recently used as a tool to investigate the German Shiga-toxigenic *E.coli*:H4 0104 outbreak of 2011 (Loman *et al.*, 2013). Using metagenomics, the outbreak strain was detected and a draft of the complete genome was obtained directly from faecal samples submitted during the outbreak, without the need to culture the organism at all. These results indicate the huge potential of metagenomics as a culture-independent approach for the identification of bacterial pathogens during an outbreak (Relman, 2013). Metagenomic-based applications have also been used for the culture-independent detection of urinary (Lu *et al.*, 2011) and respiratory (Zhou *et al.*, 2010) bacterial pathogens direct from samples, along with detection of *Clostridium difficile* direct from stool samples (Wroblewski *et al.*, 2009). In addition to providing information on which pathogens are present, pyrosequencing also has the ability to provide information on heterogeneous populations of the same species that might be present within the same sample and has useful epidemiological applications. Future challenges include improving diagnostic sensitivity, developing robust methods to distinguish between true infection and colonization and speeding up and simplifying workflows so that the methods can be easily translated into routine diagnostic laboratories in a cost effective manner.

## Rapid result changes management

### Blood culture

The speed of molecular-based techniques compared to conventional culture-based methods has resulted in the development of many assays that can be used in settings where a rapid result can change clinical management and therefore directly benefit the outcome for a patient. Examples of this include the diagnosis of serious infections, such as sepsis, endocarditis and central nervous system (CNS) infections, and identification of resistant organisms where detection of the causative agent can guide administration of appropriate antibiotic therapy earlier than if culture alone is used.

Molecular techniques have been used to improve the diagnosis of infective endocarditis, for which conventional culture and serological methods, which are often very slow and can be unreliable, do not always provide an aetiological diagnosis. 16S PCR and sequencing, performed directly on heart valve tissue, has been shown to be a highly sensitive and specific method for identification of causative organisms directly from the site of infection (Miyazato *et al.*, 2012). This method is of particular value for diagnosing culture-negative endocarditis caused by difficult-to-grow organisms, such as *Mycoplasma hominis* and *Bartonella henselae*.

The rapid identification of organisms that cause sepsis and the subsequent administration of correct antimicrobial therapy is correlated with favourable patient outcome, with morbidity and mortality being reduced. Conventionally, empirical therapy is initially guided by a Gram stain that is performed directly on a positive blood culture, with confirmed identification of the bacterial species by culture then taking up to 48 hours to achieve. There are now a number of different commercial molecular assays available that enable identification of organisms directly from positive blood cultures within a number of hours, speeding up the time to identification by up to 24–48 hours compared to conventional culture-based methods. The Prove-it™ Sepsis assay (Mobidiag) allows the identification of 50 bacterial pathogens and antibiotic resistance markers (discussed in more detail below) such as mecA, gyrB, and parE directly from positive blood cultures. Based on PCR and microarray technology, the assay has been shown to give highly sensitive and specific results 18 hours earlier than culture after a blood culture flagging positive on an automated instrument (Tissari *et al.*, 2010).

MALDI-TOF MS is also proving effective in diagnosis from blood culture (Schubert *et al.*, 2011; Lagace-Wiens *et al.*, 2012). The direct identification of microorganisms from blood cultures has a success rate around 80–85% (Schubert *et al.*, 2011; Romero-Gomez *et al.*, 2012). Enterobacteriaceae are most successfully identified whereas Gram-positive bacteria and non-fermentative Gram-negative bacilli together with yeasts are less successfully identified with reported success rates of 95% (Ashelford *et al.*, 2005; Hrabak *et al.*, 2011), 70–75% (Ashelford *et al.*, 2005; Bullman *et al.*, 2011) and 56% respectively (Meex *et al.*, 2012). There are technical limitations, but despite these identification of bacteria and yeasts from a positive blood culture 24 hours in advance of conventional methods has obvious clinical benefits and contributes to improved clinical management of patients (Gorton *et al.*, 2011; Clerc *et al.*, 2013).

### Detection of resistance

In terms of guiding patient therapy, rapid detection of antibiotic resistance can be as important, if not more important, than the rapid identification of the aetiological agent. This is because genes encoding antibiotic resistance determinants are often carried on plasmids or other mobile genetic elements, so identification of an organism from a clinical specimen such as *S. aureus* gives no indication of whether it is a methicillin-resistant or -sensitive strain.

It is well documented that early antimicrobial therapy with antibiotics that are effective *in vitro* is a powerful predictor of a favourable outcome in patients with Gram-negative sepsis (Schwaber *et al.*, 2006). Therapeutic options to treat some multi-resistant Gram-negative organisms such as those that produce ESBLs, AmpC or carbapenemases are often very limited, with temocillin, tigecycline, colistin, carbapenems and fosfomycin being amongst the few agents likely to remain active in systemic sepsis. However, as there is understandable reluctance to use these agents indiscriminately as empiric therapy, patients can be placed on inappropriate antimicrobial therapy until results from culture-based sensitivity testing are known (24–48 hours in most cases). A number of molecular-based assays have been developed that can detect common genetic resistant determinants, such as *mecA* in *Staphylococcus aureus* (indicating MRSA, rather than MSSA) (Carroll, 2008) and genes encoding for extended-spectrum β-lactamases (ESBL) (Vanstone *et al.*, 2012), AmpC and carbapenemases from Enterobacteriaceae directly from positive blood cultures. Owing to the limitations of the number of targets real-time PCR-based assays can simultaneously detect, it is impossible to design a PCR assay that looks for all clinically relevant resistant genes. Laboratories using PCR-based assays have to decide, depending on factors such as local prevalence rates and patient demographics, which assays will provide the most clinically relevant results. Recent advances in laboratory-friendly microarray assays that are able to detect multiple resistance associated genes in a single assay is promising, although much work remains to be performed before such techniques are used routinely across the world. The Check-MDR CT103 assay (Check-Points, Wageningen, The Netherlands) which detects the most prevalent ESBL, AmpC and Carbapenemase genes has recently been evaluated and was shown to be a powerful high-throughput tool, that can provide rapid results that are valuable for both epidemiological and infection control studies (Cuzon *et al.*, 2012). MALDI-TOF may also have a role to play here, with the biggest advances made in the detection of carbapenemase enzymes. Studies have focused on the detection of carbapenem hydrolysis, for example imipenem, by the carbapenemase. Success rates of 96–100% have been demonstrated using this approach (Burckhardt and Zimmermann, 2011; Hrabak *et al.*, 2011; Wang *et al.*, 2013).

Now that diagnostic laboratories are beginning to gain technology that can provide rapid identification of genes that confer resistance to antimicrobials – the next question that will need to be addressed is 'how easy is it to predict the *in vivo* antibiotic resistance profile from this information?' In a number of cases the answer to this will be straight forward – as there will be a direct correlation with presence or absence of the gene, and *in vivo* resistance. For example, the presence of *mecA* in *S. aureus* is sufficient for that strain to be methicillin resistant, and the presence of particular, well-characterized point mutations in the *rpo*B of *M. tuberculosis* is enough to confer resistance to rifampicin. However, in many cases the precise minimum inhibitory concentration (MIC) of an organism to a particular drug can be affected by many different properties within the bacterial cell – the degree to which pores or efflux pumps are expressed on the surface of the cell for example, or the ability to hyper produce particular proteins or even the tendency of the particular strain to form a biofilm. As a result, there will always be cases where the presence or absence of one or two genes will not give a clear indication to the *in vivo* antibiogram. Clinicians are used to dealing with MICs and there are vast amounts of peer-reviewed data and evidence relating MIC to the success (or failure) of particular antimicrobial treatment regimes. The challenges of linking detailed knowledge of a bacterial genome to the *in vivo* biological and clinical phenotypes of an individual bacterial strain need to be addressed as this field moves further forward.

## Detection of fastidious and difficult to grow organisms

In addition to improving our ability to diagnose infections of low bacterial load, an often overlooked advantage to molecular methods is that they allow us to detect organisms that are difficult or sometimes even impossible to grow using conventional culture methods; they can even enable detection of previously unknown pathogens. In 2011, a previously undescribed gastrointestinal pathogen, *Campylobacter ureolyticus* was identified by a diagnostic laboratory that had switched from routine culture techniques to a commercial multiplex-based PCR detection system for the detection of gastrointestinal bacterial pathogens (Bullman *et al.*, 2011). *C. ureolyticus* is unable to grow on routinely used *Campylobacter* culture media, and had, until this report, never been reported in the faeces of patients with gastrointestinal symptoms. In Ireland, *C. ureolyticus* has now surpassed *C. coli* as the second most common *Campylobacter* species isolated from patients with acute gastrointestinal symptoms. This is an excellent example of molecular-based techniques enhancing detection of fastidious organisms, which could otherwise be under-reported as false-negative results from diagnostic laboratories that use culture alone.

Traditional culture methods for slow-growing organisms such as *Actinomyces* spp. or *Mycobacteria* spp. can take days to weeks for sufficient growth of the organism to allow identification antibiotic susceptibility determination (McHugh, 2013). This can result in suboptimal management of some patients if initial, empirical therapy is not sufficient; molecular methods can increase the speed of identification of such organisms. In addition, molecular techniques allow for the detection of uncultivable or difficult to grow bacteria, such as *Mycoplasma* spp. and for the identification of aetiological agents in infections of low pathogen load, which would remain undetected using traditional culture methods.

Fungal infections present similar challenges; culture from clinical specimens lacks sensitivity with reported rates of approximately 40–60% (Passqualotto and Denning, 2005) and some fungal species are uncultivable in the laboratory, with diagnosis often made on the basis of histology. Using DNA extracted from clinical samples including wax embedded tissue sensitivity mirrors that of histopathology (Paterson *et al.*, 2003). Although PCR is sensitive enough for fungal detection, the choice of the ribosomal target is important. 18S analysis is limited in its specificity often only identifying the genus of fungi, as the homology of the 18S region is highly conserved within genera. A move towards using the internal transcribed spacer (ITS) rRNA 1 and 2 regions as appropriate targets has been initiated as these regions facilitate identification at the species level for the most commonly encountered fungi (Balajee *et al.*, 2009).

Species specific molecular diagnostic assays are also being utilized in the diagnosis of invasive fungal infections. A good example of this is the use of real-time PCR assays (both in-house and commercial) for *Pneumocystis jirovecii* detection as a rapid alternative method to conventional silver or immunofluorescence staining (Huggett *et al.*, 2008; McTaggart *et al.*, 2012). In a similar approach real-time PCR assays specific for *Aspergillus* spp. have been developed for detection of *Aspergillus* in clinical specimens. Currently global efforts are under way investigating the detection of *Aspergillus* spp. from blood samples to aid the diagnosis of invasive aspergillosis (Romero-Gomez *et al.*, 2012; Springer *et al.*, 2013). The most prevalent fungal infections within the population are superficial mycoses such as onychomycosis and tinea infection caused by dermatophytes. Commercial assays such as the Fast Track Diagnostics Dermatophyte assay (Mikrogen Diagnostik, Germany) are available. However, the implementation of PCR in this setting requires clinical justification as the cost implications are considerable compared with culture; therefore, PCR is best suited to high-throughput laboratories.

### Caveats and conclusions

There are a number of settings in the diagnostic bacteriology laboratory where the speed and sensitivity of molecular detection can enhance our ability to diagnose bacterial infections. The ability to rapidly identify organisms and/or resistance markers in patients with serious infections allows for the targeted, evidence-based treatment that can reduce morbidity and mortality. It is important that the limitations of molecular assays are not forgotten when applying molecular techniques to these situations – new genetic variants of organisms and/or resistance genes may not be detected and current techniques do not provide clinicians with an MIC. Therefore, molecular assays that are available today are a definitive improvement to our currently used culture-based methods, but until these limitations are addressed they are not quite a replacement.

## Molecular typing in control of infection

Molecular tools have facilitated the detailed analysis of microbial population structures and these tools are now being applied to the control of infections, not only in tracking the dissemination of strains with enhanced virulence or unique antibiotic resistance patterns, but also in the design of vaccines. There is considerable literature in this field but we describe two specific examples to illustrate the role of molecular techniques in these settings.

### Identifying and typing *Streptococcus* spp.

The current vaccines for pneumococcal disease are based on the capsular polysaccharide structure, which also determines the serotype. Thus continued surveillance of invasive serotypes is essential to monitor vaccine efficacy and the emergence of vaccine escape serotypes. Recently, a more accurate and cheaper serotyping test that could be used routinely has been developed that involves a single PCR reaction and nucleotide sequencing of the amplicon, termed 'sequetyping' (Leung *et al.*, 2012a). The method is able to distinguish the majority of the 23 pneumococcal polysaccharide vaccine serotypes known to cause invasive disease and all of the 13-valent conjugate vaccine and replacement serotypes, e.g. 19A. An added advantage of this 'sequetyping' method is that it is specific for *Streptococcus pneumoniae*, with no amplification observed for other closely related species (Leung *et al.*, 2012a). Furthermore, as the method is DNA based it has the potential for identifying new serotypes that may evolve in the future.

In 2004 a new, closely related, species, *S. pseudopneumoniae* (the pseudopneumococcus), in the viridans group streptococci was designated (Arbique *et al.*, 2004). This made classification within the 'mitis' streptococcal group even more difficult. The pseudopneumococcus exhibits Optochin resistance with no inhibition zone or indeterminate with an inhibition zone 8–13 mm when incubated in $CO_2$. In contrast, when incubated in air it shows optochin susceptibility (Arbique *et al.*, 2004). Further, pseudopneumococci do not react with pneumococcal typing antisera, indicating a lack of pneumococcal polysaccharide antigens. As identification of the pseudopneumococcus is recent little is known about its pathogenicity in humans (Keith *et al.*, 2006; Keith and Murdoch, 2008; Johnston *et al.*, 2010; Leegaard *et al.*, 2010; Shahinas *et al.*, 2011; Leung *et al.*, 2012b). Thus, there is a public health need for accurate differentiation of this species to evaluate its true disease potential and ensure appropriate management of patients. Recently, the pseudopneumococcus has been identified using its unique quorum-sensing 'pherotype' based on the allelic variation of the *comC* gene encoding the competence-stimulating peptide (CSP) (Leung *et al.*, 2012b). Furthermore, this method can be used to differentiate between other oral streptococci as well as the pneumococcus.

For enabling transmission chains to be established genotyping by multilocus sequencing typing (MLST) has been employed to generate a 'sequence type' (Enright and Spratt, 1998; Maiden *et al.*, 1998). This is based on the nucleotide sequences of fragments of seven 'housekeeping' genes. The sequence data are submitted to the on-line MLST pneumococcal database (www.mlst.net). The database is interrogated with the uploaded sequences and any identical matches are assigned that number; new sequences are assigned with the next numerical number for that gene fragment. The seven gene fragment numbers are compared with existing strains and assigned with either a previous number or a new number if not previously identified.

### Typing of *Staphylococcus* spp.

Linking possible cases of infection that are related is critical for the epidemiological study of Staphylococcus spp. in community settings and healthcare. Several methods are used to monitor possible transmission events. Currently, the principal method used for typing *S. aureus* is spa typing which is reliable and accurate (Grundmann *et al.*, 2010). This involves DNA sequencing of short sequence repeats in the polymorphic X region of the protein A gene (*spa*) of *S. aureus*. The region consists of a variable number of short repeats, which are around 24 bp in length. Each new base composition of a repeat is assigned an alpha-numerical code (r01, r02 etc.) and the repeat succession determines the *spa* type (e.g. t001, t002, etc.). *Spa* typing is used throughout Europe and world-wide for typing of both Methicillin sensitive and resistant *S. aureus*. Over 7000 *spa* types have been described to date; *spa* non-typable strains are rare (less than 0.1%). There is a standard international nomenclature which is web-enabled (http://www.spaserver.ridom.de) and the data are directly comparable between centres and countries. For some *S. aureus* lineages, the technique has a discriminatory index approaching that of PFGE. To discriminate between indistinguishable or closely related spa types Pulsed-Field Gel Electrophoresis (PFGE) can be used.

### Toxin gene profiling of *S. aureus*

In clinical isolates from suspected toxin-mediated infection (non-enteric disease) the presence of the following genes can be detected by multiplex PCR: the Toxic shock syndrome toxin gene, *tst*; the enterotoxin genes *sea, seb, sec, sed, see, seg, seh, sei* and *sej*; the exfoliatins (*eta, etb* and *etd*) that are associated with scalded skin syndrome and impetigo; the Panton-Valentine leukocidin (lukS-PV/lukF-PV) which is associated with primary pyogenic skin infections (abscesses, boils) especially where these are recurrent. Occasionally PVL-positive strains of *S. aureus* are associated with more invasive disease such as bone and joint infections, purpura fulminans, or necrotizing pneumonia. Panton–Valentine leukocidin (PVL) is a toxic substance produced by some strains of *S. aureus*, which is associated with an increased ability to cause disease and can include PVL-MRSA strains. Despite several countries encountering widespread problems with PVL-related disease, infections caused by PVL are still rare in the UK, and most are from strains that are sensitive to antibiotics.

In infections caused by *S. aureus* from foods and/or cases of suspected food poisoning (enteric disease) the enterotoxin genes, *sea, seb, sec, sed, see, seg, seh, sei* and *sej* are detected by multiplex PCR.

## Molecular tools in screening/high-throughput analysis

The speed and high sensitivity of molecular assays means that they are well suited to screening and high-throughput applications in the clinical bacteriology laboratory. Many aspects involved with the set up and analysis of molecular assays can now be fully automated, allowing for high-throughput to be achieved with reduced hands-on time required from laboratory staff.

Common high-throughput examples in the clinical bacterial laboratory where molecular methods are used include screening for methicillin-resistant *S. aureus* (MRSA) colonization and STI testing.

## MRSA molecular testing in the clinical laboratory

Colonization with MRSA is a prerequisite not only for the development of infection, but also for cross-transmission to other patients and staff within a healthcare facility. The rapid determination of a patient's MRSA status can therefore guide early patient management, antibiotic stewardship and allows for prompt infection control inventions such as barrier-nursing and decolonisation treatment to be implemented. Guidelines now advise that all patients admitted to hospitals across the UK must be screened for MRSA colonization, resulting in large numbers of samples that must be tested (NICE, 2011).

MRSA strains are defined by the carriage of a staphylococcal cassette chromosome *mec* (SCCmec) genetic element. This cassette, which integrates within the orfX region of the *S. aureus* genome, carries the *mec*A, which codes for the penicillin binding protein (PBP2) that is resistant to the action of most of the β-lactam agents. To increase the turn-around time to detection, MRSA screening often takes place directly from the swab (nasal and/or perineum) rather than an isolated colony, so the assays used need to be able to detect MRSA from poly-microbial biota (that might include MSSA and coagulase-negative staphylococci) taken from the skin of the patient. An assay that detects *mecA* alone would be unsuitable as coagulase-negative staphylococci (CoNS) can also carry *mecA* – leading to positive results that might not necessarily reflect the presence of MRSA. Most of the commercial real-time PCR assays that are currently available detect either the SCCmec–orfX bridging region (single-locus assay) or simultaneously detect the mecA gene and *S. aureus* specific target such as the nuclease gene, *nuc* (double-locus assay).

Both single-locus and double-locus assays are associated with limitations when used for screening samples for MRSA. The single-locus SCCmec–orfX assays can occasionally lead to false-positive reports as a result of 'mecA-dropouts' (Snyder *et al.*, 2010): where most of the SCCmec has excised from the genome (so it is no longer a MRSA strain) but enough of the sequence remains at the SCCmec–orfX junction to give a positive result by PCR. The double locus assays will not detect '*mecA*-dropouts' but require intelligent software that is able distinguish between *mecA* and *nuc* amplification from the genuine presence of MRSA and that from mixed populations containing MSSA (which would provide the *nuc*) and CoNS harbouring the *mecA* gene.

## STI testing using molecular methods

The testing for sexually transmitted infections (STIs) is an area within the clinical microbiology laboratory for which molecular testing has been used routinely for quite some time. Molecular assays have revolutionized the detection of *Chlamydia trachomatis* over the past 20 years, which was originally achieved through highly specialized and time consuming tissue culture techniques. Various commercial assays have been developed since the 1990s that are based on techniques such as real-time-PCR and ligase chain reaction (LCR). They have been shown to be comparable to tissue culture in terms of sensitivity and specificity (Schachter, 1997) and have since become well established in diagnostic laboratories across the world. In addition to improving laboratory workflow, the use of sensitive NAAT assays allows specimens from relatively non-invasive specimens to be tested, such a urine and self collected samples. This can be of particular benefit for some hard-to-reach groups of patients, in remote regions where sexual health services may not be available and for special populations where religious or cultural beliefs might restrict opportunities for specimen collection.

Some of the current NAAT assays in routine diagnostic service allow for the simultaneous detection of *C. trachomatis* and *Neisseria gonorrhoeae* from a range of patient specimens including urine, rectal and self-collected vulvo-vaginal samples. The increased sensitivity of NAATs compared to previously used culture methods enables accurate diagnosis of both symptomatic and asymptomatic gonococcal infections, that have been shown to be critical to control the disease (Whiley *et al.*, 2006). In addition, specimens collected for NAAT assays do not require the organism to be viable for detection; this is often seen as a limitation of molecular assays, but in the case of *N. gonorrhoeae* this allows for less stringent transport conditions compared with those required for culture of this particularly fastidious organism.

## Molecular assays for screening – a word of warning

A major limitation of molecular assays, which should be considered particularly when using them for high-throughput applications, is that they may be unable to detect genetic variants that might arise in a given population. Both examples described in detail above (screening for MRSA colonization and STI infections) are associated with well documented examples:

In 2011 a MRSA strain that was shown to have a novel *mecA* homologue (MecALGA251, since renamed as *mecC*) in a newly described SCCMec element was reported (Garcia-Alvarez *et al.*, 2011). This strain, which was originally discovered in bovine populations in Britain and Denmark and subsequently described in human populations, was sufficiently divergent in the SCCMec region that most commercial molecular assays available at the time were unable to detect it. As a result, false-negative results were issued from laboratories which potentially led to patients not being started on the correct decolonisation pathway and potential subsequent onward transmission events and/or complications.

Similarly, after a continuous annual increase across Sweden in the reported incidence of *C. trachomatis* during 1996–2005, an unexpected decrease of approximately 2% was observed in 2006. Investigations into the decreased incidence uncovered an outbreak of a new variant of *C. trachomatis* (serovar E) which was not detected by some of the molecular assays used in diagnostic laboratories at the time (Herrmann, 2008). This strain had a 377-bp deletion in a cryptic plasmid which is the target region of some commercial assays. Assays that amplified this particular region resulted in false negative results being issued to patients, who then failed to be given the correct treatment and advice regarding preventing onward transmission. In 2007, when all laboratories across Sweden had switched to assays that were able to detect the new-variant strain there was a sharp increase in the diagnosis of *C. trachomatis* compared with 2006 (Velicko *et al.*, 2007).

These examples highlight the importance that diagnostic laboratories are aware of the limitations of molecular assays and remain vigilant for unexpected negative results and/or periods of unusually low prevalence. Users and manufacturers of molecular assays need to ensure good surveillance methods are in place and that they have the ability to quickly adapt assay formulations in response to the discovery of new strains or substrains.

## Considerations in selecting a molecular tool

One of the perceived hurdles associated with introducing molecular-based assays is that they are expensive compared to currently used culture-based methods. In general, it is true that a direct comparison of cost per test is more expensive for NAAT-based assays than for routine culture-based methods. In addition, the initial cost of instrumentation and training can be expensive. However, when additional factors such as long-term staffing costs (molecular assays are often less hands-on) and generation of clinical waste (culture-based methods can generate tonnes of clinical waste per year, which is associated with large decontamination, landfill and associated licensing costs) are considered the two assay techniques are often more comparable. Even if overall costs to the laboratory still rise with the use of molecular assays, implications to the health care setting as a whole need to be considered, rather than costs to the diagnostic service alone. The use of rapid molecular assays has been shown to help enable earlier implementation of infection control measures which can prevent outbreaks of nosocomial pathogens, prevent the unnecessary use of antibiotics that can lead to the development of resistant strains and ensure the swift administration of correct antibiotics which can reduce the length hospital stays for patients. All of these can have a dramatic cost saving to the overall healthcare facility which is much greater than that spent by the laboratory to perform the tests.

It is an irony that whilst molecular tests are some of the most sophisticated in design and underlying technology, they often do not need a high degree of training to operate. The 'black box' technology makes many of the NAATs suitable for point of care testing with non specialist operators; for example we are evaluating the Xpert system for TB diagnosis on our mobile X-ray unit in central London and this is operated by nursing staff rather than healthcare scientists (Shorten *et al.*, 2012). However, it is critical that there are experienced staff available to oversee the service, provide quality assurance and trouble shoot issues that arise.

## Conclusion

The diagnostic microbiology laboratory is entering a new era; increasing implementation of molecular methods is drastically improving our ability to diagnose infectious disease. This presents organizational, technical and cultural challenges as conventional microbiological methods replaced by highly sensitive molecular techniques that can provide timely results to clinicians. Training needs for laboratory staff are changing and strong links between diagnostic laboratories, research institutions and commercial groups will have to be formed to maintain and ensure the continual development of ever refined, clinically relevant assays that can be automated and performed easily in busy laboratory settings. Newly introduced assays must be validated with appropriate clinical studies that correlate clinical findings with assay results. There will be a need for extensive peer-reviewed and maintained, international bioinformatics databases that are freely accessible to all. As we enter the new era, it is imperative that both laboratory staff and clinicians remain aware of the limitations of the molecular assays, and consider the biological properties of the organism and clinical presentation to ensure the correct interpretation of molecular-based results.

In our title we posited the idea that the horizon is drawing near: in our daily clinical practice and research we experience the ever changing diagnostic environment. Molecular tools have the potential to make significant improvements to the management of patients with infections and also to the control of microbial infections at the population level. There is a danger that the technology blinds us to the underlying biology, but it is essential that we consider the new technologies with the same critical rigour that we would apply to conventional assays and in the context of the clinical setting. The horizon is close and full of opportunity.

## References

Arbique, J.C., Poyart, C., Trieu-Cuot, P., Quesne, G., Carvalho Mda, G., Steigerwalt, A.G., Morey, R.E., Jackson, D., Davidson, R.J., and Facklam, R.R. (2004). Accuracy of phenotypic and genotypic testing for identification of *Streptococcus pneumoniae* and description of Streptococcus *pseudopneumoniae* sp. nov. J. Clin. Microbiol. 42, 4686–4696.

Ashelford, K.E., Chuzhanova, N.A., Fry, J.C., Jones, A.J., and Weightman, A.J. (2005). At least 1 in 20 16S rRNA sequence records currently held in public repositories is estimated to contain substantial anomalies. Appl. Environ. Microbiol. *71*, 7724–7736.

Balajee, S.A., Borman, A.M., Brandt, M.E., Cano, J., Cuenca-Estrella, M., Dannaoui, E., Guarro, J., Haase, G., Kibbler, C.C., Meyer, W., *et al.* (2009). Sequence-based identification of *Aspergillus*, *Fusarium*, and *Mucorales* species in the clinical mycology laboratory: where are we and where should we go from here? J. Clin. Microbiol. *47*, 877–884.

Bourbeau, P.P., and Ledebrei, N.A. (2013). Automation in Clinical Microbiology. J. Clin. Microbiol. *51*, 1658–1665.

Bullman, S., Corcoran, D., O'Leary, J., Lucey, B., Byrne, D., and Sleator, R.D. (2011). *Campylobacter ureolyticus*: an emerging gastrointestinal pathogen? FEMS Immunol. Med. Microbiol. *61*, 228–230.

Burckhardt, I., and Zimmermann, S. (2011). Using matrix-assisted laser desorption ionization-time of flight mass spectrometry to detect carbapenem resistance within 1 to 2.5 hours. J. Clin. Microbiol. *49*, 3321–3324.

Carroll, K.C. (2008). Rapid diagnostics for methicillin-resistant *Staphylococcus aureus*: current status. Mol. Diagn. Ther. *12*, 15–24.

Clark, A.E., Kaleta, E.J., Arora, A., and Wolk, D.M. (2013). Matrix-assisted laser desorption ionization-time of flight mass spectrometry: a fundamental shift in the routine practice of clinical microbiology. Clin. Microbiol. Rev. *26*, 547–603.

Clerc, O., Prod'hom, G., Vogne, C., Bizzini, A., Calandra, T., and Greub, G. (2013). Impact of matrix-assisted laser desorption ionization time-of-flight mass spectrometry on the clinical management of patients with Gram-negative bacteremia: a prospective observational study. Clin. Infect. Dis. *56*, 1101–1107.

Creer, D.D., Dilworth, J.P., Gillespie, S.H., Johnston, A.R., Johnston, S.L., Ling, C., Patel, S., Sanderson, G., Wallace, P.G., and McHugh, T.D. (2006). Aetiological role of viral and bacterial infections in acute adult lower respiratory tract infection (LRTI) in primary care. Thorax *61*, 75–79.

Cuzon, G., Naas, T., Bogaerts, P., Glupczynski, Y., and Nordmann, P. (2012). Evaluation of a DNA microarray for the rapid detection of extended-spectrum beta-lactamases (TEM, SHV and CTX-M), plasmid-mediated cephalosporinases (CMY-2-like, DHA, FOX, ACC-1, ACT/MIR and CMY-1-like/MOX) and carbapenemases (KPC, OXA-48, VIM, IMP and NDM). J. Antimicrob. Chemother. *67*, 1865–1869.

Dantas, G., and Sommer, M.O. (2012). Context matters – the complex interplay between resistome genotypes and resistance phenotypes. Curr. Opin. Microbiol. *15*, 577–582.

El-Bouri, K., Johnston, S., Rees, E., Thomas, I., Bome-Mannathoko, N., Jones, C., Reid, M., Ben-Ismaeil, B., Davies, A.R., Harris, L.G., *et al.* (2012). Comparison of bacterial identification by MALDI-TOF mass spectrometry and conventional diagnostic microbiology methods: agreement, speed and cost implications. Br. J. Biomed. Sci. *69*, 47–55.

Enright, M.C., and Spratt, B.G. (1998). A multilocus sequence typing scheme for Streptococcus pneumoniae: identification of clones associated with serious invasive disease. Microbiology *144*, 3049–3060.

Espy, M.J., Uhl, J.R., Sloan, L.M., Buckwalter, S.P., Jones, M.F., Vetter, E.A., Yao, J.D., Wengenack, N.L., Rosenblatt, J.E., Cockerill, F.R., 3rd, *et al.* (2006). Real-time PCR in clinical microbiology: applications for routine laboratory testing. Clin. Microbiol. Rev. *19*, 165–256.

Garcha, D.S., Thurston, S.J., Patel, A.R., Mackay, A.J., Goldring, J.J., Donaldson, G.C., McHugh, T.D., and Wedzicha, J.A. (2012). Changes in prevalence and load of airway bacteria using quantitative PCR in stable and exacerbated COPD. Thorax *67*, 1075–1080.

Garcia-Alvarez, L., Holden, M.T., Lindsay, H., Webb, C.R., Brown, D.F., Curran, M.D., Walpole, E., Brooks, K., Pickard, D.J., Teale, C., *et al.* (2011). Meticillin-resistant *Staphylococcus aureus* with a novel mecA homologue in human and bovine populations in the UK and Denmark: a descriptive study. Lancet Infect. Dis. *11*, 595–603.

Gorton, R.L., Ramnarain, P., Stone, N.R.H., Barker, K., McHugh, T.D., and Kibbler, C.C. (2011). A comparison of three rapid identification techniques for the identification of yeasts from positive blood cultures: Gram's stain, PNA-FISH and MALDI-TOF. Poster TIMM.

Grundmann, H., Aanensen, D.M., van den Wijngaard, C.C., Spratt, B.G., Harmsen, D., and Friedrich, A.W. (2010). Geographic distribution of Staphylococcus aureus causing invasive infections in Europe: a molecular-epidemiological analysis. PLoS Med *7*, e1000215.

Herrmann, B. (2008). Update on the new variant of Chlamydia trachomatis: prevalence and diagnostics. Euro. Surveill. *13*, 18913.

Honeyborne, I., McHugh, T.D., Phillips, P.P., Bannoo, S., Bateson, A., Carroll, N., Perrin, F.M., Ronacher, K., Wright, L., van Helden, P.D., *et al.* (2011). Molecular bacterial load assay, a culture-free biomarker for rapid and accurate quantification of sputum Mycobacterium tuberculosis bacillary load during treatment. J. Clin. Microbiol. *49*, 3905–3911.

Hrabak, J., Walkova, R., Studentova, V., Chudackova, E., and Bergerova, T. (2011). Carbapenemase activity detection by matrix-assisted laser desorption ionization-time of flight mass spectrometry. J. Clin. Microbiol. *49*, 3222–3227.

Huggett, J.F., Taylor, M.S., Kocjan, G., Evans, H.E., Morris-Jones, S., Gant, V., Novak, T., Costello, A.M., Zumla, A., and Miller, R.F. (2008). Development and evaluation of a real-time PCR assay for detection of *Pneumocystis jirovecii* DNA in bronchoalveolar lavage fluid of HIV-infected patients. Thorax *63*, 154–159.

Jenkins, C., Ling, C.L., Ciesielczuk, H.L., Lockwood, J., Hopkins, S., McHugh, T.D., Gillespie, S.H., and Kibbler, C.C. (2012). Detection and identification of bacteria in clinical samples by 16S rRNA gene sequencing: comparison of two different approaches in clinical practice. J. Med. Microbiol. *61*, 483–488.

Johnston, C., Hinds, J., Smith, A., van der Linden, M., Van Eldere, J., and Mitchell, T.J. (2010). Detection of large numbers of pneumococcal virulence genes in streptococci of the mitis group. J. Clin. Microbiol. *48*, 2762–2769.

Keith, E.R., and Murdoch, D.R. (2008). Antimicrobial susceptibility profile of *Streptococcus pseudopneumoniae* isolated from sputum. Antimicrob. Agents Chemother. *52*, 2998.

Keith, E.R., Podmore, R.G., Anderson, T.P., and Murdoch, D.R. (2006). Characteristics of *Streptococcus pseudopneumoniae* isolated from purulent sputum samples. J. Clin. Microbiol. *44*, 923–927.

Lagace-Wiens, P.R., Adam, H.J., Karlowsky, J.A., Nichol, K.A., Pang, P.F., Guenther, J., Webb, A.A., Miller, C., and Alfa, M.J. (2012). Identification of blood culture isolates directly from positive blood cultures by use of matrix-assisted laser desorption ionization-time of flight mass spectrometry and a commercial extraction system: analysis of performance, cost, and turnaround time. J. Clin. Microbiol. *50*, 3324–3328.

Lawn, S.D., Mwaba, P., Bates, M., Piatek, A., Alexander, H., Marais, B.J., Cuevas, L.E., McHugh, T.D., Zijenah, L., Kapata, N., *et al.* (2013). Advances in tuberculosis diagnostics: the Xpert MTB/RIF assay and future prospects for a point-of-care test. Lancet Infect. Dis. *13*, 349–361.

Leegaard, T.M., Bootsma, H.J., Caugant, D.A., Eleveld, M.J., Mannsaker, T., Froholm, L.O., Gaustad, P., Hoiby, E.A., and Hermans, P.W. (2010). Phenotypic and genomic characterization of pneumococcus-like streptococci isolated from HIV-seropositive patients. Microbiology *156*, 838–848.

Leung, M.H., Bryson, K., Freystatter, K., Pichon, B., Edwards, G., Charalambous, B.M., and Gillespie, S.H. (2012a). Sequetyping: serotyping Streptococcus pneumoniae by a single PCR sequencing strategy. J. Clin. Microbiol. *50*, 2419–2427.

Leung, M.H., Ling, C.L., Ciesielczuk, H., Lockwood, J., Thurston, S., Charalambous, B.M., and Gillespie, S.H. (2012b). Streptococcus pseudopneumoniae identification by pherotype: a method to assist understanding of a potentially emerging or overlooked pathogen. J. Clin. Microbiol. *50*, 1684–1690.

Ling, C.L., and McHugh, T.D. (2013). Rapid detection of atypical respiratory bacterial pathogens by real-time PCR. Methods Mol. Biol. *943*, 125–133.

Loman, N.J., Constantinidou, C., Christner, M., Rohde, H., Chan, J.Z., Quick, J., Weir, J.C., Quince, C., Smith, G.P., Betley, J.R., *et al.* (2013). A culture-independent sequence-based metagenomics approach to the investigation of an outbreak of Shiga-toxigenic *Escherichia coli* O104:H4. JAMA *309*, 1502–1510.

Lu, J., Yu, R., Yan, Y., Zhang, J., and Ren, X. (2011). Use of Pyromark Q96 ID pyrosequencing system in identifying bacterial pathogen directly with urine specimens for diagnosis of urinary tract infections. J. Microbiol. Methods *86*, 78–81.

McHugh, T.D. ed. (2013). Tuberculosis: Diagnosis and Treatment (CABI Publishing, Wallingford, UK).

McHugh, T.D., Ramsay, A.R., James, E.A., Mognie, R., and Gillespie, S.H. (1995). Pitfalls of PCR: misdiagnosis of cerebral nocardia infection. Lancet *346*, 1436.

McHugh, T.D., Pope, C.F., Ling, C.L., Patel, S., Billington, O.J., Gosling, R.D., Lipman, M.C., and Gillespie, S.H. (2004). Prospective evaluation of BDProbeTec strand displacement amplification (SDA) system for diagnosis of tuberculosis in non-respiratory and respiratory samples. J. Med. Microbiol. *53*, 1215–1219.

McTaggart, L.R., Wengenack, N.L., and Richardson, S.E. (2012). Validation of the MycAssay Pneumocystis kit for detection of Pneumocystis jirovecii in bronchoalveolar lavage specimens by comparison to a laboratory standard of direct immunofluorescence microscopy, real-time PCR, or conventional PCR. J. Clin. Microbiol. *50*, 1856–1859.

Maiden, M.C., Bygraves, J.A., Feil, E., Morelli, G., Russell, J.E., Urwin, R., Zhang, Q., Zhou, J., Zurth, K., Caugant, D.A., *et al.* (1998). Multilocus sequence typing: a portable approach to the identification of clones within populations of pathogenic microorganisms. Proc. Natl. Acad. Sci. U.S.A. *95*, 3140–3145.

Marin, M., Garcia-Lechuz, J.M., Alonso, P., Villanueva, M., Alcala, L., Gimeno, M., Cercenado, E., Sanchez-Somolinos, M., Radice, C., and Bouza, E. (2012). Role of universal 16S rRNA gene PCR and sequencing in diagnosis of prosthetic joint infection. J. Clin. Microbiol. *50*, 583–589.

Marko, D.C., Saffert, R.T., Cunningham, S.A., Hyman, J., Walsh, J., Arbefeville, S., Howard, W., Pruessner, J., Safwat, N., Cockerill, F.R., *et al.* (2012). Evaluation of the Bruker Biotyper and Vitek MS matrix-assisted laser desorption ionization-time of flight mass spectrometry systems for identification of nonfermenting Gram-negative bacilli isolated from cultures from cystic fibrosis patients. J. Clin. Microbiol. *50*, 2034–2039.

Meex, C., Neuville, F., Descy, J., Huynen, P., Hayette, M.P., De Mol, P., and Melin, P. (2012). Direct identification of bacteria from BacT/ALERT anaerobic positive blood cultures by MALDI-TOF MS: MALDI Sepsityper kit versus an in-house saponin method for bacterial extraction. J. Med. Microbiol. *61*, 1511–1516.

Miyazato, A., Ohkusu, K., Tabata, M., Uwabe, K., Kawamura, T., Tachi, Y., Ezaki, T., Niinami, H., and Mitsutake, K. (2012). Comparative molecular and microbiological diagnosis of 19 infective endocarditis cases in which causative microbes were identified by PCR-based DNA sequencing from the excised heart valves. J. Infect. Chemother. *18*, 318–323.

NICE (2011). NICE prevention and control of healthcare associated infections (PH36). Available at: http://guidance.nice.org.uk/PH36 (accessed 7 November 2013).

Passqualotto, A., and Denning, D.W. (2005). Diagnosis of invasive fungal infections – current limitations of classical and new diagnostic methods. Eur. Oncol. Rev. *2005*, 1–11.

Paterson, P.J., Seaton, S., McLaughlin, J., and Kibbler, C.C. (2003). Development of molecular methods for the identification of aspergillus and emerging moulds in paraffin wax embedded tissue sections. Mol. Pathol. *56*, 368–370.

Relman, D.A. (2013). Metagenomics, infectious disease diagnostics, and outbreak investigations: sequence first, ask questions later? JAMA *309*, 1531–1532.

Romero-Gomez, M.P., Gomez-Gil, R., Pano-Pardo, J.R., and Mingorance, J. (2012). Identification and susceptibility testing of microorganism by direct inoculation from positive blood culture bottles by combining MALDI-TOF and Vitek-2 Compact is rapid and effective. J. Infect. *65*, 513–520.

Saiki, R.K., Scharf, S., Faloona, F., Mullis, K.B., Horn, G.T., Erlich, H.A., and Arnheim, N. (1985). Enzymatic amplification of beta-globin genomic sequences and restriction site analysis for diagnosis of sickle cell anemia. Science *230*, 1350–1354.

Schachter, J. (1997). DFA, EIA, PCR, LCR and other technologies: what tests should be used for diagnosis of chlamydia infections? Immunol. Invest. *26*, 157–161.

Schmid-Hempel, P., and Frank, S.A. (2007). Pathogenesis, virulence, and infective dose. PLoS Pathog. *3*, 1372–1373.

Schubert, S., Weinert, K., Wagner, C., Gunzl, B., Wieser, A., Maier, T., and Kostrzewa, M. (2011). Novel, improved sample preparation for rapid, direct identification from positive blood cultures using matrix-assisted laser desorption/ionization time-of-flight (MALDI-TOF) mass spectrometry. J. Mol. Diagn. *13*, 701–706.

Schwaber, M.J., Navon-Venezia, S., Kaye, K.S., Ben-Ami, R., Schwartz, D., and Carmeli, Y. (2006). Clinical and economic impact of bacteremia with extended- spectrum-beta-lactamase-producing Enterobacteriaceae. Antimicrob. Agents Chemother. *50*, 1257–1262.

Shahinas, D., Tamber, G.S., rya, G., Wong, A., Lau, R., Jamieson, F., Ma, J.H., Alexander, D.C., Low, D.E., and Pillai, D.R. (2011). Whole-genome sequence of Streptococcus pseudopneumoniae isolate IS7493. J. Bacteriol. *193*, 6102–6103.

Shorten, R., Story, A., Tacou, J., Patel, S., Gant, V., Lipman, M., McHugh, T., and Watson, J. (2012). The implementation of Xpert MTB/RIF on the mobile x-ray unit to improve the diagnosis of pulmonary TB in hard to reach patients. Available at: https://molecularhub.org/resources/708/download/P2355-1.pdf (accessed 09 July 2013).

Snyder, J.W., Munier, G.K., and Johnson, C.L. (2010). Comparison of the BD GeneOhm methicillin-resistant *Staphylococcus aureus* (MRSA) PCR assay to culture by use of BBL CHROMagar MRSA for detection of MRSA in nasal surveillance cultures from intensive care unit patients. J. Clin. Microbiol. *48*, 1305–1309.

Springer, J., Morton, C.O., Perry, M., Heinz, W.J., Paholcsek, M., Alzheimer, M., Rogers, T.R., Barnes, R.A., Einsele, H., Loeffler, J., *et al.* (2013). Multicenter comparison of serum and whole-blood specimens for detection of *Aspergillus* DNA in high-risk hematological patients. J. Clin. Microbiol. *51*, 1445–1450.

Tissari, P., Zumla, A., Tarkka, E., Mero, S., Savolainen, L., Vaara, M., Aittakorpi, A., Laakso, S., Lindfors, M., Piiparinen, H., *et al.* (2010). Accurate and rapid identification of bacterial species from positive blood cultures with a DNA-based microarray platform: an observational study. Lancet 375, 224–230.

Vanstone, G.L., Yorgancioglu, A., Wilkie, L., Mouskos, K., Charalambous, B.M., and Balakrishnan, I. (2012). A real-time multiplex PCR assay for the rapid detection of CTX-M-type extended spectrum beta-lactamases directly from blood cultures. J. Med. Microbiol. *61*, 1631–1632.

Velicko, I., Kuhlmann-Berenzon, S., and Blaxhult, A. (2007). Reasons for the sharp increase of genital chlamydia infections reported in the first months of 2007 in Sweden. Euro. Surveill. *12*, E5–6.

Wang, L., Han, C., Sui, W., Wang, M., and Lu, X. (2013). MALDI-TOF MS applied to indirect carbapenemase detection: a validated procedure to clearly distinguish between carbapenemase-positive and carbapenemase-negative bacterial strains. Anal. Bioanal. Chem. *405*, 5259–5266.

Wang, Z., Gerstein, M., and Snyder, M. (2009). RNA-Seq: a revolutionary tool for transcriptomics. Nat. Rev. Genet. *10*, 57–63.

Wassenberg, M., Kluytmans, J., Erdkamp, S., Bosboom, R., Buiting, A., van Elzakker, E., Melchers, W., Thijsen, S., Troelstra, A., Vandenbroucke-Grauls, C., *et al.* (2012). Costs and benefits of rapid screening of methicillin-resistant *Staphylococcus aureus* carriage in intensive care units: a prospective multicenter study. Crit. Care *16*, R22.

Whiley, D.M., Tapsall, J.W., and Sloots, T.P. (2006). Nucleic acid amplification testing for Neisseria gonorrhoeae: an ongoing challenge. J. Mol. Diagn. *8*, 3–15.

Williams, K.J., Ling, C.L., Jenkins, C., Gillespie, S.H., and McHugh, T.D. (2007). A paradigm for the molecular identification of *Mycobacterium* species in a routine diagnostic laboratory. J. Med. Microbiol. *56*, 598–602.

Wroblewski, D., Hannett, G.E., Bopp, D.J., Dumyati, G.K., Halse, T.A., Dumas, N.B., and Musser, K.A. (2009). Rapid molecular characterization of *Clostridium difficile* and assessment of populations of *C. difficile* in stool specimens. J. Clin. Microbiol. 47, 2142–2148.

Zhou, Y., Lin, P., Li, Q., Han, L., Zheng, H., Wei, Y., Cui, Z., Ni, Y., and Guo, X. (2010). Analysis of the microbiota of sputum samples from patients with lower respiratory tract infections. Acta Biochim. Biophys. Sin. (Shanghai) *42*, 754–761.

# Viral Diagnostics and qPCR-based Methodologies

6

Sophie Collot-Teixeira, Philip Minor and Robert Anderson

## Abstract

Molecular techniques, based on the detection of viral genomes, are revolutionizing the field of diagnostic virus detection. They are currently replacing more traditional detection methods, such as cell culture, because they offer the multiple advantages of speed, sensitivity and specificity, which together, are not possible with other established techniques. Molecular methods afford the opportunity to detect viral genome(s), quantify viral load and detect different viral genotypes or mutations. Instruments and chemistries are becoming reasonably affordable and enclosed automated nucleic acid extraction/detection techniques are becoming more and more available increasing the throughput, the reproducibility and confidence in the results.

## Introduction

The aim of infectious disease diagnostics is the early detection of infection allowing for administration of appropriate therapy. This will have the effect of reducing disease spread, leading to an improvement in public health. Correct diagnosis is underpinned by the fidelity of the laboratory diagnostic test in question. In the United Kingdom, over 300 000 assays are performed annually in cases of suspected viral infection.

Traditionally, diagnostic virology has relied on cell culture systems to grow virus sufficiently to observe cytopathic effect. These techniques are generally costly, time consuming, subjective to limiting factors and can require a large volume of body fluid necessitating, in some cases, invasive procedures for the patients.

It is essential that the most accurate result possible is delivered to the patient. Molecular techniques, based on the detection of viral genomes, are replacing more traditional detection methods such as cell culture. These techniques offer the possibility to provide faster and more accurate results in clinical diagnostics, leading to earlier and more adapted therapies for the patients.

The principal means of viral detection is quantitative real-time PCR (qPCR). qPCR has many advantages over cell culture-based methods as it is faster, more specific and relies on a smaller, less invasive patient sample. Additionally, by multiplexing, qPCR can simultaneously be used to detect several different targets within the same sample.

qPCR is, however, not without problems. Different laboratories use different assays for the same target (e.g. different primer sequences and different instruments). When compared, these assays will often give different endpoint or quantitative results for the same sample. Some assays are more sensitive than others, which may lead to differing ideas of whether a sample is positive or not.

This chapter will review the different molecular techniques used in the field of diagnostic virology. These techniques can essentially be broken into two main groups: the PCR-based methods and the non-PCR-based methods.

## PCR-based methods

### Classic PCR

The polymerase chain reaction (PCR), which was first invented by Kary Mullis in 1983 (Mullis *et al.*, 1986; Bartlett and Stirling, 2003), has become the most widely used technique in laboratory-based diagnostic virology. This is because PCR is inexpensive, rapid and specific. The exponential amplification of the reaction means that PCR is exquisitely sensitive, a characteristic, which, together with those noted, means that PCR adequately fulfils the requirements of a robust diagnostic test.

Although there are many variations of PCR, which are described in detail below, all PCR reactions are fundamentally very similar. A segment of the genome of interest is amplified by the addition of five separate components that are required for all PCR reactions. These are the nucleic acid target, primers complementary to the sequence being amplified, a suitable buffer containing magnesium, deoxyribonucleotide triphosphates (dNTPs) and a thermostable DNA polymerase. The resulting amplified DNA is called amplicon.

PCR requires thermostable DNA polymerases to carry out the 5′ to 3′ synthesis of a complementary strand of DNA on a single stranded template but from a double-stranded region. The double-stranded regions are provided by the addition of primers each complementary to opposite strands of the region of DNA to be amplified (see Fig. 6.1A). The double stranded starting target molecule is heated to make it single stranded, cooled to allow the annealing of the primers and then heated again to allow the thermostable DNA polymerase to synthesize the new complementary DNA strand. By repeating the heating and cooling cycles, DNA will continue to accumulate in an exponential manner until either one of the components is expended or the DNA polymerase, being damaged by the repeated temperatures cycles, is unable to synthesize new DNA.

Thus, a basic PCR run of many cycles can be broken up into four distinct phases. These are the baseline, the exponential phase where the amplicon quantity is doubled at each cycle, the linear phase where the amplification slows down due the reaction components being used and the plateau phase where the reaction stopped with no more accumulation of products (see Fig. 6.2). The number of cycles required to provide optimum amplification is dependent on a number of factors including the amount of input material and the efficiency of the reaction, but most reactions are completed in less than two hours.

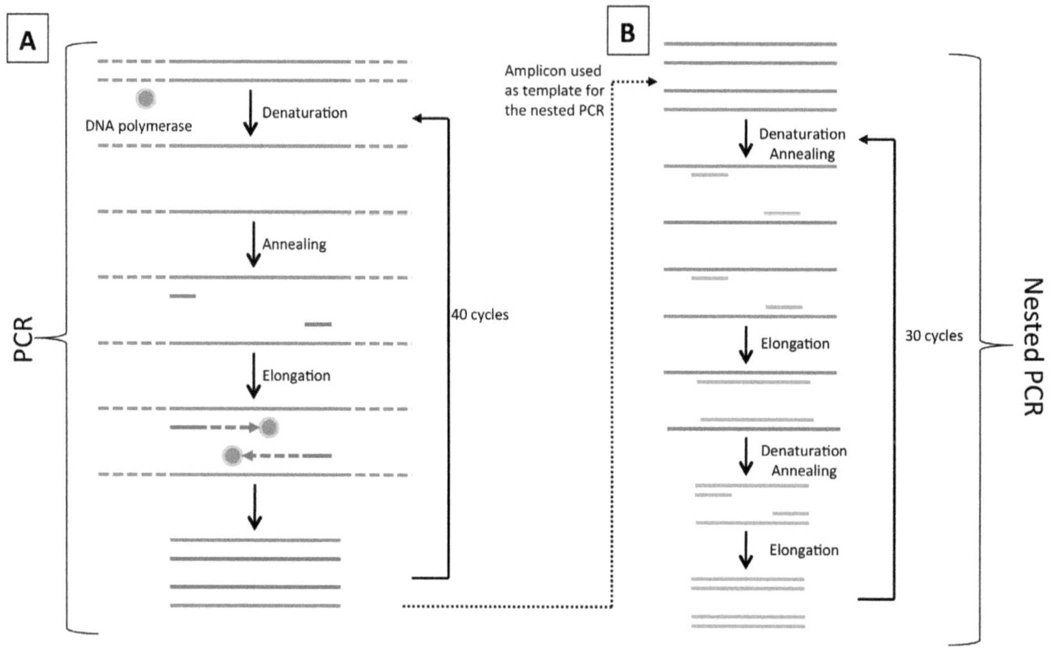


**Figure 6.1** (A) PCR and (B) nested PCR.

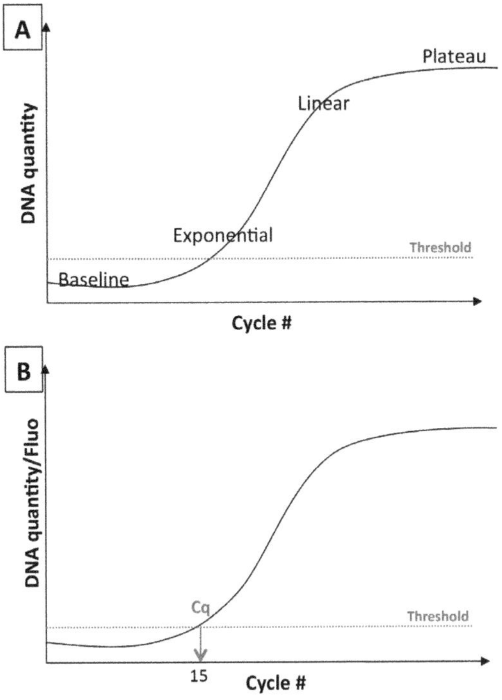


**Figure 6.2** (A) Kinetic of a PCR reaction and (B) quantification cycle (Cq).

Once the reaction has finished, the DNA target is detected by loading the reaction mixture and electrophoresing the reaction through an agarose gel containing an intercalating agent such as ethidium bromide and visualizing with UV light.

The PCR 'revolution' began with the discovery of the thermostable DNA polymerase from bacterium *Thermus aquaticus*. This enzyme functions optimally at 72°C but can sustain higher temperatures such as 95°C allowing the denaturation of double stranded DNA. Nowadays, DNA polymerases used for PCR are no longer isolated from *Thermus aquaticus* but are genetically engineered to fit certain criteria, such as proofreading, long fragment amplification or fast amplification, and expressed in other microorganisms such as *Escherichia coli.*

The detection of RNA viruses may be achieved using PCR, but an additional step is required. This step is called reverse transcription (RT), where, prior the amplification, the input RNA is converted to complementary DNA (cDNA). This cDNA strand of the resultant RNA/DNA hybrid serves as the PCR template. The conversion of RNA into complementary DNA is performed using reverse transcriptases initially extracted from retro viruses such AMV (*Avian myeloblastosis virus* reverse transcriptase) or MMLV (*Moloney murine leukemia virus* reverse transcriptase).

More recently, manufacturers have developed PCR mixes containing both reverse transcriptase and DNA polymerase, suitable for the detection of DNA and/or RNA. This simplifies the detection of DNA and/or RNA viruses in the same clinical sample (Koppelman *et al.*, 2012).

## Nested PCR

Occasionally the problem of low sensitivity arises when performing classic PCR. This is especially likely when dealing with either poor quality or low copy number template or both. Often the nucleic acid target becomes degraded when samples have been inappropriately stored or transported. By employing the technique of nested PCR, the sensitivity and specificity of the amplification is considerably enhanced.

The principle of nested PCR is to use the amplicon as a template for a second PCR. Two new primers located within the primary amplicon are used to amplify the primary amplicon template in a second PCR reaction (see Fig. 6.1B). This technique can be used when no signal or low signals are obtained with classical PCR. The signal is detectable in the same way as classical PCR – on an agarose gel containing an intercalating agent.

A major drawback with nested PCR is the issue of potential contamination, as it relies on using amplified product as a starting point. Owing to the exponential nature of the PCR process, it is essential that the amplified final product is unable to contaminate any subsequent reactions of the same target. This method presents a major drawback as it does create dilemma in a one-way workflow molecular laboratory. Tubes, plates or capillary containing amplicons should only be opened in the post-PCR room. In this context, the amplicons will be used as template therefore they could be opened in the PCR room but with huge risks to contaminate the subsequent assays. The other option is to load the nested PCR in the post-PCR room but this time with high risk to contaminate the reaction itself (Nie *et al.*, 2012).

## Quantitative real-time PCR (qPCR)

Once PCR became an established diagnostic tool, the issue of quantification of the product was addressed. There are several methods available to quantify PCR products. One of the most accurate techniques to quantify amplicons involves the incorporation of radio-labelled nucleotides or primers into the PCR product. However, for diagnostic use this is impractical.

An alternative method is to compare the fluorescence (from an intercalant dye) from the PCR product on an agarose gel with known standards. However, this technique can be insensitive with low amounts of product and the relationship between the fluorescence and the amount of product loaded is frequently non-linear.

These shortcomings led to the development of quantitative real-time PCR (qPCR). The principle of qPCR is that detection is in the early phase of amplification rather than the end of the reaction like classical PCR (Chiang *et al.*, 1996; Gibson *et al.*, 1996; Heid *et al.*, 1996). Quantification of amplicons at the saturation/plateau phase of PCR cannot be accurately related to the amount of starting template, as each reaction will plateau at a different point due to the different reaction kinetics for each sample. In qPCR, the cycle at which the fluorescence increases above threshold (Cq) can be directly related to the amount of starting target nucleic acid in the sample. Two samples containing different quantities of target will be detected at different Cqs (see Fig. 6.3A) in qPCR, but might show similar band intensity at end-point detection on an agarose gel (see Fig. 6.3B). On the contrary, two replicates of the same samples run by qPCR that have the same Cq (see Fig. 6.3C) might show different band intensity on an agarose gel (see Fig. 6.3D).

qPCR is performed without the need of gel electrophoresis due to the inclusion of a fluorescent reporter into the reaction and use of a qPCR machine that combines a thermal cycler with a

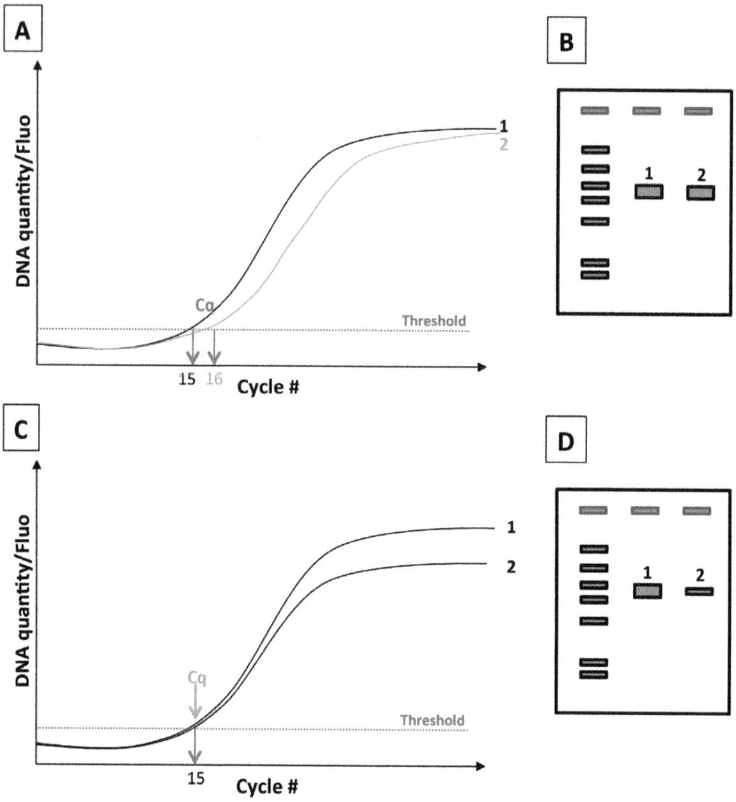


**Figure 6.3** (A, C) Real-time and (B, D) end-point detection of amplicons.

fluorometer. There are a number of dyes or probes that are used as reporters.

Key to qPCR is the setting of a threshold. This is an arbitrary level of fluorescence, which might be different for each reaction, chosen on the basis of the baseline variability. When a signal crosses the threshold, the sample is considered positive. This point is then used to define the Cq (Bustin *et al.*, 2009) for a sample (see Fig. 6.2B).

### Intercalating dye-based qPCR

Initially, qPCR was performed using the dye SYBR Green. As for classical PCR, qPCR uses exactly the same components as described above, but the intercalating dye SYBR Green is also added to the reaction. SYBR Green is a cyanine dye which binds non-specifically to the minor groove of double stranded (ds) DNA (see Fig. 6.4). In its unbound form, SYBR Green fluoresces poorly in the presence of blue light at 497 nm. However when bound to ds DNA, the dye fluoresces strongly and emits green light at about 520 nm. The amount of fluorescence is proportional to the amount of ds DNA present in the reaction. By reading the amount of fluorescence at the end of each elongation cycle, the amount of ds DNA present in the reaction can be determined.

The main disadvantage of qPCR using SYBR Green is that the dye is non-specific. This means that it binds any ds DNA present in the reaction such as primer dimers or unspecific amplicons. However, the specificity of the amplicon can be checked by analysis the melting curve at the end of the qPCR run (see Fig. 6.5). This melting/dissociation curve is obtained by adding an extra step at the end of the normal run. During this step, the temperature will be increased slowly (from 50°C to 95°C) denaturing the amplicons and releasing the dye, which reduces the fluorescence. By background subtraction and differentiation ($-dF/dT$), the melting curves are converted into melting peaks (Ririe *et al.*, 1997). The melting

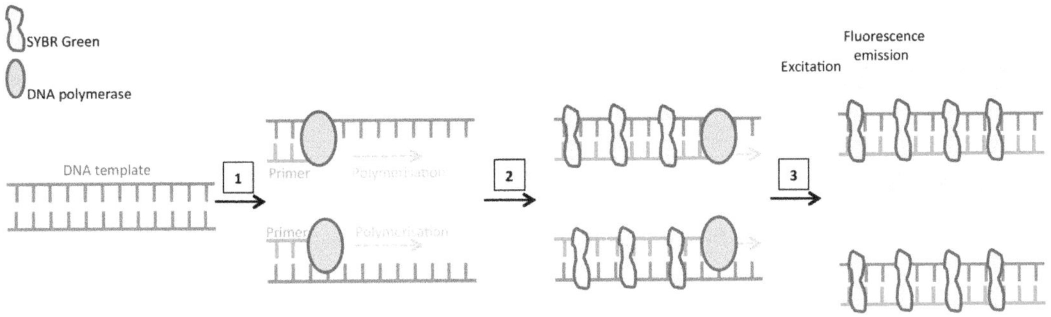


**Figure 6.4** qPCR using SYBR Green.

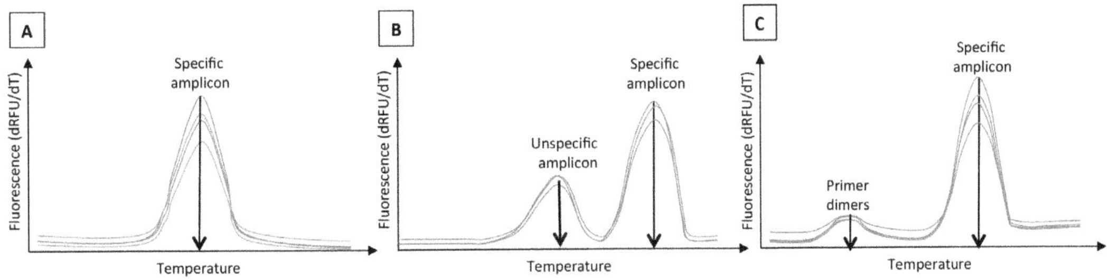


**Figure 6.5** Melting curve obtained after amplification of a specific target (A). Non-optimized assays may result in the presence of unspecific amplicons (B) or primer dimers (C) that will be detected and quantified using SYBR Green.

temperature ($T_m$) of each product is defined as the temperature at which the corresponding peak maximum (see Fig. 6.5). A specific amplification will give only one PCR product and therefore will display only one melting curve (see Fig. 6.5A). A non-specific amplification can produce different amplicons and display several melting curves (see Fig. 6.5B). A similar situation is observed for primer dimers (see Fig. 6.5C). The fluorescence coming from the unspecific target and/or the primer dimers will be quantified in addition to the fluorescence from the specific target giving an overestimation of the results. This underlines the absolute requirement for developing very specific assays when using SYBR Green.

As qPCR is performed in an enclosed system, this dramatically reduces the possibility of contamination, as there is no requirement to open the tubes containing amplified products for subsequent analysis.

SYBR Green has been the most common intercalant dye used for qPCR. However, it has been reported to inhibit PCR in a concentration-dependent manner (Nath *et al.*, 2000). The next generation of DNA-binding dyes, such as Eva Green (Eischeid, 2011), LC Green (Wittwer *et al.*, 2003; Nakagawa *et al.*, 2008) and SYTO9 (Monis *et al.*, 2005) are thought to give better results for melting curve analysis.

### Hydrolysis/TaqMan® probe-based qPCR

Although SYBR Green technology was a major advance over classical PCR, the problems of specificity sometimes remained. To overcome these issues, an additional specific component was added to the reaction. Together with specific primers for the target of interest, a specific probe complementary to a sequence within the amplified region is included. The probe is labelled with a reporter fluorophore covalently attached to the 5′-end and a quencher molecule at the 3′-end. The proximity of the quencher to the reporter prevents emission from the fluorescent reporter. As the PCR reaction progresses, the DNA polymerase meets the labelled probe which is bound to the target DNA. The 5′ exonuclease activity of the DNA polymerase hydrolyses the probe, releasing the reporter from the quencher. Thus, after excitation, the reporter is free to fluoresce (see Fig. 6.6). The quantity of released fluorescence is directly proportional to the quantity of amplified target. The amplification process is generally performed in two steps, a denaturation step at 95°C and a annealing/elongation step at approximately 55–65°C. The single annealing/elongation step is designed to optimize the exonuclease activity of the polymerase and to prevent the probe dissociating from the target before the hydrolysis.

### Hybridisation/FRET probe-based qPCR

Fluorescence resonance energy transfer (FRET) probe chemistry is based on the use of two probes (see Fig. 6.7). With FRET probes, one is labelled with a donor fluorophore and the second probe with an acceptor fluorophore. Both probes are designed to hybridize closely, head to tail, on the target to be amplified. When the two probes are hybridized to the single strand target, and after excitation of the donor, the fluorescence will transfer to the acceptor and this will be will be monitored. The use of two probes increases the specificity but also increases the complexity of the assay design. As the probes remain intact, melt peak analysis can also be performed for additional specificity.

### Molecular beacon probe-based qPCR

Molecular beacon probes are single probes which are labelled with a reporter and a quencher similar to TaqMan® probes. However the difference between TaqMan® probes and beacon probes is that the beacon probes are oligonucleotides with a hairpin-loop structure (Tyagi and Kramer, 1996). The hairpin structure brings the 3′ quencher close to the 5′ reporter preventing the reporter fluorescing when excited in the absence of target nucleic acid. When the beacon probe hybridizes to the target, the reporter is separated from the quencher and can fluoresce after excitation (see Fig. 6.8). Beacon probes are highly specific for the target even at low annealing temperatures and can be used for single nucleotide polymorphism (SNP) detection (Broude, 2002) and in isothermal nucleic acid amplification systems. They are less frequently used as their design is more complex than hydrolysis probes.

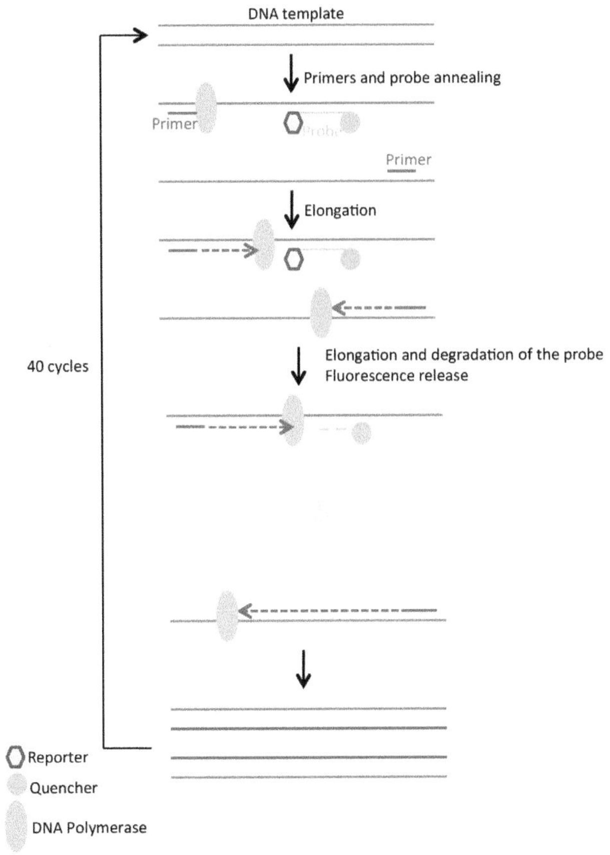


**Figure 6.6** Hydrolysis probe.

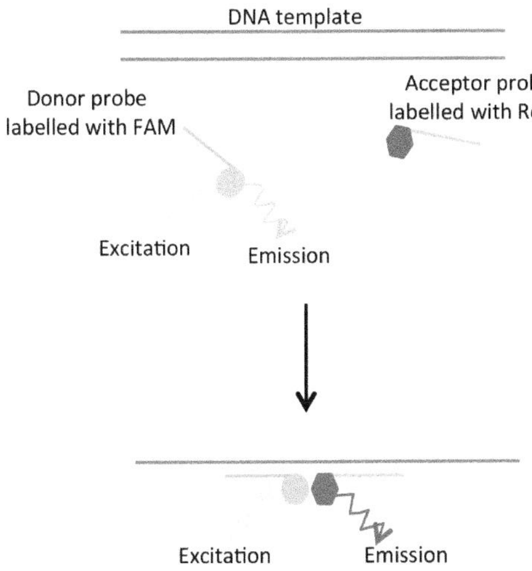


**Figure 6.7** Hybridization/FRET probes.

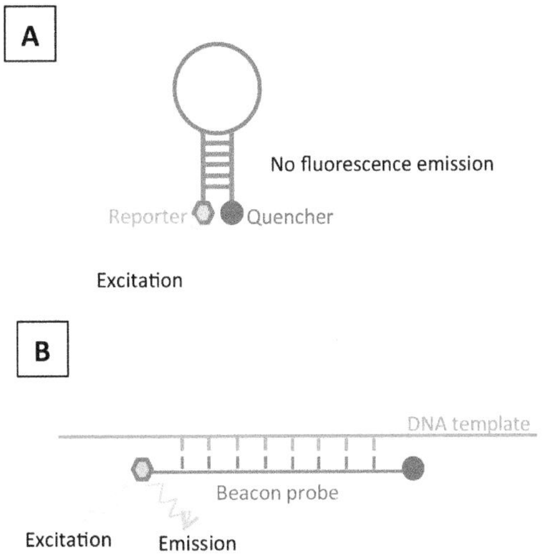


**Figure 6.8** Beacon probe. (A) Unhibridized probe does not emit any fluorescence even after excitation due to the proximity of the quencher. (B) Hybridization of the probe on the DNA allows the reporter to be free from the quencher and to emit some fluorescence (after excitation).

## Scorpion probe-based qPCR

Scorpion probes (Dreier *et al.*, 2005) are also dual labelled oligonucleotides with a hairpin shape but with a covalently bound primer. The probe is complementary to the extension product of the primer it is attached to. The amplification is a monomolecular reaction as the probe attaches to the newly synthesized DNA target allowing the reporter to fluoresce after excitation (see Fig. 6.9). These probe assays are very versatile and allow different options, such as the genotyping or detection of mutations (by the use of two different probes in the same tube, one carrying the wild-type sequence and the other one carrying the mutation) or simultaneous detection of several viruses. Each probe is labelled with a different fluorophore allowing the detection of either or both (Alby *et al.*, 2013).

## Absolute and relative quantification using qPCR

As mentioned previously, qPCR is capable of quantifying the pathogen of interest in a sample. Quantification may take several forms. At its most basic, quantification is relative, e.g. there is more of the pathogen of interest in sample A than sample B. This type of quantification may be performed when the samples are run at the same time. However, this is not always practical, for example, when following a course of treatment

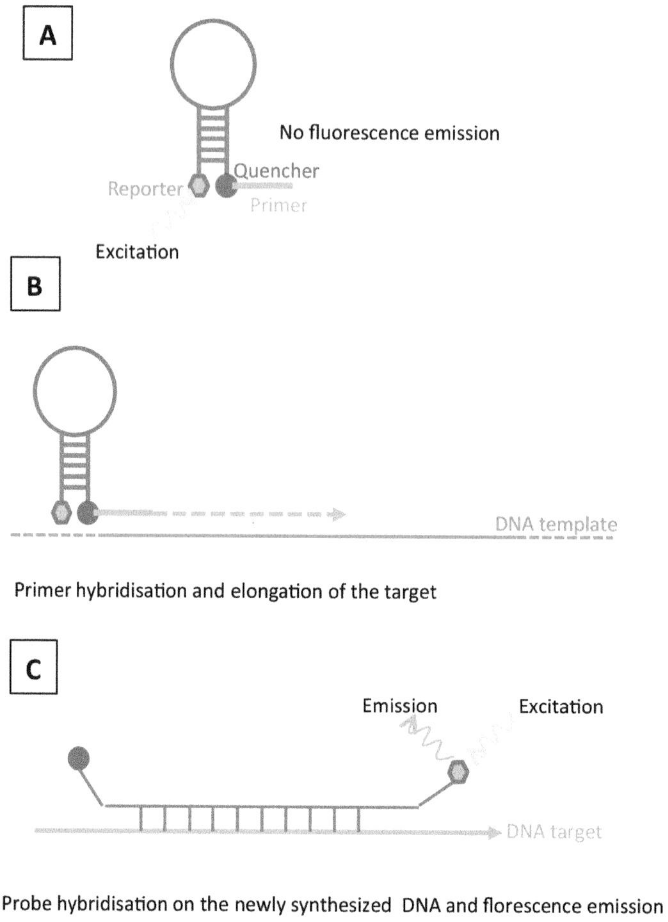


**Figure 6.9** Scorpion probe. (A) Unhibridized probe does not emit any fluorescence even after excitation due to the proximity of the quencher. (B) The primer located on the probe hybridizes on the DNA and allows elongation of the DNA. (C) The probe hybridizes on the newly synthesised DNA which allows the reporter to be free from the quencher and to emit some fluorescence (after excitation).

over several days or weeks for a particular disease. The other option is 'absolute' quantification. It is called absolute because it refers to an 'absolute' quantity expressed in, for example, copy number, ng of DNA/RNA or international units (IU) as opposed to a fold change used in 'relative' quantification. 'Absolute' quantification is made relative to a standard curve. For each run, serial dilutions of a known amount of target are assayed in parallel with the clinical samples. A linear regression is calculated based on the standards and this is used to quantify the amount of target nucleic acid in the clinical samples (see Fig. 6.10). It is important to note that inter laboratory comparisons expressing an absolute value are only valid when the same control material is used in all cases.

## The international standard (IS)

In an effort to standardize the quantification assays across different labs, an approach could be to use an International Standard when available. International Standards are reference materials provided by the World Health Organization (WHO) which serve as reference sources of defined biological activity expressed in an internationally agreed unit: the International Unit (IU). Thirteen WHO International Standards are available for viral PCR: hepatitis A, B, C and E, HIV-1, and 2, HPV16 and 18, CMV, EBV and parvovirus B19 (see Table 6.1).

International standards are not designed to be used in every assay but instead for an individual laboratory to assign a value to a batch of secondary calibrators. It is important to point out that the IS and the secondary calibrator should be diluted in the same matrix as the clinical samples investigated, e.g. whole blood or plasma, so the extraction and the amplification steps will be the same between the IS, secondary calibrator and clinical samples. Each new batch of secondary calibrators must therefore be re-evaluated against the appropriate IS. Serial 10-fold dilutions of the secondary calibrator are subsequently used in every run to create the standard curve (as described above) with a known value in IU. When this method is employed, all results are traceable back to the IU. Thus, results generated under these circumstances become comparable on both an intra- and inter-laboratory basis. Accurate viral load determination is paramount for some infections, such as HIV, as it is used to decide when to administer intervention, for

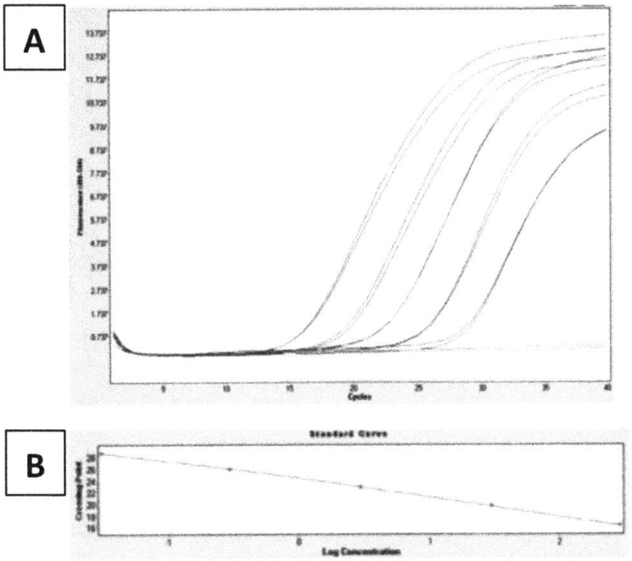

**Figure 6.10** Standard curve for absolute quantification. (A) Amplification curves of a series of dilutions of a standard. (B) Standard curve obtained by plotting the Cq values against the known concentrations (in Log) of the dilutions.

**Table 6.1** List of WHO international standards (IS) for virus detection using nucleic acid detection techniques (NAT)

| Pathogens | WHO IS | Cat no | Repository | Reference |
|---|---|---|---|---|
| B19 | 2nd WHO international standard for parvovirus B19 DNA for NAT | 99/802 | NIBSC, UK | |
| B19 | 1st WHO international reference panel for parvovirus B19 genotypes for NAT | 09/110 | NIBSC, UK | Baylis *et al.*, Vox. Sang., 2012 |
| CMV | 1st WHO international standard for human cytomegalovirus for NAT | 09/162 | NIBSC, UK | Fryer *et al.*, WHO ECBS Report 2010; WHO/BS/10.2138 |
| EBV | 1st WHO international standard for Epstein–Barr virus for NAT | 09/260 | NIBSC, UK | Fryer *et al.*, WHO ECBS Report 2011; WHO/BS/11.2172 |
| HAV | First WHO international standard for hepatitis A virus RNA for NAT | 00/560 | NIBSC, UK | Saldanha *et al.* Vox. Sang., 2005 |
| HBV | 3rd WHO international standard for hepatitis B virus for NAT | 10/264 | NIBSC, UK | Fryer *et al.*, WHO ECBS report 2011; WHO/BS/2011.2170 |
| HBV | 1st WHO international reference panel for hepatitis B virus genotypes for NAT | 5086/08 | PEI, Germany | Chudy *et al.*, WHO Report 2009, WHO/BS/09.2121 |
| HCV | 4th WHO international standard for hepatitis C virus for NAT | 06/102 | NIBSC, UK | Fryer *et al.*, WHO ECBS report 2011; WHO/BS/2011.2173 |
| HEV | 1st WHO international standard for hepatitis E virus RNA for NAT | 6329/10 | PEI, Germany | Baylis *et al.*, J. Clin. Microbiol., 2011 |
| HIV-1 | 3rd WHO international standard for HIV-1 NAT assays | 10/152 | NIBSC, UK | |
| HIV-2 | HIV-2 RNA international standard | 08/150 | NIBSC, UK | |
| HPV16 | 1st WHO international standard for human papillomavirus (HPV) type 16 DNA | 06/202 | NIBSC, UK | Quint *et al.*, J. Clin. Microbiol., 2006/Wilkinson *et al.*, Int. J. Cancer, 2010 |
| HPV18 | 1st WHO International Standard for Human Papillomavirus (HPV) Type 18 DNA | 06/206 | NIBSC, UK | Quint *et al.*, J. Clin. Microbiol., 2006/Wilkinson *et al.*, Int. J. Cancer, 2010 |

determining the efficacy of drugs, and for assessing drug resistance.

The situation is more complex when no IS is available. Different laboratories therefore use different standards that are defined differently in each case. This makes comparisons impossible. Owing to the limit in the number of International Standards available, most viral pathogens, such as the viruses responsible for respiratory or gastroenteritis symptoms, do not yet have a common International Standard.

## qPCR use outside pathogen quantification

As described in the above paragraphs, qPCR is commonly used to quantify viral load in clinical samples using standards. However in some cases, detecting and typing the viruses is more useful than a very accurate viral load. This is exemplified by the Influenza viruses. The common circulating influenza viruses are influenza type B virus and two subtypes of type A virus, H1N1 and H3N2. Therefore, multiplex assays using type specific primers and/or probes have been developed to diagnose and identify these viruses, with no accurate quantification required (Suwannakarn *et al.*, 2008; Wu *et al.*, 2008). Another example would be the detection and typing of HSV in cervico-vaginal lavage fluid of HIV-infected subjects. Anti-HSV therapies are efficient but require rapid, efficient, reliable and type specific diagnosis (Aryee *et al.*, 2005).

Another example of the use of qPCR for applications other than quantification of viral pathogens, specifically called allele-specific real-time PCR, is the detection or monitoring of antiretroviral drug resistance mutations in the HIV genome (Metzner *et al.*, 2011). This technique is based on the use of several primers or probes bearing the mutations of interest. It allows clinicians to choose or change specific therapeutic regimens based on the patient's drug-resistance profile, thus improving patient outcomes.

### Multiplexing

The detection of several targets simultaneously in clinical samples, called multiplexing, is frequently used in clinical diagnostics. Simultaneous detection can be achieved by using probes labelled with fluorophores that emit at different wavelengths. Most qPCR devices are able to differentiate between four and six wavelength ranges, therefore, several reactions can take place in at the same time. Multiplexing is not without its challenges, as initial assay design, optimization and set-up are costly and time consuming. In practice, multiplexing is limited by the spectral overlap of the dyes, the number of detector channels available on the common qPCR instruments and the competition for reagents of the simultaneous amplifications. These factors can reduce the sensitivity and dynamic range of the assays in multiplex format when compared to singleplex reactions.

## Non-PCR-based methods

### Strand displacement amplification (SDA)

SDA requires primers specific to the target of interest and a high fidelity, 5′–3′ exonuclease-deficient DNA polymerase. In most cases this is Phi29 DNA polymerase (Walker *et al.*, 1992). This technique is based upon the known ability of the 5′–3′ exonuclease-deficient DNA polymerase to extend the 3′ end of the primer and displace the downstream strand. As synthesis proceeds, strand displacement of complementary DNA generates new single-stranded DNA. The subsequent priming and strand displacement replication of this DNA results in the formation of double-stranded DNA (see Fig. 6.11). This procedure is isothermal except for denaturation. It can be applied to either single or double-stranded DNA and it is very sensitive (Hellyer and Nadeau, 2004).

### Ligase chain reaction (LCR)

The LCR uses a DNA ligase and two adjacent synthetic primers for each target DNA strand. The primers are ligated together to form a single probe that will be subsequently amplified. The junction of the two primers can be positioned so that the nucleotide at the 3′ end of the upstream primer coincides with a potential SNP (see Fig. 6.12). The mismatch will prevent the ligation of the two probes and therefore prevent amplification. This technique allows the discrimination of DNA sequences differing in only a single base pair (Wiedmann *et al.*, 1994). Variations of this technique were also developed involving a DNA polymerase (Barany, 1991; Wiedmann *et al.*, 1994).

### Nucleic acid sequence-based amplification (NASBA) and transcription-mediated amplification (TMA)

NASBA and TMA are single-step isothermal processes used to amplify RNA targets(Compton, 1991) that can be used for virus detection (Kamisango *et al.*, 1999; Lanciotti and Kerst, 2001). A NASBA reaction needs several different

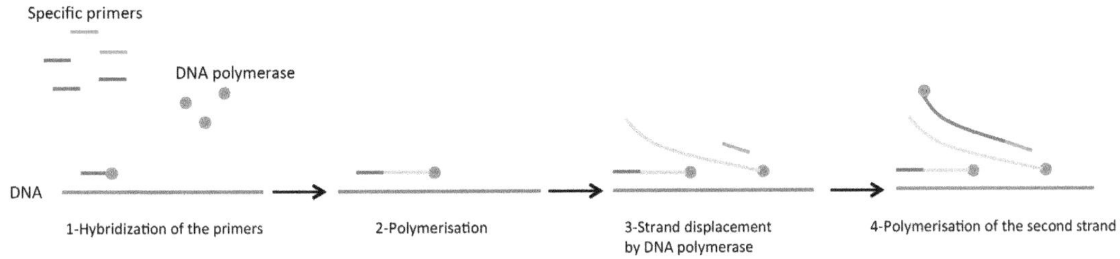


**Figure 6.11** Strand displacement amplification (SDA).

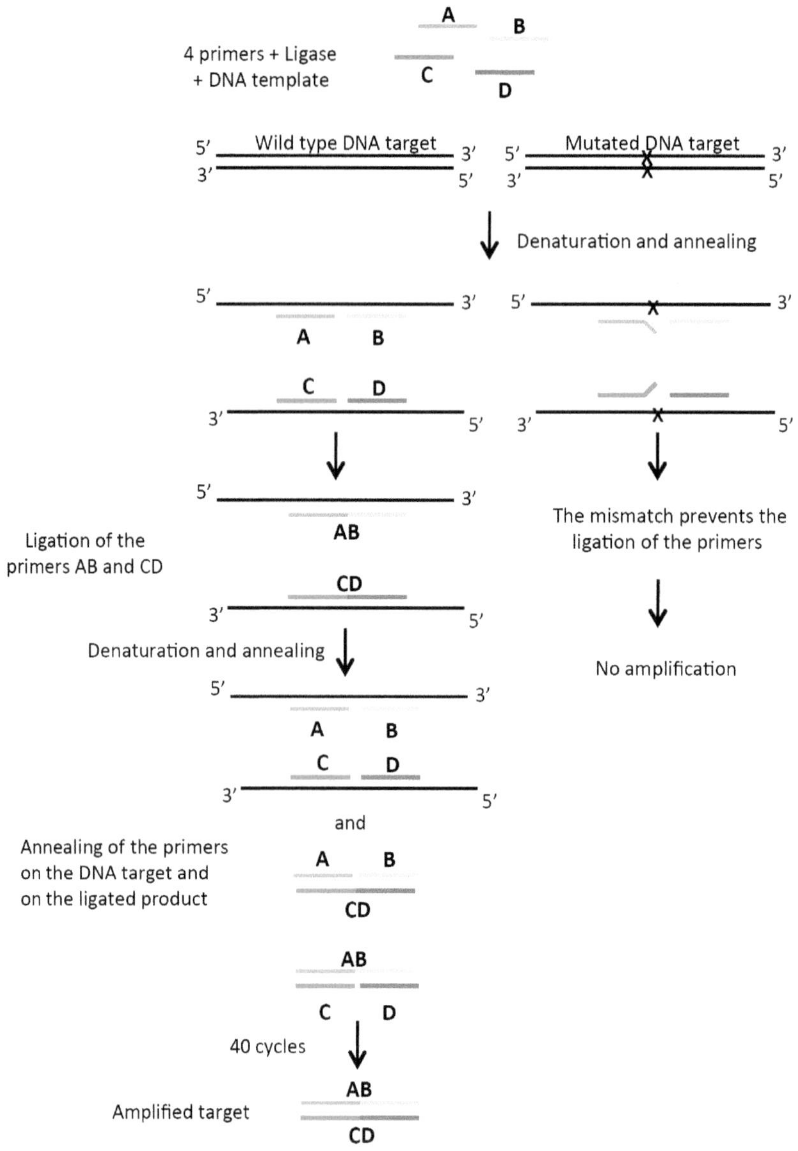


**Figure 6.12** Ligase chain reaction (LCR).

enzymes, namely reverse transcriptase, T7 RNA polymerase and RNase H, as well as target specific primers. The forward primer must contain a T7 RNA polymerase promoter sequence. In the first step, the reverse transcriptase, together with the two two primers, synthesizes a double stranded DNA amplicon containing a T7 RNA polymerase promoter sequence. This sequence is then used as template for the synthesis of large amount of RNA by the T7 RNA polymerase (see Fig. 6.13). Initially, the method relied on gel-based detection of products but more recently real-time fluorescence detection has also been applied using probe technology such as molecular beacons (Niesters, 2001). TMA is very similar to NASBA but uses a reverse transcriptase (AMV) with an endogenous RNase H activity (Langabeer *et al.*, 2002).

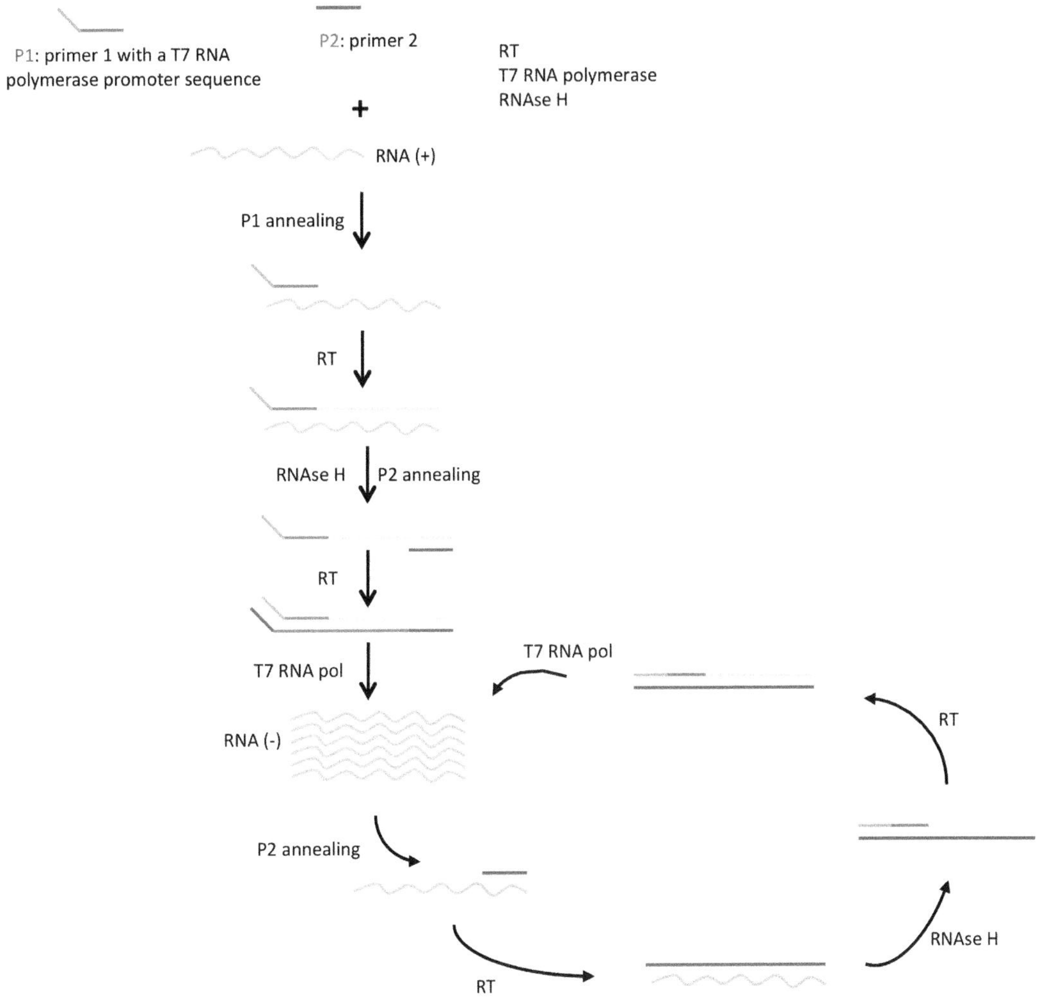


**Figure 6.13** Nucleic acid sequence-based amplification (NASBA).

## Loop-mediated isothermal amplification (LAMP)

Another isothermal nucleic acid amplification technique used to detect viruses (Okafuji *et al.*, 2005; Parida *et al.*, 2005; Toriniwa and Komiya, 2006) is the loop-mediated isothermal amplification (LAMP). This technique requires the use of a DNA polymerase with a strand displacement activity and four primers (F, forward outer primer; B, backward outer primer; FIP, forward inner primer; and BIP, backward inner primer) (see Fig. 6.14). The simultaneous initiation of polymerization by the different primers allows a quick synthesis of large amount of DNA. Amplification starts when FIP anneals to the complementary region of the target DNA, then 'F' hybridizes and displaces the first strand that subsequently forms a loop structure at one end. The newly synthesized DNA strand serves as template for the BIP hybridization and DNA synthesis. Finally, 'B' starts the strand displacement by annealing to the target, leading to the formation of loop-shaped DNA structure (Notomi *et al.*, 2000). The F and B primers are used mainly at the beginning of the reaction for strand displacement then the FIP and BIP primers are used for the loop structure amplification. It requires simple equipment rendering the method particularly suitable for field settings where sophisticated equipment or skilled personnel are lacking (Parida *et al.*, 2008).

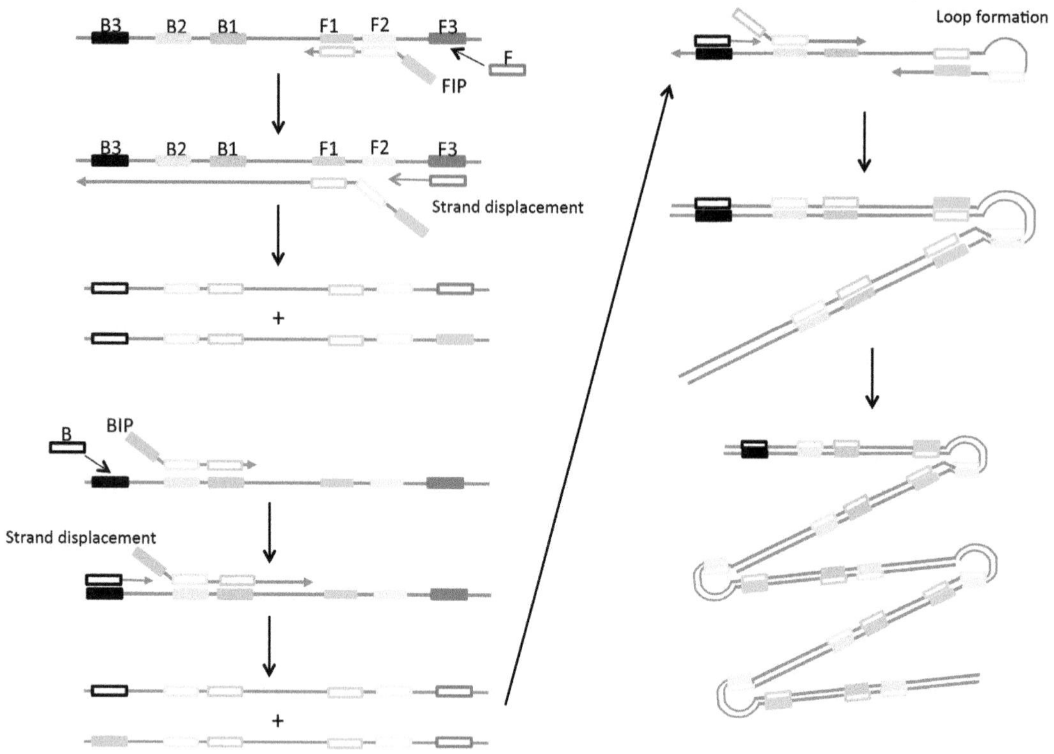


**Figure 6.14** Loop-mediated isothermal amplification (LAMP).

### Amplification-free hybridization-based assays

This technique is based on the use of UPT-reporter (Up-converting Phosphor Technology) which are inorganic microcrystals that emit visible light upon excitation with IR light (Zuiderwijk *et al.*, 2003). Advantages of UPT reporters compared to more conventionally used fluorescent dyes are higher sensitivity (~10- to 100-fold better detection), long shelf life, lower cost and the signals generated do not fade (Corstjens *et al.*, 2001). Thanks to the higher sensitivity of this reporter, amplification of the nucleic acids is not required. The technique is based on the use of oligonucleotides labelled with the UPT-reporter that will anneal on the viral targeted sequence (Corstjens *et al.*, 2003).

## Perspectives

In parallel with the introduction of qPCR has been the introduction of automated analysers, which are common in diagnostic laboratories. In these 'sample to answer' analysers, clinical samples are loaded into the device followed by nucleic acid extraction, amplification and detection. This means that the technician handles the sample for the minimum amount of time possible, reducing the risk of cross contamination. The nature of the automated devices means that they are highly reproducible which allows a high level of confidence in the results obtained. All data analysis is performed by device specific software. Dependent upon the device in question, a qualitative or quantitative result may be provided. Quantitative results may be provided in IU values, e.g. when a diagnostic kit is calibrated using the international standard, or in copy number when a standard curve is applied by the operator.

### Emerging technologies

PCR is an area of constant development and new technologies arise that move forward the state-of-the-art. One such technique is digital PCR (dPCR). dPCR is similar to qPCR but it relies on the principle that the sample being tested is separated into many fractions each with ≤1 copy of the target. Each fraction is then analysed separately and the results are featured as negative or positive. The final result is quantitative but based on a qualitative analysis of each fraction. In theory, dPCR offers several advantages over qPCR. Data published by Zhong *et al.* (2011) have shown that dPCR is able to simultaneously amplify multiple targets without competition as only single reactions are carried out in each cell of the instrument. Providing manufacturers develop instruments capable of differentiating many different wavelengths, dPCR should allow the multiplexing of many more targets than qPCR, aiding throughput and turnaround time to results. The results generated by dPCR should be operator independent allowing for higher intra- and inter- laboratory standardization.

Another area of rapid development is that of next generation sequencing (NGS) also known as massively paralleled sequencing. NGS is a name that covers many different variations of the same process. NGS is a high-throughput technology where the sequencing process is performed in parallel and produces tens of thousands to millions of sequences in one reaction, depending on instrument. Key steps towards routine use of NGS in diagnostic virology laboratories is the reduction in the amount of time required for the library preparation steps, standardization of this highly variable process and the development of the bioinformatics software necessary to rapidly make sense of the derived sequences. These factors currently make NGS unsuitable for routine diagnostics. However, several examples of its potential future use in molecular microbial diagnostics have been published (Palacios *et al.*, 2008).

Microarrays have been available for many years. A microarray is a collection of gene sequences (probes) that are synthesized onto a microscope slide, known as a DNA chip. Nucleic acid is hybridized to these sequences and probe–target hybridization, usually detected by fluorescence, indicated the specific detection of a particular target. Although microarrays offers huge advantages in throughput, e.g. more than 40,000 targets may be screened at once for gene expression, the inherent lack of sensitivity and high cost have led to a limited uptake of microarray technology for diagnostic use. Recently, the combination of PCR and microarray technology has led to the development of highly multiplexable diagnostics technology for the detection of infectious diseases (Chen *et al.*, 2011; Leveque *et al.*, 2013; Perotin *et al.*, 2013). However, so far, the use of microarrays in the screening of clinical specimens has mainly been for epidemiological purposes.

## Validation/standardization/ harmonization

Any new assay introduced by a diagnostic laboratory must be validated as fit for purpose. This validation should ensure that the assay is sensitive, accurate and precise. For 'in-house' assays, this responsibility lies with the individual laboratory (Dimech *et al.*, 2004). Commercial assays must comply with international standards and regulations (ISO 13485, Directive 98/79/EC, FDA regulations, etc.) (Fryer and Minor, 2009). Guidance on the development and validation of diagnostic tests that depends on nucleic acid amplification and detection techniques is currently being written/reviewed by the Health Protection Agency (HPA) (http://www.hpa.org.uk/webc/HPAwebFile/HPAweb_C/1317137954762). For all assays, validating performance on a day to day basis is crucial. Internal quality control (IQC) is the responsibility of individual laboratories. Appropriate quality controls (QC) should be included in all the runs. An overview of the ideal QC is summarized in Table 6.2 (Fryer *et al.*, 2008; Holden *et al.*, 2011).

Possibly the biggest problem associated with multiple different assays and instrumentation for specific analytes, is the issue of standardization. This is both intra- laboratory standardization, when comparing one operator to another in the same laboratory, and inter-laboratory standardization when comparing one laboratory to another. This is particularly important where viral load assays are performed and the results are used to inform on treatment. Inter- laboratory comparison is essential when, for example, new treatment strategies are being developed and refined using the results for different laboratories. Owing to their clinical importance, most assays for the common blood viruses, such as Hepatitis B and C

**Table 6.2** List of quality controls (QC)

| QC name | Description | Role |
| --- | --- | --- |
| Positive extraction/ amplification control | Whole virus solution that should be extracted and amplified alongside unknown samples. Commercially available controls must comply with the ISO 17511 | Part of a continuing quality control programme used to monitor assay and inter-assay performance. This control must be amplified at the same level (Cq, IU, copy numbers...) in all the runs |
| Negative extraction/ amplification control | Target virus-free clinical matrix that should be extracted and amplified alongside unknown samples | Validate the absence of target contamination. Must be negative for the virus target and positive for the IC |
| Internal amplification control (IC) | DNA spiked in all the samples (clinical and QC), extracted and amplified ideally simultaneously | Validate the extraction and amplification processes and demonstrate the absence of inhibitors |
| Positive amplification control | Known amount of target nucleic acids to be amplified. The IC can be added as well | Validate the amplification process. Must be positive for the virus target and the IC |
| Negative amplification control | Contains all the reagents necessary for the amplification except the nucleic acids that are replaced by molecular grade water. The IC should also be added | Demonstrate the absence of contamination at the amplification process. Must be negative for the virus target and positive for the IC |
| International Standard (IS) | Whole virus in solution (after reconstitution) or DNA solution that was assigned an arbitrary value (International Units IU) used to calibrate a second calibrator | The IS are intended to be used in the standardization of nucleic acid amplification technique (NAT)-based assays. They are produced by NIBSC and other collaborating centres commissioned by the World Health Organization (WHO) |
| Standards (second calibrator) | Serial 10-fold dilutions of the virus target that was calibrated against the IS if available or evaluated by spectrophotometry, that must be extracted and amplified along with the samples | Allows the assay to be quantitative. The results obtained for the standards will be used by the software to calculate the quantity of target in the unknown samples by linear regression |

and HIV are performed using commercially available viral detection kits and platforms. These are calibrated against the International Standard (as described above), thus allowing easy inter- and intra-laboratory comparison. This is not possible where no International Standard exits. Assays for respiratory pathogens, for example, have tended to be locally developed to meet a specific need. Most have been thoroughly designed, displaying good efficiency, sensitivity and specificity. However, owing to the lack of a unifying standard it is often very difficult to compare results between laboratories.

## Quality assurance schemes

A reliable means of ensuring continued high standards in laboratory testing is to ensure that a laboratory participates in an external quality assurance scheme (EQA). Indeed, to adhere to the laboratory ISO standard 17025, it is mandatory to participate in an EQA scheme. Examples of such schemes are those run by UK National External Quality Assessment Service (NEQAS). Such schemes send out blinded samples to many different laboratories and the results are collated and distributed to the participants. In this way each laboratory can see how they compare with colleagues in other laboratories. Data obtained from such schemes indicate a large variability in quantitative results between laboratories for their molecular assays. This indicates that there is great need for harmonization and standardization in PCR for infectious diseases and as a result, the IVD directive 98/79/EC is currently being reviewed to improve harmonization. The new regulations are scheduled to be published in 2014. Similarly, the Joint Committee on Traceability in Laboratory Medicine (JCTLM) (http://www.bipm.org/jctlm/home.do) was created to meet the need for a worldwide platform to promote and give guidance on internationally recognized and accepted equivalence of measurements in laboratory medicine and traceability to appropriate measurement standards.

## Conclusions

The aim of viral disease diagnostics is the detection of infection allowing for administration of appropriate therapy. During the last decade or so, PCR and more recently qPCR, have replaced the traditional diagnostic tests for viral pathogens, resulting in both improved detection rates and quicker sample turnaround. qPCR is a versatile, sensitive, fast and cost-effective method to detect/quantify viral targets. As qPCR has been introduced, most diagnostic laboratories have witnessed an increase in the number of tests requested, necessitating a multiplexing approach to maintain throughput. In parallel, automated analysers have become common in diagnostic laboratories increasing the throughput, the reproducibility and the confidence in the results, and reducing the risk of cross contamination. However, an associated problem with multiple different assays and instrumentation for specific analytes remains that makes inter-laboratory comparison impossible.

## Key regulatory documents

Code of Federal Regulations, Food and Drugs. 21 CFR 820 (Revised as of 1 April 2008). Published by the office of the Federal Register, National Archives and Records Administration (USA).

Commission decision of 7 May 2002 on common technical specifications for *in vitro* diagnostic medical devices. 2002. Off. J. Eur. Commun. *131*, 17–30.

Directive 98/79/EC of the European Parliament and of the Council of 27 October 1998 on *in vitro* diagnostic medical devices. 1998. Off. J. Eur. Commun. *331*, 1–37.

ISO 13485:2003. Medical devices – Quality management systems – Requirements for regulatory purposes. Geneva, Switzerland: ISO; 2003.

## References

Alby, K., Popowitch, E.B., and Miller, M.B. (2013). Comparative evaluation of the Nanosphere Verigene RV+ assay and the Simplexa Flu A/B & RSV kit for detection of influenza and respiratory syncytial viruses. J. Clin. Microbiol. *51*, 352–353.

Aryee, E.A., Bailey, R.L., Natividad-Sancho, A., Kaye, S., and Holland, M.J. (2005). Detection, quantification and genotyping of Herpes Simplex Virus in cervicovaginal secretions by real-time PCR: a cross sectional survey. Virol. J. *2*, 61.

Barany, F. (1991). The ligase chain reaction in a PCR world. PCR Methods Appl. *1*, 5–16.

Bartlett, J.M., and Stirling, D. (2003). A short history of the polymerase chain reaction. Methods Mol. Biol. *226*, 3–6.

Broude, N.E. (2002). Stem–loop oligonucleotides: a robust tool for molecular biology and biotechnology. Trends Biotechnol. *20*, 249–256.

Bustin, S.A., Benes, V., Garson, J.A., Hellemans, J., Huggett, J., Kubista, M., Mueller, R., Nolan, T., Pfaffl, M.W., Shipley, G.L., *et al.* (2009). The MIQE guidelines: minimum information for publication of quantitative real-time PCR experiments. Clin. Chem. *55*, 611–622.

Chen, E.C., Miller, S.A., DeRisi, J.L., and Chiu, C.Y. (2011). Using a pan-viral microarray assay (Virochip) to screen clinical samples for viral pathogens. J. Vis. Exp. *50*, 2536.

Chiang, P.W., Song, W.J., Wu, K.Y., Korenberg, J.R., Fogel, E.J., Van Keuren, M.L., Lashkari, D., and Kurnit, D.M. (1996). Use of a fluorescent-PCR reaction to detect genomic sequence copy number and transcriptional abundance. Genome Res. *6*, 1013–1026.

Compton, J. (1991). Nucleic acid sequence-based amplification. Nature *350*, 91–92.

Corstjens, P., Zuiderwijk, M., Brink, A., Li, S., Feindt, H., Niedbala, R.S., and Tanke, H. (2001). Use of up-converting phosphor reporters in lateral-flow assays to detect specific nucleic acid sequences: a rapid, sensitive DNA test to identify human papillomavirus type 16 infection. Clin. Chem. *47*, 1885–1893.

Corstjens, P.L., Zuiderwijk, M., Nilsson, M., Feindt, H., Sam Niedbala, R., and Tanke, H.J. (2003). Lateral-flow and up-converting phosphor reporters to detect single-stranded nucleic acids in a sandwich-hybridization assay. Anal Biochem. *312*, 191–200.

Dimech, W., Bowden, D.S., Brestovac, B., Byron, K., James, G., Jardine, D., Sloots, T., and Dax, E.M. (2004). Validation of assembled nucleic acid-based tests in diagnostic microbiology laboratories. Pathology *36*, 45–50.

Dreier, J., Stormer, M., and Kleesiek, K. (2005). Use of bacteriophage MS2 as an internal control in viral reverse transcription-PCR assays. J. Clin. Microbiol. *43*, 4551–4557.

Eischeid, A.C. (2011). SYTO dyes and EvaGreen outperform SYBR Green in real-time PCR. BMC Res. Notes *4*, 263.

Fryer, J.F., and Minor, P.D. (2009). Standardisation of nucleic acid amplification assays used in clinical diagnostics: a report of the first meeting of the SoGAT Clinical Diagnostics Working Group. J. Clin. Virol. *44*, 103–105.

Fryer, J.F., Baylis, S.A., Gottlieb, A.L., Ferguson, M., Vincini, G.A., Bevan, V.M., Carman, W.F., and Minor, P.D. (2008). Development of working reference materials for clinical virology. J. Clin. Virol. *43*, 367–371.

Gibson, U.E., Heid, C.A., and Williams, P.M. (1996). A novel method for real time quantitative RT-PCR. Genome Res. *6*, 995–1001.

Heid, C.A., Stevens, J., Livak, K.J., and Williams, P.M. (1996). Real time quantitative PCR. Genome Res. *6*, 986–994.

Hellyer, T.J., and Nadeau, J.G. (2004). Strand displacement amplification: a versatile tool for molecular diagnostics. Expert Rev. Mol. Diagn. *4*, 251–261.

Holden, M.J., Madej, R.M., Minor, P., and Kalman, L.V. (2011). Molecular diagnostics: harmonization through reference materials, documentary standards and proficiency testing. Expert Rev. Mol. Diagn. *11*, 741–755.

Kamisango, K., Kamogawa, C., Sumi, M., Goto, S., Hirao, A., Gonzales, F., Yasuda, K., and Iino, S. (1999). Quantitative detection of hepatitis B virus by transcription-mediated amplification and hybridization protection assay. J. Clin. Microbiol. *37*, 310–314.

Koppelman, M.H., Cuijpers, H.T., Wessberg, S., Valkeajarvi, A., Pichl, L., Schottstedt, V., and Saldanha, J. (2012). Multicenter evaluation of a commercial multiplex polymerase chain reaction test for screening plasma donations for parvovirus B19 DNA and hepatitis A virus RNA. Transfusion *52*, 1498–1508.

Lanciotti, R.S., and Kerst, A.J. (2001). Nucleic acid sequence-based amplification assays for rapid detection of West Nile and St. Louis encephalitis viruses. J. Clin. Microbiol. *39*, 4506–4513.

Langabeer, S.E., Gale, R.E., Harvey, R.C., Cook, R.W., Mackinnon, S., and Linch, D.C. (2002). Transcription-mediated amplification and hybridisation protection assay to determine BCR-ABL transcript levels in patients with chronic myeloid leukaemia. Leukaemia *16*, 393–399.

Leveque, N., Renois, F., and Andreoletti, L. (2013). The microarray technology: facts and controversies. Clin. Microbiol. Infect. *19*, 10–14.

Metzner, K.J., Rauch, P., Braun, P., Knechten, H., Ehret, R., Korn, K., Kaiser, R., Sichtig, N., Ranneberg, B., van Lunzen, J., *et al.* (2011). Prevalence of key resistance mutations K65R, K103N, and M184V as minority HIV-1 variants in chronically HIV-1 infected, treatment-naive patients. J. Clin. Virol. *50*, 156–161.

Monis, P.T., Giglio, S., and Saint, C.P. (2005). Comparison of SYTO9 and SYBR Green I for real-time polymerase chain reaction and investigation of the effect of dye concentration on amplification and DNA melting curve analysis. Anal Biochem. *340*, 24–34.

Mullis, K., Faloona, F., Scharf, S., Saiki, R., Horn, G., and Erlich, H. (1986). Specific enzymatic amplification of DNA *in vitro:* the polymerase chain reaction. Cold Spring Harb. Symp. Quant. Biol. *51*, 263–273.

Nakagawa, T., Higashi, N., and Nakagawa, N. (2008). Detection of antigenic variants of the influenza B virus by melting curve analysis with LCGreen. J. Virol. Methods *148*, 296–299.

Nath, K., Sarosy, J.W., Hahn, J., and Di Como, C.J. (2000). Effects of ethidium bromide and SYBR Green I on different polymerase chain reaction systems. J. Biochem. Biophys. Methods *42*, 15–29.

Nie, J.J., Sun, K.X., Li, J., Wang, J., Jin, H., Wang, L., Lu, F.M., Li, T., Yan, L., Yang, J.X., *et al.* (2012). A type-specific nested PCR assay established and applied for investigation of HBV genotype and subgenotype in Chinese patients with chronic HBV infection. Virol. J. *9*, 121.

Niesters, H.G. (2001). Quantitation of viral load using real-time amplification techniques. Methods *25*, 419–429.

Notomi, T., Okayama, H., Masubuchi, H., Yonekawa, T., Watanabe, K., Amino, N., and Hase, T. (2000). Loop-mediated isothermal amplification of DNA. Nucleic Acids Res. *28*, E63.

Okafuji, T., Yoshida, N., Fujino, M., Motegi, Y., Ihara, T., Ota, Y., Notomi, T., and Nakayama, T. (2005). Rapid diagnostic method for detection of mumps virus genome by loop-mediated isothermal amplification. J. Clin. Microbiol. *43*, 1625–1631.

Palacios, G., Druce, J., Du, L., Tran, T., Birch, C., Briese, T., Conlan, S., Quan, P.L., Hui, J., Marshall, J., *et al.* (2008). A new arenavirus in a cluster of fatal transplant-associated diseases. N. Engl. J. Med. *358*, 991–998.

Parida, M., Horioke, K., Ishida, H., Dash, P.K., Saxena, P., Jana, A.M., Islam, M.A., Inoue, S., Hosaka, N., and Morita, K. (2005). Rapid detection and differentiation of dengue virus serotypes by a real-time reverse transcription-loop-mediated isothermal amplification assay. J. Clin. Microbiol. *43*, 2895–2903.

Parida, M., Sannarangaiah, S., Dash, P.K., Rao, P.V., and Morita, K. (2008). Loop mediated isothermal amplification (LAMP): a new generation of innovative gene amplification technique; perspectives in clinical diagnosis of infectious diseases. Rev. Med. Virol. *18*, 407–421.

Perotin, J.M., Dury, S., Renois, F., Deslee, G., Wolak, A., Duval, V., De Champs, C., Lebargy, F., and Andreoletti, L. (2013). Detection of multiple viral and bacterial infections in acute exacerbation of chronic obstructive pulmonary disease: a pilot prospective study. J. Med. Virol. *85*, 866–873.

Ririe, K.M., Rasmussen, R.P., and Wittwer, C.T. (1997). Product differentiation by analysis of DNA melting curves during the polymerase chain reaction. Anal Biochem. *245*, 154–160.

Suwannakarn, K., Payungporn, S., Chieochansin, T., Samransamruajkit, R., Amonsin, A., Songserm, T., Chaisingh, A., Chamnanpood, P., Chutinimitkul, S., Theamboonlers, A., *et al.* (2008). Typing (A/B) and subtyping (H1/H3/H5) of influenza A viruses by multiplex real-time RT-PCR assays. J. Virol. Methods *152*, 25–31.

Toriniwa, H., and Komiya, T. (2006). Rapid detection and quantification of Japanese encephalitis virus by real-time reverse transcription loop-mediated isothermal amplification. Microbiol. Immunol. *50*, 379–387.

Tyagi, S., and Kramer, F.R. (1996). Molecular beacons: probes that fluoresce upon hybridization. Nat. Biotechnol. *14*, 303–308.

Walker, G.T., Fraiser, M.S., Schram, J.L., Little, M.C., Nadeau, J.G., and Malinowski, D.P. (1992). Strand displacement amplification--an isothermal, *in vitro* DNA amplification technique. Nucleic Acids Res. *20*, 1691–1696.

Wiedmann, M., Wilson, W.J., Czajka, J., Luo, J., Barany, F., and Batt, C.A. (1994). Ligase chain reaction (LCR)

– overview and applications. PCR Methods Appl. *3*, S51–64.

Wittwer, C.T., Reed, G.H., Gundry, C.N., Vandersteen, J.G., and Pryor, R.J. (2003). High-resolution genotyping by amplicon melting analysis using LCGreen. Clin. Chem. *49*, 853–860.

Wu, C., Cheng, X., He, J., Lv, X., Wang, J., Deng, R., Long, Q., and Wang, X. (2008). A multiplex real-time RT-PCR for detection and identification of influenza virus types A and B and subtypes H5 and N1. J. Virol. Methods *148*, 81–88.

Zhong, Q., Bhattacharya, S., Kotsopoulos, S., Olson, J., Taly, V., Griffiths, A.D., Link, D.R., and Larson, J.W. (2011). Multiplex digital PCR: breaking the one target per color barrier of quantitative PCR. Lab. Chip *11*, 2167–2174.

Zuiderwijk, M., Tanke, H.J., Sam Niedbala, R., and Corstjens, P.L. (2003). An amplification-free hybridization-based DNA assay to detect Streptococcus pneumoniae utilizing the up-converting phosphor technology. Clin. Biochem. *36*, 401–403.

# XMRV: A Cautionary Tale

7

Jeremy A. Garson

## Abstract

Xenotropic murine leukaemia virus-related virus (XMRV) was discovered in 2006 in tumour tissues from patients with a familial form of prostate cancer. The discovery was made using a powerful molecular detection technique involving random PCR amplification and a high density microarray composed of oligonucleotides from the most conserved sequences of all known viral families. XMRV was claimed to be the first example of a *Gammaretrovirus* to infect humans and its discovery was followed by several studies that confirmed its presence in prostate cancer patients. The publication in 2009 of a highly controversial paper in Science associating the virus with chronic fatigue syndrome (CFS) brought XMRV to the attention of the worldwide public. Its detection in 3.7% of healthy controls sparked fears that millions might be infected with a novel retrovirus of unknown pathogenic potential. However, almost all recent efforts to confirm the association between XMRV, CFS and prostate cancer have failed and XMRV is now thought much more likely to represent a contamination related laboratory artefact rather than a genuine human infection. This chapter reviews the spectacular rise and fall of XMRV and attempts to draw useful lessons from this painful episode in the history of molecular diagnostics.

## Introduction

The development and application of highly sensitive molecular biological techniques over the past two decades has led to significant advances in many areas of medicine including the discovery of novel human pathogens such as hepatitis C virus, human herpesvirus 8, hantavirus, human metapneumovirus and the SARS coronavirus (van den Hoogen *et al.*, 2001; Jones *et al.*, 2005). Initially, it looked as though xenotropic murine leukaemia virus-related virus would represent another triumphant discovery to add to the growing list of human pathogens identified by applying molecular techniques but for reasons to be explained in this chapter, the triumph proved to be short-lived.

Xenotropic murine leukaemia virus-related virus (XMRV) is so named because its genetic sequence is closely related to sequences of the xenotropic murine leukaemia viruses (MLV-Xs). Murine leukaemia viruses (MLVs) are retroviruses of the *Gammaretrovirus* genus that cause cancers and other diseases, including neurological disease, in mice. They are divided into the ecotropic, amphotropic, polytropic and xenotropic classes on the basis of their host ranges. MLV-Xs are endogenous retroviruses (ERVs) carried by most strains of mice as proviral genomes integrated into the chromosomes of every cell and inherited as Mendelian traits through the germline (Weiss, 2006). Under certain conditions, endogenous retroviral proviruses may become activated to yield potentially infectious viral particles. The term 'xenotropic' is applied to murine ERVs which have the peculiar characteristic of being able to infect the cells of foreign species, such as human, but are unable to infect the cells of many inbred mouse strains. This apparently paradoxical tropism of MLV-Xs is determined primarily by the cell surface receptor molecules to which the retroviral envelope glycoproteins bind prior to cell entry.

## Discovery of XMRV

XMRV was discovered in 2006 in the tumour tissues of patients suffering a familial form of prostate cancer associated with mutations that impair the function of the antiviral defence protein RNase L (Urisman *et al.*, 2006). The hypothesis underlying the discovery was that patients with prostate cancer, who had an inherited mutation in the RNase L gene, would be more likely to harbour a chronic and potentially oncogenic virus infection (Silverman, 2007). The discovery was dependent on a molecular detection technique which entailed random amplification by reverse transcription PCR of RNA extracted from human prostatic tumour tissues. Products of random amplification were hybridized to a DNA ViroChip microarray composed of oligonucleotides corresponding to the most conserved sequences of all known human, animal, plant, and bacterial viruses. A positive hybridization signal suggestive of a *Gammaretrovirus* was detected in 7 of 11 tumours from patients homozygous for the R462Q RNase L variant QQ. DNA eluted from the positive microarray spots was re-amplified and plasmid libraries constructed and screened by colony hybridization using the spot's oligonucleotides as probes. Further characterization of the viral nucleic acid obtained from the patients' tumour tissues enabled the entire sequence of the novel *Gammaretrovirus* genome to be obtained. XMRV was thus discovered and claimed to be the first example of a *Gammaretrovirus* to infect humans. An expanded survey of 86 tumours using an 'XMRV-specific' RT-PCR detected the virus in 8 of 20 RNase L variant QQ cases (40%), compared with only one (1.5%) among 66 RQ and RR cases. Using *in situ* hybridization and immunohistochemistry Urisman and colleagues (2006) reported that XMRV nucleic acid and protein were both present in the prostatic tumour biopsies, although unexpectedly they were located in stromal cells (i.e. fibroblasts and haematopoietic cells) rather than in the carcinoma cells themselves.

Two overlapping partial cDNAs of XMRV from prostate tumour sample VP62 (Urisman *et al.*, 2006) were joined to generate a full-length XMRV molecular clone (Dong *et al.*, 2007). This full length clone, designated XMRV VP62, was shown to be infectious by transfecting the human prostate tumour cell line LNCaP. The XMRV particles released by transfected LNCaP cells were in turn shown to be infectious to another human prostate tumour cell line, DU145. LNCaP-derived XMRV viral particles also efficiently infected and replicated in human ovarian carcinoma cells and in the human cervical carcinoma cell line, HeLa. In addition to this evidence that XMRV is capable of infecting human cells *in vitro* it has also been reported that XMRV can infect non-human primates *in vivo* (Onlamoon *et al.*, 2011). Rhesus macaques inoculated intravenously with extremely high doses of DU145-derived XMRV established a persistent chronic disseminated infection. XMRV in the macaques exhibited organ specific tissue tropism; CD4 T-cells in lymphoid organs and alveolar macrophages in lung became infected, as well as epithelial/interstitial cells in other organs, including the reproductive tract. It is noteworthy that XMRV replication was observed in the macaque prostate during acute but not chronic infection (Onlamoon *et al.*, 2011). However, Del Prete *et al.*, (2012) in a more recent study of XMRV in pigtailed macaques found only very limited and transient replication following intravenous XMRV inoculation, with minimal if any spread to prostate tissue. They concluded that XMRV replication was severely restricted by intrinsic defence mechanisms, predominantly by APOBEC-mediated hypermutation.

## Reports of XMRV in prostate cancer and chronic fatigue syndrome

Following its discovery in 2006, three independent groups reported the presence of XMRV in a proportion of human prostate tumours using a variety of 'XMRV-specific' PCR assays, *in situ* hybridization and immunological techniques. However, the percentage of positive tumours varied considerably between different studies and, puzzlingly, the linkage to polymorphisms of the RNase L gene was not confirmed (Schlaberg *et al.*, 2009; Arnold *et al.*, 2010; Danielson *et al.*, 2010). The epithelial cell location of XMRV reported by Schlaberg and colleagues (2009) within tumours was also inconsistent with the stromal location originally described by Urisman *et al.* (2006). As well as these reports of XMRV being detected in human prostatic tumour biopsy tissues, XMRV was also found to be produced at high levels by the long-established and widely used human prostate cancer cell line 22Rv1 (Knouf *et al.*, 2009). The special significance of this observation will be discussed later in this chapter. Importantly, integration of XMRV sequences into the chromosomal DNA of human prostate tumour biopsies was reported by Silverman's group (Dong *et al.* 2007; Kim *et al.* 2008). Chromosomal integration of the cDNA copy of genomic viral RNA to form the provirus is essential for retroviruses to establish productive infection and its demonstration was therefore regarded as the most direct evidence that bona fide, naturally occurring XMRV infections of humans had actually occurred.

For the first three years after its discovery, publications on XMRV accumulated slowly at around four per year but the rate increased spectacularly to ~90 papers per year following the high-profile publication of a study linking XMRV to chronic fatigue syndrome (CFS) (Lombardi *et al.*, 2009). RNase L dysfunction had earlier been reported in patients with CFS, a disorder of unknown aetiology affecting millions worldwide in which diverse viral infections had already been inconclusively implicated (Suhadolnik *et al.*, 1997; Devanur and Kerr, 2006). It therefore seemed a rational extrapolation from the work on prostate cancer to look for evidence of XMRV infection in CFS. The Lombardi study, published by the journal Science in October 2009 received extensive media coverage and brought XMRV to worldwide public attention. Undertaken by Mikovits and co-workers at the newly established Whittemore Peterson Institute in Nevada, in collaboration with Silverman, co-discoverer of XMRV, and Ruscetti (US National Cancer Institute), co-discoverer of the first human retrovirus HTLV-1, the study reported detection of XMRV sequences in 67% of 101 CFS patients and 3.7% of healthy controls by nested PCR on DNA extracted from peripheral blood mononuclear cells (PBMCs). The PCR findings were supported by flow cytometry and Western blotting demonstrating XMRV antigens in PBMCs, and by cell culture experiments demonstrating the presence of infectious XMRV in CFS patients' blood (Lombardi *et al.*, 2009). Ominously the paper concluded with the following words, 'Finally it is worth noting that 3.7% of the healthy donors in our study tested positive for XMRV sequences. This suggests that several million Americans may be infected with a retrovirus of as yet unknown pathogenic potential'. Understandably, this portentous message elicited fearful memories of the early days of the global HIV pandemic and triggered urgent efforts to replicate the findings by laboratories in the USA and across the world. The Lombardi study also caused Blood Transfusion Services in several countries to exclude those with a history of CFS from donating blood (Dodd, 2011).

A second high profile paper reporting the presence of retroviral sequences in patients with CFS was published by PNAS a few months later (Lo *et al.*, 2010). The paper, co-authored by the renowned Harvey J. Alter and colleagues from the National Institutes of Health, described the detection by nested PCR of MLV-like sequences in blood from 32 out of 37 (86.5%) CFS patients and in 6.8% of healthy volunteer blood donors. This result was erroneously interpreted by many as confirming the findings of Lombardi *et al.* but in fact the sequences amplified were not of XMRV itself but of closely related agents including polytropic and modified polytropic murine endogenous retroviruses (the term 'polytropic' in this context means capable of infecting many species including mice). Nevertheless, the study by Lo *et al.* did provide independent support for the belief that retroviruses related to murine leukaemia virus might be infecting a significant proportion of the general population and be associated with CFS.

## Mounting negative evidence ignites XMRV controversy

The first study on CFS patients to be published after the Lombardi *et al.* (2009) Science paper was by McClure's group at Imperial College London (Erlwein *et al.*, 2010). They studied a cohort of 186 patients from the UK all of whom met the international consensus criteria for CFS (Fukuda *et al.*, 1994). XMRV sequences were not found in any of the patients by nested PCR with either 'XMRV-specific' or consensus MLV primer sets. Similar negative findings from two other European research groups quickly followed (Groom *et al.*, 2010; van Kuppeveld *et al.*, 2010) but were dismissed by Mikovits and Ruscetti on the grounds that, like McClure's group, they had failed to use exactly the same PCR methodology as that described by Lombardi *et al.* (Mikovits and Ruscetti, 2010).

It was also asserted that the discrepant findings could be due to geographic differences in the distribution of XMRV between Europe and the USA, or due to sequence divergence between different strains of XMRV in different locations, or due to different patient selection criteria (Mikovits *et al.*, 2010). However, negative data continued to accumulate and by mid 2011 approximately 20 negative CFS studies had been published by independent research groups on several continents, including several studies from the USA (Switzer *et al.*, 2010). A recent review by Robinson *et al.* (2011a) notes that two of these negative studies (Knox *et al.*, 2011; Shin *et al.*, 2011) used exactly the same methodology and the same patient selection criteria as described in the original study by Lombardi *et al.* and actually tested a subset of patients that had previously been reported as being XMRV-positive in the Lombardi study. Even more disturbingly, a large multi-laboratory study by the Blood XMRV Scientific Research Working Group reported their findings on blind testing of coded replicate blood samples from 15 subjects previously reported to be XMRV/MLV-positive (14 with CFS) and from 15 healthy donors previously determined to be negative for the viruses (Simmons *et al.*, 2011). The results showed that none of the nine participating laboratories, including the Mikovits laboratory at the Whittemore Peterson Institute, were able to reliably distinguish between samples from CFS subjects and controls by using PCR, virus isolation or serological assay. They concluded, 'These results indicate that current assays do not reproducibly detect XMRV/MLV in blood samples and that blood donor screening is not warranted'.

The controversy surrounding the accumulating evidence against XMRV being a cause of CFS sparked an extraordinary and disturbing social phenomenon. Virologists, physicians and psychiatrists associated with these negative research studies were subjected to a campaign of intimidation, attacks and death threats from extreme activists within the CFS patient community (Hawkes, 2011). Professor Myra McClure, head of infectious diseases at Imperial College London and co-author of the first of the negative studies published in *PLoS One*, was one of those subjected to extreme abuse and had to withdraw from a US collaboration because she was warned she might be shot (McKie, 2011). In response to the intimidation of McClure and other scientists working in the field, the editors of the journal *PLoS One* issued a statement 'In Support of XMRV Researchers' on 2 September 2011. The statement concluded with the following: 'Those who threaten researchers' safety above all do themselves a major disservice by dissuading other researchers from entering the field, chasing away the very people who may be able to help them. It is bad both for science and for patients, and should absolutely not be tolerated' (*PLoS One*, 2011).

Encouraged by the initial reports of XMRV in prostate cancer and CFS, studies have also looked for evidence of XMRV infection in a number of other diseases of unknown aetiology. Cohorts of patients with amyotrophic lateral sclerosis (McCormick *et al.*, 2008), multiple sclerosis (Hohn *et al.*, 2010), autism (Satterfield *et al.*, 2010), rheumatoid arthritis (Henrich *et al.*, 2010), fibromyalgia and paediatric idiopathic disease (Jeziorski *et al.*, 2010) have all been investigated but no association with XMRV has emerged. Large numbers of healthy blood donors and patients considered at increased risk of infection, including HIV-1-positive individuals, have also been tested for XMRV but all with negative results (Arredondo *et al.*, 2011; Grey *et al.*, 2011; Mi *et al.*, 2011; Robinson *et al.*, 2011b; Tang *et al.*, 2011).

Whilst most of the prevalence studies over the past 2 years have been directed towards testing the claimed association with chronic fatigue syndrome, work on XMRV in prostate cancer has continued. Once again, the pattern that has emerged is that of a few early positive studies (Urisman *et al.*, 2006; Schlaberg *et al.*, 2009; Arnold *et al.*, 2010; Danielson *et al.*, 2010) being overwhelmed by negative studies from independent research groups in many countries (for example,

Hohn *et al.*, 2009; Aloia *et al.*, 2010; Sakuma *et al.*, 2011; Stieler *et al.*, 2011; Zhang *et al.*, 2011). These negative studies have employed a variety of high-sensitivity techniques, such as co-culture infectivity assays and immunohistochemistry, in addition to PCR. Aloia and colleagues employed HPLC purified proteins to raise the antisera for immunostaining, and were unable to find any trace of XMRV in ~800 prostate tumours analysed. They suggested that the positive immunostaining described in earlier studies may have been due to the use of non-specific antisera exhibiting cross-reactivity with human cellular proteins (Aloia *et al.*, 2010).

Central to the hypothesis that XMRV is a genuine human pathogen is the observation that it integrates into the chromosomal DNA of prostate tumour tissues (Dong *et al.* 2007; Kim *et al.* 2008). Given the importance of this observation we sought to examine the authenticity of the integration sites that had been described (Garson *et al.*, 2011). The only research group that had reported patient-derived XMRV integration sites provided sequence data from 14 sites cloned from the prostatic tumour tissues of nine patients. To our surprise, nucleotide BLAST searches using each of the 14 integration site sequences against the GenBank nr database revealed that 2 of the 14 integration sites, obtained from two different patients, were absolutely identical to XMRV integration sites which had previously been cloned from the human tumour cell line DU145 (Fig. 7.1), which had been experimentally infected with XMRV in the same laboratories (Kim *et al.*, 2008, 2010). Since identical integration sites had only extremely rarely been described in any retrovirus infection and since both patient and cell line XMRV integration sites were PCR amplified and cloned in the same laboratories, we proposed that the patient-derived sites were very probably the result of PCR contamination (Garson *et al.*, 2011). This seemed particularly likely given the unusual technique used for cloning the prostate tissue-derived integration sites, which involved an extraordinary degree of PCR amplification with 80 preliminary amplification cycles followed by nested PCR consisting of 29 cycles and then an additional 18 cycles. PCR tubes were opened during the procedure for the addition of fresh DNA polymerase after 40 cycles. Using such a technique would have entailed a significant risk of direct or indirect contamination from experimentally infected DU145 cells, cellular DNA, plasmids or PCR products that had been handled in the same environment. No negative controls were mentioned in the published method. Independent evidence for contamination during amplification and cloning of XMRV integration sites was subsequently provided by Rusmevichientong *et al.* (2011) based on comparative analysis of single-nucleotide polymorphisms.

## Human pathogen or laboratory contaminant?

In addition to the aforementioned studies that raise serious doubts about the authenticity of the chromosomal integration site data, an increasing number of other investigations have also suggested that XMRV is more likely to represent a laboratory contaminant than a genuine human pathogen. Hué *et al.* (2010) demonstrated that published PCR primers previously regarded as XMRV-specific could in fact amplify endogenous retroviral sequences present in the genome of many mouse strains. Some of these murine endogenous retroviral sequences are known to be present in the mouse genome at very high copy number and thus minuscule traces of mouse DNA contaminating patient samples would potentially lead to false-positive PCR results and confound specific XMRV detection. Concrete evidence of false-positive XMRV detection due to such contamination was provided by Oakes *et al.* (2010), who demonstrated that all human samples that tested positive for XMRV and/or MLV were also PCR positive for the highly abundant murine intracisternal A-type particle (IAP) long terminal repeat and most were positive for murine mitochondrial cytochrome oxidase sequences. Oakes and colleagues (2010) therefore concluded that it is vital that contamination by mouse DNA be monitored with adequately sensitive assays in all samples tested for XMRV and related viruses. Similar conclusions were reached by Robinson *et al.* (2010) who reported that XMRV-like sequences were found in 4.8% of prostate cancers but that all positive samples were also positive

A)

```
EU981808     CTCCTCAGAGTGATTGACTACCCAGCTCGGGGGTCTTTCAaaagcacaca
GU816103     ------------ATTGACTACCCAGCTCGGGGGTCTTTCAaaagcacaca
                         **************************************

EU981808     gatataagtgctgtcatatagtaaatgcctaaataaaagtgttttgtgta
GU816103     gatataagtgctgtcatatagtaaatacctaaataaaagtgttttgtgta
             ************************* ************************

EU981808     gttttaatttatattctatttttcagaaacacaactaccatataaactga
GU816103     gttttaatttatattctatttttcagaaacacaactaccatataaactga
             **************************************************

EU981808     gagagtatttttatttctttgggattttacaaagagcaatttaccatttt
GU816103     gagagtatttttatttctttgggattttacaaagagcaatttaccatttt
             **************************************************

EU981808     tgaaaatcaggccattcacgggaacttgtagttccagctaatcgggaggc
GU816103     tgaaaatcaggccattcacgggaacttgtagttccagctaatcgggaggc
             **************************************************

EU981808     tgaggcaggagaatgacgtgaacctgggacgtgaacccatgagcttgcag
GU816103     tgaggcaggagaatgacgtgaacctgggacgtgaacccatgagcttgcag
             **************************************************

EU981808     tgagccagatcatgcctctgcactccagcctgggcaacagagcaagactc
GU816103     tgagccagatcatgcctctgcactccagcctgggcaacagagcaagactc
             **************************************************

EU981808     catctcaaaaaaaaaaaaaaaaaaaaaaaaaaaaaa
GU816103     catctcaaaaaaaaaaaaaaaaaaaaaaaaaa----
             ********************************
```

B)

```
EU981810     CTCCTCAGAGTGATTGACTACCCAGCTCGGGGGTCTTTCAatatgtttgg
EU981678     CTCCTCAGAGTAATTAACTACCCAGCTCGGGGGTCTTTCAatatgtttgg
             *********** *** **********************************

EU981810     ttaacacccttatcg
EU981678     ttaacancccttatcg
             ****** ********
```

**Figure 7.1** Nucleotide alignments of XMRV integration site sequences derived from patients' prostate cancer tissues and the experimentally infected human tumour cell line DU145. (A) Alignment of sequence EU981808 (patient 122-derived) and sequence GU816103 (DU145 cell line-derived). (B) Alignment of sequence EU981810 (patient VP268-derived) and sequence EU981678 (DU145 cell line-derived). The initial 169 nt segment of sequence EU981678 is not shown as it includes a repeat within the XMRV sequence which is not covered by the much shorter EU981810 sequence and is therefore redundant for purposes of alignment. Upper case letters represent the XMRV LTR sequence and lower case letters represent the flanking human chromosomal sequence. Note that the viral 3′ ends terminate with a conserved CA dinucleotide (reproduced from Garson *et al.*, 2011, with permission).

for murine IAP sequences and many but not all of these were also positive for mouse mitochondrial DNA sequences. These and similar studies establish that contamination of human samples with mouse DNA is widespread and can give rise to false-positive XMRV PCR results unless highly sensitive IAP control assays are included as part of a process of rigorous experimental control and validation.

Other sources of laboratory contamination that may have contributed to invalid and irreproducible findings in the early publications on XMRV have also been identified. Several groups have independently discovered that commercial enzymes (reverse transcriptases and thermostable DNA polymerases) and other reagents used in PCR assays may be contaminated with traces of mouse DNA and give rise to intermittent false positives in XMRV studies (Sato *et al.*, 2010; Tuke *et al.*, 2011; Zheng *et al.*, 2011). Mouse DNA contamination of murine monoclonal antibodies used in certain 'hot start' Taq polymerases may be responsible for some of these observations but often it is not possible to identify the exact origin of the contamination in PCR reagents. In addition, the DNA-binding columns used in many commercially available kits to extract and purify DNA from tissues and cells have also been shown to be responsible for introducing contaminating mouse DNA into test samples, thereby potentially generating spurious XMRV PCR results (Erlwein *et al.*, 2011). Plasmid DNA is another potential source of laboratory contamination. The recent disclosure by Silverman and colleagues that XMRV plasmid VP62 had apparently contaminated some of the CFS patients' DNA samples that had been used in the seminal Science publication by Lombardi *et al.* (2009) resulted in a partial retraction of the paper (Silverman *et al.*, 2011). Unaccountably, the plasmid DNA was found to have selectively contaminated patient samples whilst the negative control samples remained unaffected. Although the reason for this apparently selective contamination is unclear, it is possible that the patient samples were handled more frequently than the controls and were, therefore, more exposed to contamination risk.

It has also become clear that XMRV and XMRV-like viruses released from infected cell lines represent yet another potentially crucial source of contamination. Several studies, including those by Hué *et al.* (2010) and Sfanos *et al.* (2011) have shown that in addition to the prostate cancer cell line 22Rv1 mentioned above (Knouf *et al.*, 2009), a significant proportion of other widely used human tumour cell lines produce replication competent murine xenotropic gammaretroviruses (MLV-Xs). Sfanos and colleagues (2011) were surprised to observe the ease with which uninfected human cell lines became infected through contamination with XMRV from 22Rv1 cells, despite the uninfected cell lines and 22Rv1 cells not being handled simultaneously in the same tissue culture hood. Similarly, Shin *et al.* (2011) reported that a DNA extraction robot that had previously been used to process an XMRV infected cell line, subsequently contaminated DNA extracted from patient samples and led to false positive XMRV PCR results, despite an interval of several months between the two extractions.

Using detailed phylogenetic analysis, Hué *et al.* (2010) demonstrated that XMRV sequences, reportedly derived from unlinked patients with prostate cancer or CFS, displayed much less genetic diversity than would be expected from a retrovirus being transmitted from person to person by a process of natural infectious transmission (Fig. 7.2). In fact, the genetic distance between different 'patient-derived' XMRV sequences was found to be less than that observed between XMRV clones derived from the XMRV producing human prostate cancer cell line 22Rv1 (Knouf *et al.*, 2009). Furthermore, Hué and colleagues' phylogenetic analysis showed that the 22Rv1 cell line-derived sequences were probably *ancestral* to the patient-associated sequences, suggesting that XMRV from 22Rv1 cells or 22Rv1-derived DNA may somehow have contaminated the patient samples. In view of these findings, Hué *et al.* (2010) concluded that XMRV probably represented a laboratory contaminant rather than a genuine human infection.

## The recombinant origin of XMRV

Although Hué *et al.* (2010) had observed that XMRV derived from the 22Rv1 cell line appeared to be phylogenetically ancestral to all known 'patient-derived' XMRV sequences, the true origin of human prostate cancer-associated and CFS-associated XMRV remained uncertain until the decisive contribution of Paprotka and colleagues (2011). Two hypotheses concerning how the 22Rv1 cell line became infected with XMRV had been proposed (Sfanos *et al.*, 2011). One hypothesis (Knouf *et al.*, 2009) was that the virus was in the human prostate tumour from which the cell line was originally established. The second hypothesis (Hué *et al.*, 2010; Yang *et al.*, 2011) was that the 22Rv1 cell line was infected with the virus after the prostate tumour was removed

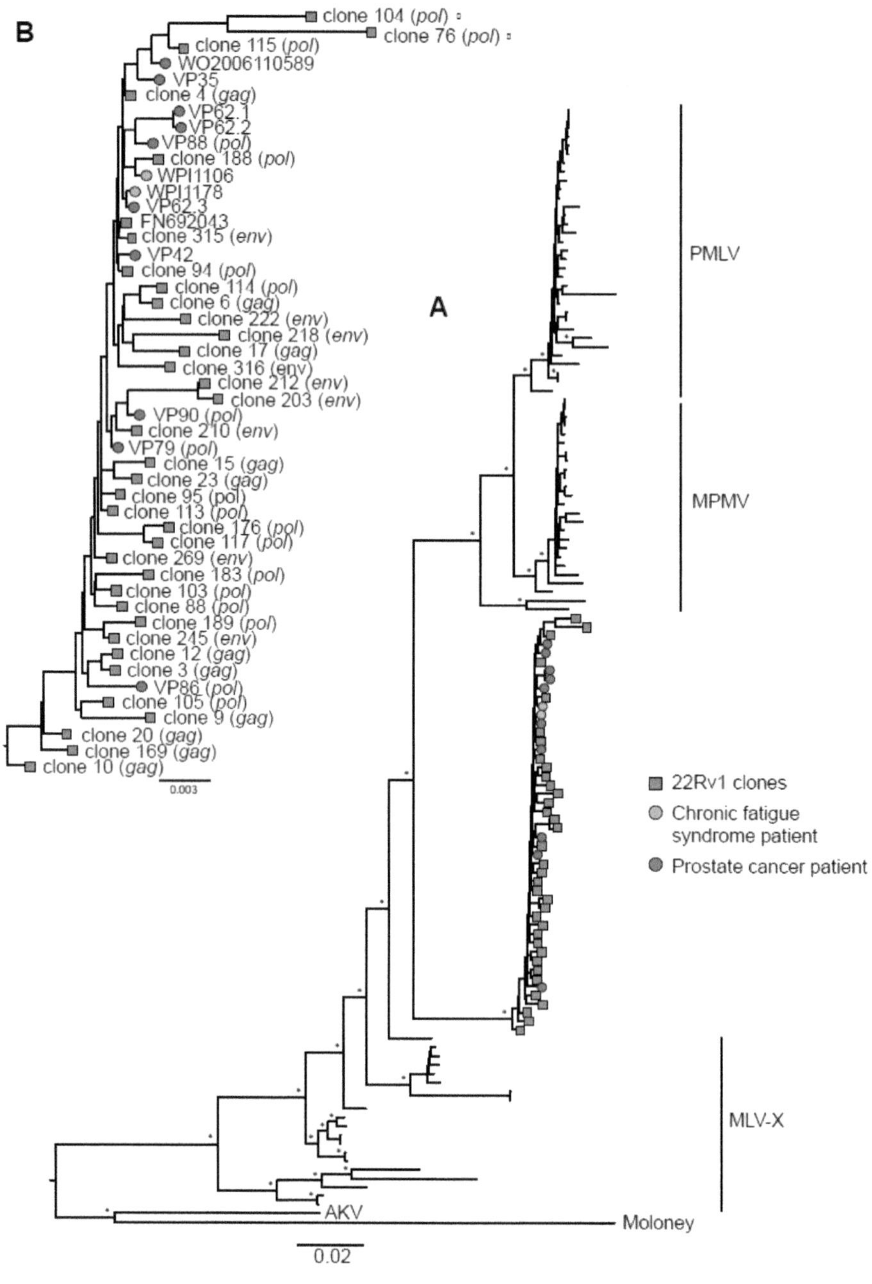


**Figure 7.2** Phylogeny of 22Rv1 and patient-derived XMRV sequences and other murine leukaemia viruses. Bayesian maximum clade credibility phylogeny of 22Rv1 cell line-derived XMRV clones, patient-derived XMRV sequences and other murine leukaemia viruses (A). The monophyletic cluster formed by 22Rv1 cell line-derived XMRV clones and patient-derived XMRVs is magnified in (B). Xenotropic MLV (MLV-X), polytropic MLV (PMLV), and modified polytropic MLV (MPMLV) were added as controls. Sequences derived from prostate cancer patients (VP and WO) and chronic fatigue syndrome patients (WPI) are indicated by red and yellow circles respectively. Gene sequences derived from 22Rv1 clones are indicated by blue squares. When full-length genomes were not available, the loci of the sequence used in the phylogenetic reconstruction are shown in brackets. APOBEC-3G/F hypermutated clones are labelled with a closed circle. The tree is rooted against AKV and Moloney MLVs. Bayesian posterior probabilities of 1.00 are indicated on the corresponding branches by a star. The scale bar represents the number of nucleotide substitutions per site (reproduced from Hué *et al.*, 2010, with permission).

from the patient, probably during the process of xenotransplantation and serial passage in immunodeficient mice. Note that xenotransplantation has, in the past, routinely been employed to facilitate the establishment of human prostate tumour cell lines.

The final definitive proof that the second hypothesis was correct was provided by Paprotka *et al.* (2011), who investigated the 22Rv1 cell line and stored samples of early to late passages of its progenitor human prostate tumour xenograft (CWR22) that had been serially passaged through nude (immunodeficient) mice during the period 1993–1996. They detected XMRV infection in 22Rv1 cells and in the later passage xenografts, but *not in the early passages.* Furthermore, they found that the host mice contained two proviruses, PreXMRV-1 and PreXMRV-2, which shared >99.9% identity with XMRV and therefore concluded that XMRV was not present in the original CWR22 tumour but was generated by recombination of the two proviruses during passaging of the xenografted tumour in mice. On the basis of their findings Paprotka and colleagues (2011) deduced that the association of XMRV with human disease is due to contamination of human samples with virus originating from this recombination event. In a recent review on the XMRV/CFS controversy, Robinson *et al.*, (2011a) point out that because XMRV was created by a laboratory recombination event in the mid 1990s, it is impossible (in the absence of time travel) that it caused CFS in patients diagnosed in the 1980s.

## Dénouement and conclusions

There is also circumstantial evidence pointing towards the 22Rv1 cell line having been the actual source of the ViroChip signal that led to the XMRV discovery in patients' prostate tissues by Silverman's group (Urisman *et al.* 2006). Silverman recently admitted that, '22RV1 cells were once previously (more than a year earlier) grown in my lab but were being stored in a liquid nitrogen freezer at the time, and not the same freezer used to store prostate tissues – at the time it was unknown that 22RV1 cells were infected with XMRV' (Tsouderos, 2011). In view of this, it may be relevant to note that viral contamination of liquid nitrogen tanks has been reported previously. Tedder *et al.* (1995) described a case in which hepatitis B virus (HBV) from a stored frozen sample contaminated a cryopreservation tank, and resulted in the infection of several bone marrow transplant patients with HBV.

The unravelling of the XMRV story accelerated dramatically in late 2011 with the sacking of Mikovits from the Whittemore Peterson Institute, followed by her arrest and brief imprisonment in relation to charges of 'unlawful taking of computer data, equipment, supplies, or other computer related property' (Cohen, 2011). The editors of the journal Science fully retracted the already partially retracted Lombardi *et al.* (2009) paper after having earlier issued an Editorial Expression of Concern (Alberts, 2011a,b; Silverman *et al.*, 2011). The grounds for the retraction included poor quality control, plasmid contamination and the mislabelling of figure legends, issues that led Science to lose 'confidence in the Report and the validity of its conclusions.' Only days after the retraction of Lombardi *et al.* (2009), the other key paper (Lo *et al.*, 2010) that had supported a link between CFS and XMRV-like viruses was also retracted (PNAS, 2012). The chief reasons for the retraction of the paper were that its findings had not been confirmed by other studies, and that Katzourakis *et al.* (2011) had demonstrated by phylogenetic analysis that retroviral sequences from the longitudinally sampled CFS patients described by Lo *et al.* were probably the result of PCR contamination rather than of viral evolution.

Therefore, it seems that the discovery of XMRV, although initially regarded as a triumph of molecular diagnostic technology, is destined to be viewed in retrospect as a colossal waste of time, money and scientific resources. It raised false hope amongst patients and spread unnecessary fear of a major threat to public health. There are many painful lessons, highly relevant to all those involved in molecular diagnostics, that can be learnt from the spectacular rise and fall of XMRV. Most of them are obvious but worth listing nevertheless:

- The strength of PCR, that is its exquisite sensitivity, is also its Achilles heel.
- Contamination is an ever present threat and can come from many sources including PCR

amplicons, plasmids, cell lines, DNA derived from and viruses released from cell lines, environmental mouse DNA, enzymes and other PCR reagents, DNA purification columns, DNA extraction robots, cryopreservation tanks, etc., etc.

- Xenotransplantation of human tumours using immunodeficient mice carries a significant risk of infection with xenotropic MLVs.
- To minimize the risk of falsely claiming that a particular nucleic acid sequence is associated with a particular disease, it is essential that patient and healthy control samples are collected, stored and handled in exactly the same manner and subjected to rigorous 'blind' testing.
- Publication of studies in highly ranked journals such as *Science* and *PNAS* is, of course, no guarantee of their veracity.
- Extraordinary claims require extraordinary evidence.

It is now certain that XMRV arose in the laboratory during the mid-1990s as the result of a rare recombination event. As Sfanos *et al.* (2012) state in their recent review, there is no reason to believe that XMRV actively circulates in humans or is present in the environment outside the laboratory. In conclusion, it transpires that XMRV is not a new human tumour virus after all, but rather the latest unfortunate example of a human 'rumour' virus (Voisset *et al.*, 2008; Hohn and Bannert, 2011).

## Note added in proof

On 18 September 2012, the editors of the journal *PLOS Pathogens* retracted the original paper (Urisman *et al.*, 2006) describing XMRV and its association with prostate cancer. The editors write in their retraction statement that, 'The association of XMRV with prostate cancer has now been thoroughly refuted.' On the same day, the long awaited results of a large multicentre study funded by the US National Institutes of Health were published (Alter *et al.* 2012). The results confirm unequivocally that the theory linking XMRV with chronic fatigue syndrome is most definitely dead.

## Acknowledgements

Professor Greg Towers (University College London) is thanked for his critical reading of the manuscript and for valuable comments and suggestions.

## References

Alberts, B. (2011a). Editorial expression of concern. Science *333*, 35.

Alberts, B. (2011b). Retraction. Science *334*, 1636.

Aloia, A.L., Sfanos, K.S., Isaacs, W.B., Zheng, Q., Maldarelli, F., De Marzo, A.M., and Rein, A. (2010). XMRV: a new virus in prostate cancer? Cancer Res. *70*, 10028–10033.

Alter H.J., Mikovits J.A., Switzer W.M., Ruscetti F.W., Lo S.C., Klimas N., Komaroff A.L., Montoya J.G., Bateman L., Levine S., *et al.* (2012). A multicenter blinded analysis indicates no association between chronic fatigue syndrome/myalgic encephalomyelitis and either xenotropic murine leukaemia virus-related virus or polytropic murine leukaemia virus. MBio *3*, e00266-12.

Arnold, R.S., Makarova, N.V., Osunkoya, A.O., Suppiah, S., Scott, T.A., Johnson, N.A., Bhosle, S.M., Liotta, D., Hunter, E., Marshall, F.F., *et al.* (2010). XMRV infection in patients with prostate cancer: novel serologic assay and correlation with PCR and FISH. Urology *75*, 755–761.

Arredondo, M., Hackett, J., de Bethencourt, F.R., Trevino, A., Escudero, D., Collado, A., Qiu, X., Swanson, P., Soriano, V., and de Mendoza, C. (2011). Prevalence of XMRV infection in different risk populations in Spain. AIDS Res. Hum. Retroviruses *28*, 1089–1094.

Cohen, J. (2011). Details emerge of the criminal charges against embattled CFS researcher. ScienceInsider. Available at: http://news.sciencemag.org/scienceinsider/2011/11/details-emerge-of-the-criminal.html?ref=ra (accessed 9 September 2013).

Danielson, B.P., Ayala, G.E., and Kimata, J.T. (2010). Detection of xenotropic murine leukaemia virus-related virus in normal and tumor tissue of patients from the southern United States with prostate cancer is dependent on specific polymerase chain reaction conditions. J. Infect. Dis. *202*,1470–1477.

Del Prete, G.Q., Kearney, M.F., Spindler, J., Wiegand, A., Chertova, E., Roser, J.D., Estes, J.D., Hao, X.P., Trubey, C.M., Lara, A., *et al.* (2012). Restricted replication of xenotropic murine leukaemia virus-related virus in pigtailed macaques. J. Virol. *86*, 3152–3166.

Devanur, L.D., and Kerr, J.R. (2006). Chronic fatigue syndrome. J. Clin. Virol. *37*, 139–150.

Dodd, R.Y. (2011). Chronic fatigue syndrome, XMRV and blood safety. Future Microbiol. *6*, 385–389.

Dong, B., Kim, S., Hong, S., Das Gupta, J., Malathi, K., Klein, E.A., Ganem, D., Derisi, J.L., Chow, S.A., and Silverman, R.H. (2007). An infectious retrovirus susceptible to an IFN antiviral pathway from human prostate tumors. Proc. Natl. Acad. Sci. U.S.A. *104*, 1655–1660.

Erlwein, O., Kaye, S., McClure, M.O., Weber, J., Wills, G., Collier, D., Wessely, S., and Cleare, A. (2010). Failure to detect the novel retrovirus XMRV in chronic fatigue syndrome. PLoS One 5, e8519.

Erlwein, O., Robinson, M.J., Dustan, S., Weber, J., Kaye, S., and McClure, M.O. (2011). DNA extraction columns contaminated with murine sequences. PLoS One *6*, e23484.

Fukuda, K., Straus, S.E., Hickie, I., Sharpe, M.C., Dobbins, J.G., and Komaroff, A. (1994). The chronic fatigue syndrome: a comprehensive approach to its definition and study. International Chronic Fatigue Syndrome Study Group. Ann. Intern. Med. *121*, 953–959.

Garson, J.A., Kellam, P., and Towers, G.J. (2011). Analysis of XMRV integration sites from human prostate cancer tissues suggests PCR contamination rather than genuine human infection. Retrovirology *8*, 13.

Gray, E.R., Garson, J.A., Breuer, J., Edwards, S., Kellam, P., Pillay, D., and Towers, G.J. (2011). No evidence of XMRV or related retroviruses in a London HIV-1-positive patient cohort. PLoS One *6*, e18096.

Groom, H.C., Boucherit, V.C., Makinson, K., Randal, E., Baptista, S., Hagan, S., Gow, J.W., Mattes, F.M., Breuer, J., Kerr, J.R., *et al.* (2010). Absence of xenotropic murine leukaemia virus-related virus in UK patients with chronic fatigue syndrome. Retrovirology 7, 10.

Hawkes, N. (2011). Dangers of research into chronic fatigue syndrome. BMJ *342*, d3780.

Henrich, T.J., Li, J.Z., Felsenstein, D., Kotton, C.N., Plenge, R.M., Pereyra, F., Marty, F.M., Lin, N.H., Grazioso, P., Crochiere, D.M., *et al.* (2010). Xenotropic murine leukaemia virus-related virus prevalence in patients with chronic fatigue syndrome or chronic immunomodulatory conditions. J. Infect. Dis. *202*, 1478–1481.

Hohn, O., and Bannert, N. (2011). Origin of XMRV and its demise as a human pathogen associated with chronic fatigue syndrome. Viruses 3, 1312–1319.

Hohn, O., Krause, H., Barbarotto, P., Niederstadt, L., Beimforde, N., Denner, J., Miller, K., Kurth, R., and Bannert, N. (2009). Lack of evidence for xenotropic murine leukaemia virus-related virus (XMRV) in German prostate cancer patients. Retrovirology *6*, 92.

Hohn, O., Strohschein, K., Brandt, A.U., Seeher, S., Klein, S., Kurth, R., Paul, F., Meisel, C., Scheibenbogen, C., and Bannert, N. (2010). No evidence for XMRV in German CFS and MS patients with fatigue despite the ability of the virus to infect human blood cells *in vitro*. PLoS One 5, e15632.

van den Hoogen, B.G., de Jong, J.C., Groen, J., Kuiken, T., de Groot, R., Fouchier, R.A., and Osterhaus, A.D. (2001). A newly discovered human pneumovirus isolated from young children with respiratory tract disease. Nat. Med. 7, 719–724.

Hué, S., Gray, E.R., Gall, A., Katzourakis, A., Tan, C.P., Houldcroft, C.J., McLaren, S., Pillay, D., Futreal, A., Garson, J.A., *et al.* (2010). Disease-associated XMRV sequences are consistent with laboratory contamination. Retrovirology 7, 111.

Jeziorski, E., Foulongne, V., Ludwig, C., Louhaem, D., Chiocchia, G., Segondy, M., Rodiere, M., Sitbon, M., and Courgnaud, V. (2010). No evidence for XMRV association in pediatric idiopathic diseases in France. Retrovirology 7, 63.

Jones, M.S., Kapoor, A., Lukashov, V.V., Simmonds, P., Hecht, F., and Delwart, E. (2005). New DNA viruses identified in patients with acute viral infection syndrome. J. Virol. *79*, 8230–8236.

Katzourakis, A., Hué, S., Kellam, P., and Towers, G.J. (2011). Phylogenetic analysis of murine leukaemia virus sequences from longitudinally sampled chronic fatigue syndrome patients suggests PCR contamination rather than viral evolution. J. Virol. *85*, 10909–10913.

Kim, S., Kim, N., Dong, B., Boren, D., Lee, S.A., Das Gupta, J., Gaughan, C., Klein, E.A., Lee, C., Silverman, R.H., *et al.* (2008). Integration site preference of xenotropic murine leukaemia virus-related virus, a new human retrovirus associated with prostate cancer. J. Virol. *82*, 9964–9977.

Kim, S., Rusmevichientong, A., Dong, B., Remenyi, R., Silverman, R.H., and Chow, S.A. (2010). Fidelity of target site duplication and sequence preference during integration of xenotropic murine leukaemia virus-related virus. PLoS One 5, e10255.

Knouf, E.C., Metzger, M.J., Mitchell, P.S., Arroyo, J.D., Chevillet, J.R., Tewari, M., and Miller, A.D. (2009). Multiple integrated copies and high-level production of the human retrovirus XMRV (xenotropic murine leukaemia virus-related virus) from 22Rv1 prostate carcinoma cells. J. Virol. *83*, 7353–7356.

Knox, K., Carrigan, D., Simmons, G., Teque, F., Zhou, Y., Hackett, J., Jr., Qiu, X., Luk, K.C., Schochetman, G., Knox, A., *et al.* (2011). No evidence of murine-like gammaretroviruses in CFS patients previously identified as XMRV-infected. Science *333*, 94–97.

van Kuppeveld, F.J., de Jong, A.S., Lanke, K.H., Verhaegh, G.W., Melchers, W.J., Swanink, C.M., Bleijenberg, G., Netea, M.G., Galama, J.M., and van der Meer, J.W. (2010). Prevalence of xenotropic murine leukaemia virus-related virus in patients with chronic fatigue syndrome in the Netherlands: retrospective analysis of samples from an established cohort. BMJ *340*, c1018.

Lo, S.C., Pripuzova, N., Li, B., Komaroff, A.L., Hung, G.C., Wang, R., and Alter, H.J. (2010). Detection of MLV-related virus gene sequences in blood of patients with chronic fatigue syndrome and healthy blood donors. Proc. Natl. Acad. Sci. U.S.A. *107*, 15874–15879.

Lombardi, V.C., Ruscetti, F.W., Das Gupta, J., Pfost, M.A., Hagen, K.S., Peterson, D.L., Ruscetti, S.K., Bagni, R.K., Petrow-Sadowski, C., Gold, B., *et al.* (2009). Detection of an infectious retrovirus, XMRV, in blood cells of patients with chronic fatigue syndrome. Science *326*, 585–589.

McCormick, A.L., Brown, R.H., Jr., Cudkowicz, M.E., Al-Chalabi, A., and Garson, J.A. (2008). Quantification of reverse transcriptase in ALS and elimination of a novel retroviral candidate. Neurology *70*, 278–283.

McKie, R. (2011). Chronic fatigue syndrome researchers face death threats from militants. The Observer. Available at: http://www.guardian.co.uk/society/2011/aug/21/chronic-fatigue-syndrome-myalgic-encephalomyelitis (accessed 9 September 2013).

Mi, Z., Lu, Y., Zhang, S., An, X., Wang, X., Chen, B., Wang, Q., and Tong, Y. (2011). Absence of xenotropic murine leukaemia virus-related virus in blood donors in China. Transfusion *52*, 326–331.

Mikovits, J.A., and Ruscetti, F.W. (2010). Response to Comments on 'Detection of an Infectious Retrovirus, XMRV, in Blood Cells of Patients with Chronic Fatigue Syndrome'. Science *328*, 825.

Mikovits, J.A., Huang, Y., Pfost, M.A., Lombardi, V.C., Bertolette, D.C., Hagen, K.S., and Ruscetti, F.W. (2010). Distribution of xenotropic murine leukaemia virus-related virus (XMRV) infection in chronic fatigue syndrome and prostate cancer. AIDS Rev. *12*, 149–152.

Oakes, B., Tai, A.K., Cingoz, O., Henefield, M.H., Levine, S., Coffin, J.M., and Huber, B.T. (2010). Contamination of human DNA samples with mouse DNA can lead to false detection of XMRV-like sequences. Retrovirology *7*, 109.

Onlamoon, N., Das Gupta, J., Sharma, P., Rogers, K., Suppiah, S., Rhea, J., Molinaro, R.J., Gaughan, C., Dong, B., Klein, E.A., *et al.* (2011). Infection, viral dissemination, and antibody responses of rhesus macaques exposed to the human gammaretrovirus XMRV. J. Virol. *85*, 4547–4557.

Paprotka, T., Delviks-Frankenberry, K.A., Cingoz, O., Martinez, A., Kung, H.J., Tepper, C.G., Hu, W.S., Fivash, M.J. Jr., Coffin, J.M., and Pathak, V.K. (2011). Recombinant origin of the retrovirus XMRV. Science *333*, 97–101.

PLoS One (2011). In support of XMRV researchers. Available at: http://blogs.plos.org/everyone/2011/09/02/in-support-of-xmrv-researchers/ (accessed 9 September 2013).

PNAS (2012). Retraction for Lo *et al.* Detection of MLV-related virus gene sequences in blood of patients with chronic fatigue syndrome and healthy blood donors. Proc. Natl. Acad. Sci. U.S.A. *109*, 346.

Robinson, M.J., Erlwein, O.W., Kaye, S., Weber, J., Cingoz, O., Patel, A., Walker, M.M., Kim, W.J., Uiprasertkul, M., Coffin, J.M., *et al.* (2010). Mouse DNA contamination in human tissue tested for XMRV. Retrovirology *7*, 108.

Robinson, M.J., Erlwein, O., and McClure, M.O. (2011a). Xenotropic murine leukaemia virus-related virus (XMRV) does not cause chronic fatigue. Trends Microbiol. *19*, 525–529.

Robinson, M.J., Tuke, P.W., Erlwein, O., Tettmar, K.I., Kaye, S., Naresh, K.N., Patel, A., Walker, M.M., Kimura, T., Gopalakrishnan, G., *et al.* (2011b). No evidence of XMRV or MuLV sequences in prostate cancer, diffuse large B-Cell lymphoma, or the UK Blood donor population. Adv. Virol. Article ID 782353.

Rusmevichientong, A., Das Gupta, J., Elias, P.S., Silverman, R.H., and Chow, S.A. (2011). Analysis of single-nucleotide polymorphisms in patient-derived retrovirus integration sites reveals contamination from cell lines acutely infected by xenotropic murine leukaemia virus-related virus. J. Virol. *85*, 12830–12834.

Sakuma, T., Hue, S., Squillace, K.A., Tonne, J.M., Blackburn, P.R., Ohmine, S., Thatava, T., Towers, G.J., and Ikeda, Y. (2011). No evidence of XMRV in prostate cancer cohorts in the Midwestern United States. Retrovirology *8*, 23.

Sato, E., Furuta, R.A., and Miyazawa, T. (2010). An endogenous murine leukaemia viral genome contaminant in a commercial RT-PCR kit is amplified using standard primers for XMRV. Retrovirology *7*, 110.

Satterfield, B.C., Garcia, R.A., Gurrieri, F., and Schwartz, C.E. (2010). PCR and serology find no association between xenotropic murine leukaemia virus-related virus (XMRV) and autism. Mol. Autism *1*, 14.

Schlaberg, R., Choe, D.J, Brown, K.R., Thaker, H.M., and Singh, I.R. (2009). XMRV is present in malignant prostatic epithelium and is associated with prostate cancer, especially high-grade tumors. Proc. Natl. Acad. Sci. U.S.A. *106*, 16351–16356.

Sfanos, K.S., Aloia, A.L., Hicks, J.L., Esopi, D.M., Steranka, J.P., Shao, W., Sanchez-Martinez, S., Yegnasubramanian, S., Burns, K.H., Rein, A., *et al.* (2011). Identification of replication competent murine gammaretroviruses in commonly used prostate cancer cell lines. PLoS One *6*, e20874.

Sfanos, K.S., Aloia, A.L., De Marzo, A.M., and Rein, A. (2012). XMRV and prostate cancer – a 'final' perspective. Nat. Rev. Urol. *9*, 111–118.

Shin, C.H., Bateman, L., Schlaberg, R., Bunker, A.M., Leonard, C.J., Hughen, R.W., Light, A.R., Light, K.C., and Singh, I.R. (2011). Absence of XMRV retrovirus and other murine leukaemia virus-related viruses in patients with chronic fatigue syndrome. J. Virol. *85*, 7195–7202.

Silverman, R.H. (2007). A scientific journey through the 2–5A/RNase L system. Cytokine Growth Factor Rev. *18*, 381–388.

Silverman, R.H., Das Gupta, J., Lombardi, V.C., Ruscetti, F.W., Pfost, M.A., Hagen, K.S., Peterson, D.L., Ruscetti, S.K., Bagni, R.K., Petrow-Sadowski, C., *et al.* (2011). Partial retraction. Detection of an infectious retrovirus, XMRV, in blood cells of patients with chronic fatigue syndrome. Science *334*, 176.

Simmons, G., Glynn, S.A., Komaroff, A.L., Mikovits, J.A., Tobler, L.H., Hackett, J., Jr. Tang, N., Switzer, W.M., Heneine, W., Hewlett, I.K., *et al.* (2011). Failure to confirm XMRV/MLVs in the blood of patients with chronic fatigue syndrome: a multi-laboratory study. Science *334*, 814–817.

Stieler, K., Schindler, S., Schlomm, T., Hohn, O., Bannert, N., Simon, R., Minner, S., Schindler, M., and Fischer, N. (2011). No detection of XMRV in blood samples and tissue sections from prostate cancer patients in Northern Europe. PLoS One *6*, e25592.

Suhadolnik, R.J., Peterson, D.L., O'Brien, K., Cheney, P.R., Herst, C.V., Reichenbach, N.L., Kon, N., Horvath, S.E., Iacono, K.T., Adelson, M.E., *et al.* (1997). Biochemical evidence for a novel low molecular weight 2–5A-dependent RNase L in chronic fatigue syndrome. J. Interferon Cytokine Res. *17*, 377–385.

Switzer, W.M., Jia, H., Hohn, O., Zheng, H., Tang, S., Shankar, A., Bannert, N., Simmons, G., Hendry, R.M., Falkenberg, V.R., *et al.* (2010). Absence of evidence of xenotropic murine leukaemia virus-related virus infection in persons with chronic fatigue syndrome and healthy controls in the United States. Retrovirology 7, 57.

Tang, S., Zhao, J., Haleyur Giri Setty, M.K., Devadas, K., Gaddam, D., Viswanath, R., Wood, O., Zhang, P., and Hewlett, I.K. (2011). Absence of detectable XMRV and other MLV-related viruses in healthy blood donors in the United States. PLoS One *6*, e27391.

Tedder, R.S., Zuckerman, M.A., Goldstone, A.H, Hawkins, A.E., Fielding, A., Briggs, E.M., Irwin, D., Blair, S., Gorman, A.M., Patterson, K.G., *et al.* (1995). Hepatitis B transmission from contaminated cryopreservation tank. Lancet *346*, 137–140.

Tsouderos, T. (2011). Research casts doubt on theory of cause of chronic fatigue. Chicago Tribune. Available at: http://www.chicagotribune.com/health/ct-met-chronic-fatigue-xmrv-20110317,0,6116823.story?page=1 (accessed 9 September 2013).

Tuke, P.W., Tettmar, K.I., Tamuri, A., Stoye, J.P., and Tedder, R.S. (2011). PCR master mixes harbour murine DNA sequences. Caveat emptor! PLoS One *6*, e19953.

Urisman, A., Molinaro, R.J., Fischer, N., Plummer, S.J., Casey, G., Klein, E.A., Malathi, K., Magi-Galluzzi, C., Tubbs, R.R., Ganem, D., *et al.* (2006). Identification of a novel Gammaretrovirus in prostate tumors of patients homozygous for R462Q RNASEL variant. PLoS Pathog. *2*, e25.

Voisset, C., Weiss, R.A., and Griffiths, D.J. (2008). Human RNA 'rumor' viruses: the search for novel human retroviruses in chronic disease. Microbiol. Mol. Biol. Rev. *72*, 157–196.

Weiss, R.A. (2006). The discovery of endogenous retroviruses. Retrovirology 3, 67.

Yang, J., Battacharya, P., Singhal, R., and Kandel, E.S. (2011). Xenotropic murine leukaemia virus-related virus (XMRV) in prostate cancer cells likely represents a laboratory artifact. Oncotarget 2, 358–362.

Zhang, Z.Z., Guo, B.F., Feng, Z., Zhang, L., and Zhao, X.J. (2011). Is XMRV a causal virus for prostate cancer? Asian J. Androl. *13*, 698–701.

Zheng, H., Jia, H., Shankar, A., Heneine, W., and Switzer, W.M. (2011). Detection of murine leukaemia virus or mouse DNA in commercial RT-PCR reagents and human DNAs. PLoS One *6*, e29050.

# Ancient DNA and the Fingerprints of Disease: Retrieving Human Pathogen Genomic Sequences from Archaeological Remains Using Real-time Quantitative Polymerase Chain Reaction

8

G. Michael Taylor

**Abstract**

It is almost 20 years since the first reports of the amplification of pathogen DNA sequences from human archaeological remains. Diseases such as tuberculosis and leprosy have been productive areas of endeavour due to their association with characteristic skeletal lesions and the robust nature of the mycobacterial cell wall. A number of other pathogens apart from mycobacteria have been successfully amplified from human remains, including the microorganisms responsible for malaria, leishmaniasis, brucellosis and plague (black death). Information from dated contexts can further our understanding of past epidemics, both challenging and informing evolutionary models of microorganism phylogeny and global dissemination.

This chapter considers the practical aspects of extracting and amplifying degraded fragments of pathogen DNA that may persist in archival tissues. It includes a discussion of the measures needed to prevent contamination and to authenticate findings, in particular, considering the many advantages of using real-time quantitative PCR methods.

## Introduction

This chapter discusses the use of ancient DNA analysis in the diagnosis of infectious diseases in past human and animal populations. Since the early days of ancient DNA (aDNA) studies over 20 years ago, a growing body of work now attests to the feasibility of demonstrating surviving evidence of pathogen genomic DNA associated with archaeological remains. The majority of studies have used conventional polymerase chain reaction (PCR), sometimes supported by additional techniques, such as demonstration of the presence of associated pathogen proteins or mycolipids. An example of the latter would be confirmation of aDNA evidence for tuberculosis species, combined with demonstration of specific mycolates, derived from the mycobacterial cell wall. Over the years, a number of tests of authenticity have become established for validation of PCR data from both human and pathogen studies and, whilst these may differ slightly depending on the species of interest, the underlying principles are similar (Taylor *et al.*, 2009). In the example of mycobacterial DNA supported by mycolate analysis, the latter can be extracted and studied directly, without the need for amplification, and the associated risk for cross-contamination is therefore reduced. A comprehensive consideration of the means for validating aDNA studies will be presented below.

It would be wrong to infer that all pathogens may be detected with equal success. Whilst the greatest successes are associated with mycobacterial diseases like tuberculosis (Table 8.1) and leprosy (Table 8.2), many investigators have been

**Table 8.1** Early and key papers dealing with the identification of MTB complex DNA in archaeological remains

| Authors | Year | Region | Burials age | Period/ century[1] | Specimens | Locus | Amplicon(s) (bp) |
|---|---|---|---|---|---|---|---|
| Spigelman and Lemma | 1993 | 1. Scotland | Adult | 17th | Lumbosacral spine | IS*6110* | 123 |
| | | 2. Turkey | Child | Byzantine | Lumbosacral spine | | |
| | | 3. Borneo | Adult | Pre contact | Ulna | | |
| Salo *et al.* | 1994 | Peru | Female, 40–45 years | 11th | Hilar lymph nodes | IS*6110* | 97 |
| Taylor *et al.* | 1996 | Royal Mint site, UK | 1. Male 45 years | 14th–16th | Metacarpal | IS*6110* | 123/92 |
| | | | 2. Male 15–25 years | 14th–16th | Vertebrae | IS*6110* | 123/92 |
| Donoghue *et al.* | 1998 | Negev desert, Israel | Male, 35–45 years | 600 AD | Calcified pleura | IS*6110* | 123/92 |
| Braun and Cook | 1998 | North America | 1. Adult | 15th | Vertebrae (lumbar) | IS*6110* | 123/92 |
| | | | 2. Female, 21–22 years | 11th | Vertebra T11 | | |
| Taylor *et al.* | 1999 | Royal Mint site, UK | As above | As above | As above | IS*6110* | 123/92 |
| | | | | | | *rpoB* | 157 |
| | | | | | | *mtp40* | 152 |
| | | | | | | *oxyR*[285] | 150 |
| | | | | | | DR | Various |
| Nuorola | 1999 | Man-of-war 'Kronan' | Adult male | 17th | Long bone | IS*6110* | 123/92 |
| Mays *et al.* | 2001 | Wharram Percy, UK | Nine burials: six male, three female | 10th–15th | Vertebrae | IS*6110* | 123/92 |
| | | | | | | *rpoB* | 157 |
| | | | | | | *mtp40* | 152 |
| | | | | | | *oxyR*[285] | 150/127 |
| | | | | | | *pncA*[169] | 140 |
| | | | | | | DR | Various |
| Rothschild *et al.* | 2001 | Wyoming, USA | Long-horned bison | 17,870 years BP | Metacarpal | IS*6110* | 92 |
| | | | | | | S12 | 163 |
| Mays *et al.* | 2002a | Wharram Percy, UK | Male, 30–40 (EE062) | 10th–16th | Ribs, vertebra | IS*6110* | 123/92 |
| | | | | | | *oxyR*[285] | 127 |
| Mays *et al.* | 2002b | Wharram Percy, UK | G658 Child | 10th–16th | Ribs | IS*6110* | 123/92 |
| | | | EE002 Child | | | | |
| | | | EE005 Child | | | | |
| Spigelman *et al.* | 2002 | 1. Scotland | Adult | 17th | Lumbosacral spine | IS*6110* | 123/92 |
| | | 2. Turkey | Child | Byzantine | Lumbosacral spine | IS*6110* | 246 |
| | | 3. Borneo | Adult | Pre contact | Ulna | S12 | 204 |

| Authors | Year | Region | Burials age | Period/ century[1] | Specimens | Locus | Amplicon(s) (bp) |
|---|---|---|---|---|---|---|---|
| Fletcher *et al*. | 2003a | Vac, Hungary | 350 samples from 168 individuals | 1731–1838 AD | Multiple sites | IS*6110* | 92–372 |
| | | | | | | *19-kDa* | |
| | | | | | | $KatG^{463}$ | |
| | | | | | | $GyrA^{95}$ | |
| | | | | | | *MPB70* | |
| Fletcher *et al*. | 2003b | Vac, Hungary | Mother | d.1793 | Abdomen | IS*6110* | 92–220 |
| | | | | | | *oxyR* | |
| | | | Daughter 1 | d.1797 | Chest | *mtp40* | |
| | | | | | | $KatG^{463}$ | |
| | | | Daughter 2 | d.1795 | Chest | $KatG^{203}$ | |
| | | | | | | $GyrA^{95}$ | |
| | | | | | | DR | |
| | | | | | | *plcD* | |
| Mays and Taylor | 2003 | Tarrant Hinton, UK | Male, 30–40 years | Iron age 2200 BP | Vertebrae, ribs. | IS*6110* | 123/92 |
| Zink *et al*. | 2003 | Egypt | 25 mummies | 2050–500 BC | Various bone and soft tissues | IS*6110* | 123 |
| | | | | | | $oxyR^{285}$ | 150 |
| | | | | | | *mtp-40* | 152 and 150 |
| | | | | | | DR locus | Multiple |
| Taylor *et al*. | 2005 | Tarrant Hinton, UK | Male, 30–40 years | As above | Vertebrae, ribs | IS*1081* | 135/113 |
| | | | | | | $oxyR^{285}$ | 110/94 |
| | | | | | | *pncA* | 117/96 |
| | | | | | | TbD1 Flanking | 112/96 |
| Bouwman and Brown | 2005 | UK | Faringdon Street (four individuals) | 1730–1849 | Vertebrae, ribs | IS*6110* | 123 |
| | | | Kingston upon Hull (two individuals) | 1316–1539 | | IS*1081* | 141/113 |
| | | | Newcastle Infirmary (one individual) | 1753–1845 | | | |
| Taylor *et al*. | 2007 | Tyva, Siberia | Five individuals: two females, one male, one unknown | 360 BC–390 AD | Lumbar vertebrae, rib | IS*1081* | 135 |
| | | | | | | $oxyR^{285}$ | |
| | | | | | | $pncA^{169}$ | |
| | | | | | | RD4 | 142 |
| | | | | | | RD12 | 123 |
| | | | | | | RD13 | 136 |
| | | | | | | RD17 inner | 96 |
| | | | | | | TBD1 inner | 115 |

**Table 8.1** Continued

| Authors | Year | Region | Burials age | Period/ century[1] | Specimens | Locus | Amplicon(s) (bp) |
|---|---|---|---|---|---|---|---|
| Herschkovitz *et al.* | 2008 | Atlit Yam, eastern Mediterranean | Female infant | 9250–8126 BP | Rib, arm, long bones | IS*6110* | 123/92 |
| | | | | | | IS*1081* | 135/113 |
| | | | | | | DR | Various |
| | | | | | | TbD1 flank | 128 |
| | | | | | | CMP | 105 |
| Donoghue *et al.* | 2010 | Thebes, Egypt | Female mummy, 50 years | 600 BC | Lung, gall bladder | IS*6110* | 123/92 |
| Jaeger *et al.* | 2011 | Rio de Janeiro, Brazil | Eight males, eight females, 16 unknown | 17th | Bone and tissue samples | IS*6110* | 93 |
| | | | | | | IS*1081* | 113 |

[1]Dating and period as described in the publication. BP, before present.

**Table 8.2** Ancient DNA publications dealing with recovery and genotyping of *Mycobacterium leprae* DNA from archaeological human remains

| Authors | Year | Region | Burials, ages | Period/ century[1] | Samples | Target | Amplicon(s) (bp) |
|---|---|---|---|---|---|---|---|
| Rafi *et al.* | 1994 | Jerusalem | Mass grave | 600 AD | Metatarsal | 65 kDa antigen | 439 |
| Haas *et al.* | 2000 | South Germany | Two females | 1400–1800 | Hard palate | RLEP 1 and 3 | 320 and 370 |
| Taylor *et al.* | 2000 | Orkney | Adolescent male | 1100–1200 | Nasal region | RLEP | 153 |
| Donoghue *et al.* | 2001 | Poland | Male, 40–50 years | Mediaeval | Rhinomaxillary area | RLEP | 129/99 |
| Donoghue *et al.* | 2002 | Hungary | One male | 10th–11th | Nasal region | 18 kDa | 136/110 |
| | | | One female | | | 36 kDa antigen | 531 |
| Spigelman and Donoghue | 2001 | Israel | Female | 3rd–6th | 'Madura foot' metatarsal | RLEP | 129/99 |
| Montiel *et al.* | 2003 | Spain (Seville) | Two adults: A-43, A-120 | 12th | Metacarpal bones | RLEP | 149/97 |
| Donoghue *et al.* | 2005 | Hungary, Sweden, Egypt, Israel | Various, *n*=8 | 10th–16th, 10th–13th, 4th, 1st | Various | RLEP | 129/99 |
| Taylor *et al.* | 2006 | Wharram Percy UK | Child, 10 years | 10th–12th | Maxilla | RLEP | 133/111 |
| | | | | | | TTC Microsatellite (AGA)20 | 131 |
| | | Ipswich, UK | Male, 35–40 years | 13th–16th | Multiple sites | AGT Microsatellite (GTA)9 | 151 |
| | | | | | | ML0058 | 117/99 |

| Authors | Year | Region | Burials, ages | Period/ century[1] | Samples | Target | Amplicon(s) (bp) |
|---|---|---|---|---|---|---|---|
| Likovský *et al.* | 2006 | Czech Republic | 1. Child, 14–18 years | Early 12th | Nasal area | RLEP | 97 |
| | | | 2. Adult female? | | Vertebrae | | |
| Watson and Lockwood | 2009 | Croatia | G483 | 8th–9th | Rhinomaxillary | RLEP | 111 |
| | | Denmark | *n*=4 | 13th–16th | Palatine | SNP | 115–131 |
| | | UK | *n*=5 | 10th–11th | Various | | |
| Taylor *et al.* | 2009 | Uzbekistan | Mature female | 1st–4th | Skull, tibia | Multiple SNP and VNTR loci | 97 –136 |
| | | Ipswich | Male, 35–40 years | 13th–16th | Multiple sites | | |
| Monot *et al.* | 2009 | North Africa and European | 12 cases: eight male, four female | 4th–15th | Multiple | Multiple SNP | 97–136 |
| Suzuki *et al.* | 2010 | Japan | Male, 30–50 years | 18th–19th | Multiple | Multiple SNP, WGA | 171–465 |
| Taylor and Donoghue | 2011 | Hungary 1 | Male | 7th | Various sites. See citation and references therein | RLEP | 111 |
| | | Hungary 2 | Female | 10th–11th | | | |
| | | Hungary 3 | Male | 10th–11th | | | |
| | | UK 1 | Child? Male | 10th–12th | | | |
| | | UK 2 | Male, mature female | 13th–16th | | Multiple SNP and VNTR loci | 99–136 |
| | | Turkey | Sub-adult | 8th–9th | | | |
| | | Czech Republic | | 9th | | | |
| | | Uzbekistan | Female | 1st–4th | | | |

[1]Dating and period as described in the publication. RLEP, repetitive element; SNP, single nucleotide polymorphism; VNTR, variable nucleotide tandem repeat; WGA, whole genome amplification.

unsuccessful when it comes to finding treponemal DNA in probable cases of tertiary syphilis, leading to the conclusion that this pathogen is unsuitable for aDNA analysis (Bouwman and Brown, 2005). This may reflect the paucibacillary nature of visible lesions at this late stage of the infection. Table 8.3 lists some other pathogens that have been studied with greater success using aDNA techniques. For reviews on progress made in the study of past tuberculosis and other pathogens, readers are referred also to publications by Drancourt and Raoult (2005) and Donoghue (2008, 2011).

This chapter will briefly consider how aDNA may become modified with the passage of time and exposure to the environment and how this impacts on practical considerations for the front-end processing of archaeological samples as well as the design of PCR methods and post-PCR analysis. Samples taken from buried human or faunal remains frequently contain naturally occurring inhibitors of the Taq polymerase enzyme used in PCR, so that inhibition must be monitored and steps taken to overcome inhibitors as necessary. The chapter will also reflect on what may be learnt of bacterial pathogens using aDNA methods. Finally, as aDNA remains a controversial area of study, methods for validation of experimental findings will be considered in some depth.

**Table 8.3** Examples of pathogens other than MTB complex and *Mycobacterium leprae* identified in human archaeological remains

| Disease/pathogen | Region | Burials | Period/ century[1] | PCR locus | Amplicon (bp) | Reference |
|---|---|---|---|---|---|---|
| Chagas disease (*Trypanosoma cruzi*) | Colombia | Mummies | 4000 BP | Minicircle kinetoplast | 330 | Guhl *et al.* (1997) |
| Chagas disease (*T. cruzi*) | Chile | Mummies, *n*=4 | 2000–600 bp | As above | 121 | Ferreira *et al.* (2000) |
| Plague (*Yersinia pestis*) | France | Teeth from two adults and one child | 14th | *pla* | 148 | Raoult *et al.*, 2000) |
| Helminths (*Ascaris*) | Belgium | Human coprolites | 14th | 18SrRNA | 99–147 | Loreille *et al.*, (2001) |
| | | | | *cytb* | 142 | |
| Malaria (*Plasmodium falciparum*) | Italy | Infant skeleton | 5th | 18S rRNA | 89 | Sallares and Gomzi (2001) |
| Leishmaniasis (*Leishmania donovani*) | Egypt and Nubia | Mummies, skeletons | 4000 bp | Mt DNA | 120 | Zink *et al.* (2006) |
| | | | | Kinetoplast | | |
| Trench fever (*Bartonella quintana*) | Vilnius, Lithuania | Skeletons, dental pulp, uniforms | 1812 | *hbpE* | 429/282 | Raoult *et al.* (2006) |
| | | | | *htrA* | 192/113 | |
| Typhus (*Rickettsia prowasekii*) | | | | *dnaA* | 279/141 | |
| | | | | *dnaE* | 246/77 | |
| Malaria (*P. falciparum*) | Egypt | Bone and tissues | 1500–500 bc | 18S rDNA | | Nerlich *et al.* (2008) |
| | | | | *pfcrt* | 134 | |
| Plague (*Y. pestis*) | Netherlands, UK, France | Skeletons, bones/teeth | 14th–17th | *Pla* | 148 | Haensch *et al.* (2010) |
| | | | | *Caf 1* | 161 | |
| | | | | *16SrRNA* | 138 | |
| | | | | *rpoB* | 144–170 | |
| | | | | *glpD* | 110,167 | |
| | | | | *napA* | 113 | |
| Brucellosis (*Brucella* spp.) | Butrint, Albania | Vertebrae, ribs | 11th–13th | *IS6501 (IS711)* | 58 | Mutolo *et al.* (2011) |
| | | | | *Bcsp31* | 59 | |

[1]Dating and period as described in the publication. BP, before present.

## aDNA breakdown after deposition

DNA degradation commences immediately after the death of an individual or organism. *Post mortem* action of endonucleases and subsequent microbial activity results in fragmentation of both host and pathogen DNA. Over the passage of time, the DNA bases may be further modified by chemical processes such as hydrolysis and oxidation (Lindahl and Nyberg, 1972) and by the presence of naturally occurring fixatives. This can result in DNA polymerase blocking lesions, due to both intra and inter-strand linking. Modification or loss of nucleotide bases, particularly depurination (Lindahl, 1993) can also introduce blocking errors or changes which allow extension but which appear as nucleotide transitions when remnant DNA is subjected to PCR, cloning and sequencing. The majority of miscoding changes involve type 2 changes (C→T/U and G→A) transitions (Gilbert *et al.*, 2003; Stiller *et al.*, 2006). Only about 2–3% constitute the type 1 transitions (A→G, T→C) (Hansen *et al.*, 2001).

Smith and colleagues (2003) proposed the concept of the thermal age of specimens, with the aim of predicting the chances of successful DNA amplification from archival remains. The thermal age is defined as the time needed to produce a given degree of DNA damage when the temperature is held at a constant 10°C. In this approach, the thermal age adjusts the chronological age of sites using their individual thermal histories to give an indication of the effective depurination temperature (Smith *et al.*, 2003). However, in practice there is not always good agreement between the calculated thermal age and the ability to extract and amplify DNA by PCR (Schwarz *et al.*, 2009). This is probably due to the variety of environmental conditions that inevitably apply at different sites over time, such as pH, salt composition and oxidative free radicals. As the thermal age only considers damage due to depurination, this model is unhelpful for predicting outcome when other forms of damage, such as strand shortening or cross-linking have accrued over time (Hansen *et al.*, 2006). N-phenacylthiazolium bromide (PTB) is reportedly helpful in undoing cross-links which have occurred due to Maillard reactions, that is between sugar and amino acid residues, increasing available templates for amplification (Poinar *et al.*, 1998).

The exact conditions that have prevailed since deposition are extremely difficult to reckon or model and conditions may well have varied across the same sample. It is often worthwhile performing a pilot PCR study to check for DNA from the species or pathogen of interest. The size of locus chosen for the PCR will govern the likelihood of obtaining a product, with amplicons around 80–90 bp offering the greatest chance of success. Sequencing amplicons from successful amplifications will not normally show nucleotide changes due to damage, as these molecules usually constitute only a very minor component of the total DNA templates present (Mateiu and Rannala, 2008; Heyn *et al.*, 2010). Identification of these minority target nucleotide base changes will only be apparent if cloning and sequencing of a representative number of clones is undertaken. Conversely, consensus sequences obtained from a majority of clones will match direct sequencing (Winters *et al.*, 2011).

## Why study aDNA and what can it tell us?

### Confirmation of skeletal patterning with a particular disease

The ability to detect evidence of a particular pathogen in either human or animal remains can greatly assist with confirming impressions gained from study of the skeletal patterning of the visible lesions present. Whole skeletons, showing good preservation and with characteristic lesions of an advanced disease present fewer problems of diagnosis for the osteologist. For example, the lytic vertebral lesions of Pott's disease associated with spinal tuberculosis or the pathognomonic changes in the extremities and skull in lepromatous leprosy (LL) may allow unambiguous diagnosis. However, this may change when fewer skeletal elements have survived, the available bones are damaged or the disease process is at an early stage.

### Differential diagnoses

Brucellosis is caused by a number of the closely related species and biovars of the bacterial genus *Brucella*, which are Gram-negative, rod-shaped, facultative, intracellular organisms. Brucellosis is a zoonotic disease that can be transmitted to man from animals and sometimes vice versa. Human brucellosis may affect the skeleton in a small percentage of cases. The spine and pelvis are the most common sites of infection. In the vertebral bodies of the spine, the disease results in signs of spondylitis with cavitating lytic lesions that can be difficult to distinguish from tuberculosis. Tuberculosis is considered to result in little, if any, reactive bone growth and to affect the thoracic and cervical vertebrae more commonly that the lumbar vertebrae, whilst brucellosis is said to target the lumbar vertebrae and to be less likely to result in psoas abscess (Ortner and Putschar, 1985). Whilst these are general guidelines, individual cases must be assessed on their own merits. One case in point was an Iron Age period burial from Tarrant Hinton in Dorset, UK, of a male individual with spinal tuberculosis in which it was possible to demonstrate DNA from *M. tuberculosis* in cancellous bone sampled from lumbar vertebrae 1–3 (Taylor *et al.*, 2005). No evidence for Brucella DNA was present. Also with tuberculosis, lesions in skeletal elements other than the vertebrae may be indistinguishable from osteomyelitis due to other organisms. There are numerous other examples of the uses for aDNA in differential diagnosis. In LL, granulomatous bone lesions present in small bones of the hands and feet may resemble those seen in sarcoidosis (Ortner and Putschar, 1985). Advanced signs of LL in the skull, sometimes referred to as *facies leprosa* are usually considered pathognomonic for *Mycobacterium leprae* infection but some paleopathologists have noted that loss of the anterior nasal spine and destruction of the alveolar maxillary bone may also be seen in syphilis (Ortner and Putschar, 1985). In practice, the other signs indicative of LL, such as rounding and widening of the nasal aperture, resorption of the premaxillary alveolar process often allows distinction to be made between the two diseases.

## Association of unusual or alleged skeletal markers with a particular disease

Ancient DNA can be used to assess if particular skeletal lesions may be associated with a disease. For example, it has been proposed that the periostitic new bone deposited on the visceral side of ribs may be the result of pulmonary tuberculosis (Kelley and Micozzi, 1984). If this could be confirmed, then these readily identified lesions would provide another means for identification of the disease, even when other skeletal evidence is lacking. In a study of the Wharram Percy Mediaeval cases with periostitic rib lesions, we could find no consistent correlation with tuberculosis (Mays *et al.*, 2002). In our limited study we found evidence of tuberculosis in only 1/7 cases with rib lesions whereas in the control group, 2/7 cases showed evidence of pathogen DNA. However, our findings in the Wharram Percy material do not preclude an association in other series or population studies with visceral surface lesions. It is likely that rib periostitis is a non-specific indicator of inflammation, possibly due to a number of diseases of which tuberculosis is but one. Other diseases, which have been linked with these rib lesions, include pneumonia, bronchitis, emphysema and pleurisy.

Hypertrophic pulmonary osteoarthropathy (HPO) is a syndrome consisting of clubbing of the digits of the hands and feet, joint inflammation and diffuse subperiosteal bone deposition (Resnick and Niwayama, 1981: 2983–2996). Digital clubbing and joint changes involve soft tissue, so palaeopathological diagnosis of the condition normally relies on the pattern and nature of subperiosteal bone deposition. The condition may be either primary or secondary. Primary HPO is due to a rare genetic disease whereas secondary HPO is generally associated with thoracic (and sometimes abdominal) tumours or chronic infections. In the pre-antibiotic era it was often the latter and in the early medical literature tuberculosis was commonly recognized as a driver of HPO (Locke, 1915). Periostitis seen in a symmetrical distribution on skeletal long bones in HPO provides another example of how aDNA may be used to either confirm or refute the suggestion that these may be due to tuberculosis. In a study of two cases of HPO from the Wharram Percy collection, it was possible to demonstrate markers of *Mycobacterium tuberculosis* (MTB) complex DNA in one (burial EE062) and also to show that the disease was due to a strain of *M. tuberculosis* rather than *M. bovis*, a related and important human pathogen (Mays and Taylor 2002). In one of the few studies of tuberculosis in animal remains, Bathurst and Barta (2004) demonstrated the presence of MTB complex DNA (IS*6110*) in a fully articulated 16th-century dog skeleton displaying signs of HPO, from an Iroquoian site in Ontario, Canada.

## Identification of burials with co-infections

Ancient DNA testing can sometimes reveal when individuals have suffered from more than one infection at the same time. Probably the most documented examples come from studies of human remains showing signs of LL. Co-infection with tuberculosis has been found in a number of examples of these, in some series up to about 30% of skeletons with LL have been found to also be infected with tuberculosis (Donohue *et al.*, 2005). The remains do not always show osteological evidence of both mycobacterium infections, but pathogen DNA from both has been shown to be present. A male skeleton with clear ontological evidence for both LL and tuberculosis has recently been excavated from the site of the former St. Mary Magdalen leper hospital in Winchester (Turner and Roffey, personal communication). Co-infection of leprosy sufferers with tuberculosis has been proposed as one possible reason for the historical decline in leprosy during the mediaeval period (Manchester, 1991). Lepers, with an impaired cell-mediated immune response would have been susceptible to the faster replicating *Mycobacterium tuberculosis* whereas those who survived tuberculosis may have acquired a degree of cross-immunity to subsequent infection with leprosy.

We have also investigated an individual burial (XXXI.34, the remains of a 25- to 35-year-old female of Iron Age period) of Hunno-Sarmatian period from Tyva, south Siberia, where evidence for both *Mycobacterium bovis* and *Brucella* spp. was obtained in a specimen taken from a lumbar vertebra displaying the classic spinal lesions of tuberculosis (Murphy *et al.*, 2009). The identification of *Brucella* DNA was a late observation discovered after the completion of the aDNA analyses which had focused on the typing of *M. bovis* isolates found in a total of four burials of nomadic pastoralists living in close proximity to a number of herd species. The amplification of *Brucella* pathogen DNA was a reproducible observation found by two investigators at centre 1 but not replicated at a second laboratory, due to prior return of the material for burial. It served as a useful lesson that the situation may be more complex than originally supposed from relying on visible lesions alone. Recently, Mutolo and colleagues (2011) have reported qPCR amplification of markers of *Brucella* spp. from two adolescent male skeletons excavated from a mediaeval (11th–13th century) site in Albania, thus confirming that this organism can also be studied by aDNA methodology.

The ability to detect pathogen DNA, as shown in the Siberian pastoralists and Mediaeval Butrint mentioned above, provides examples of how aDNA can be used to study the prevalence of disease in past populations and links with human activities, in this case close contact with herd animals over a prolonged period. Given the presence of a zoonotic pathogen in human remains, it should also be possible in theory to show its presence in faunal remains from the same sites and periods. This is one of the ways proposed for validating aDNA studies of this kind.

Faunal remains, however, pose additional problems in their own right. For example, identification and study of infectious lesions in animal bones from archaeological sites is complicated by the often fragmented and disarticulated nature of the remains and the frequent lack of a consensus of reference cases for diagnosing disease. In addition, unlike most human remains, the depositional history of faunal remains is frequently complex and they have generally been subjected to cooking and scavenging. The study of animal remains is certainly an area in which more work is required, particularly when zoonotic pathogens are detected in human skeletons from the same sites.

### Genotyping and testing models of evolution and global spread of disease

The study of aDNA is valuable in reconstructing the routes taken by ancient diseases and their adaptation to different human populations. For example, by genotyping large numbers of extant strains of leprosy an association was recognized between variable loci, such as single nucleotide polymorphisms (SNPs), and geographical origin. The study showed that the original spread of a single clone of *M. leprae* could be retraced through analysis of informative SNPs. This allowed construction of a general evolutionary scheme and understanding of the global dissemination of the disease in antiquity (Monot *et al.*, 2005). The origins of the disease appear to lie in either Eastern Africa or the Near East and it likely spread with successive human migrations. Europeans or North Africans introduced leprosy into West Africa and the Americas within the past 500 years. Ancient DNA analyses can contribute to these models by allowing recovery of data from areas where the disease is no longer extant (Monot *et al.*, 2009).

Similarly, study of victims of Black Death excavated from mass graves associated with the plague from several parts of Europe has confirmed that the pathogen responsible was indeed *Yersinia pestis* (Haensch *et al.*, 2010). These landmark studies combined aDNA amplification and genotyping with demonstration of a protein marker, the F1 antigen of *Y. pestis*. This work has effectively resolved a long-standing and often controversial debate about the aetiology of Black Death. Moreover, genotyping has allowed these ancestral isolates of *Y. pestis* to be placed within the evolutionary scenario for *Yersinia* strains, showing that they lie between the ancestral *Y. pseudotuberculosis* and later biovars *Antiqua*, *Mediaevalis* and *Orientalis*. The recovery of data in cases such as *M. leprae* and *Y. pestis* provide opportunities for both testing and informing phylogenetic models of disease.

## Sampling and measures to avoid cross-contamination

Two important cornerstones in aDNA studies are (1) measures taken to avoid contamination and (2) measures taken to demonstrate authenticity of any resulting data. Relating to the first of these, an important consideration before starting analyses of precious samples is the overall strategy for avoiding contamination, not only between cases but also from earlier experiments in which PCR amplicons have been generated. Steps to avoid contamination should start from the time of sampling the material. Often, this will take place in a museum store or archaeology teaching laboratory. The provision of a suitable surface, which can be cleaned between cases, wearing of laboratory coats and disposable gloves and use of sterile blades or probes for sampling, all help prevent cross-contamination between cases. Individual bones or other remains with distinctive lesions may have been used for teaching purposes and will have been much handled in the past. Samples are therefore best taken from below the surface and preferably both near to, and at some distance from, the lesions, to compare distribution of surviving pathogen DNA. Another approach adopted by some investigators has been use of protocols that have been shown to be useful for detecting and removing contamination with modern DNA (Dissing *et al.*, 2008). If permissions can be obtained, multiple site sampling is helpful for validation as it allows a degree of replication and also comparison of DNA quantitation with respect to characteristic skeletal patterning of the disease.

Physical barriers have been shown to be one of the most effective approaches to dealing with contamination and in some instances, false negative results due to degraded templates (Dieffenbach and Dveksler, 1993; Niederhauser *et al.*, 1994). Therefore one of the most effective measures is the provision of separate laboratories for each of the main stages of the procedure, namely (1) nucleic acid extraction and PCR set-up, (2) PCR amplification and (3) post-PCR analysis (such as gel electrophoresis and purification of products for sequencing). A suitable facility would then consist of three separate laboratories with independent equipment, such as pipettes, fridge-freezers, mixers and bench-top centrifuges and separate disposables (plastic ware, filter tips and other reagents). Two-laboratory three-workstation strategies also work well but the most important provision is a clean area for initial sample processing. This must not have been used previously for PCR of the pathogens of interest. Physical separation of pre and post-PCR stages has always been advocated and used by the author and collaborators in all referenced works cited in this chapter.

Work surfaces in contact with sample tubes are easily cleaned with proprietary sprays or bleaches. Small equipment like pipettes and centrifuge rotors should be dismantled and cleaned regularly. Separate extraction controls must be processed at the same time as the samples to ensure contamination-free reagents. Another useful control is the extraction of cases likely to be negative for the microorganism under study. Ideally, pre-PCR labs should operate with a slight positive air pressure and post-PCR areas with a slight negative pressure compared to their surroundings to help prevent airborne contamination. Knapp and colleagues (2011) have recently considered key aspects of designing and working in an ancient DNA facility (Knapp *et al.*, 2011). However, having a purpose-built laboratory suite is no solution for poor housekeeping or technique, as these can quickly thwart any advantage provided by the perfect facility.

Pathogen aDNA may be extracted from a variety of animal and human tissue types. In the majority of cases the most likely material to survive is bone. Lesions characteristic of both infectious and some inherited diseases are well described in the palaeopathological literature (Ortner and Putschar, 1985). Rarely, natural or deliberate mummification allows recovery of additional tissues. These may present unique opportunities for recovering multiple lines of evidence for the signs of a particular pathogen, including conventional light microscopy and staining. For example, Donoghue and co-workers (2004), were able to show Ziehl–Neelsen staining of acid-fast bacteria in chest material taken from a 36-year-old male individual who had died in 1808 and who proved to be strongly positive for MTB complex DNA.

The extraction step must be able to recover small degraded fragments of DNA persisting in human and other remains. It must offer some degree of removal of inhibitors of the PCR reaction, which otherwise might lead to false negative results. As there are situations when intact mycobacteria may still be present, the extraction step must also be able to deal efficiently with front-end processing of these resilient organisms.

A consideration of the literature for work on 'modern' mycobacterial diseases is useful and reveals the importance of the extraction step when dealing with this group of microorganisms. Mycobacteria present some well-recognized problems not generally encountered with other bacteria or eukaryotic cells and these are related to the robust mycobacterial envelope (Brennan and Nikaido, 1995; Takade *et al.,* 2003). PCR methods should be assessed as a combination of both DNA recovery and the efficiency of PCR.

The extraction procedure must deliver 1) effective lysis of any remnant mycobacteria 2) good recovery of pathogen DNA from bone or a complex mixture of tissue debris and 3) removal of PCR inhibitors. A number of studies have addressed the problem of initial processing of modern mycobacterial samples and a number of procedures have been described. These range from boiling and centrifugation (Afghani and Stutman, 1996), trapping of DNA on Chelex resin (Heginbotham *et al.,* 2003), bead-beating (Tell *et al.,* 2003), sonication (David *et al.,* 1984), enzymatic digestion (Zhang *et al.,* 1997), sequence capture (Mangiapan *et al.,* 1996; Roring *et al.,* 2000), commercial kits with lysis reagents (Aldous *et al.,* 2005) and combinations of these various approaches (Heginbotham *et al.,* 2003). Several of these studies have compared procedures, often with differing conclusions. The literature is complex and reflects the fact that groups have compared variants of the same generic method, examined various sample types and gauged recovery using different criteria. Examples of different methods used to estimate DNA recovery in aDNA studies include the ability to obtain a PCR product (or not) on known positives (Afghani *et al.,* 1996), the absorbance (OD) of recovered DNA (Tell *et al.,* 2003) and the use of qPCR to quantify the DNA (Aldous *et al.,* 2005). It is appropriate to conclude that there are a number of different methods which can be made to work effectively if sufficient steps are included to satisfy the three main criteria listed above (Christopher-Hennings *et al.,* 2003).

The author's preferred approach has been the use of a guanidinium-based chaotropic buffer (e.g. >6 M guanidinium thiocyanate, GuSCN) for initial processing and lysis, followed by trapping of DNA on silica particles. This copes with a range of specimen types. Moreover, it is capable of inactivating any viable mycobacteria (Taylor *et al.,* 2005), which can be an added advantage in some circumstances. Should the sample consist of either fixed, mummified, or denser material, a prior incubation step in 1 × TE buffer containing proteinase K solution can be performed at 55°C. Once disrupted, the sample is transferred into GuSCN and processed as normal. Rohland and Hofreiter (2007) have published a detailed protocol for maximizing DNA release from bones and teeth. This is similar to the above method but includes an overnight step with EDTA and proteinase K for maximum demineralisation.

Silica-based DNA extraction methods (e.g. Boom *et al.,* 1990) have been widely evaluated and found to be one of the most efficient. Bouwman and Brown (2002) looked at several variations of this approach for aDNA applications, finding columns to be generally more efficient than slurries. However, the latter do provide greater freedom for incorporating additional wash steps or scaling up for larger sample volumes. If using a commercial kit that employs columns, it must be noted that many such methods do not efficiently recover smaller fragments (<150 bp) of DNA. Therefore, it is worth checking if smaller DNA fragments can be efficiently recovered, as these may be typical of the pathogen DNA that is recovered in archival material. One way of checking this is to purify DNA size markers, for example 25 or 50 bp ladders found in gel electrophoresis standards, using one of the columns from the kit in question. Recovered DNA can then be run on agarose gel or microfluidic chip system (e.g. Agilent 2100 Bioanalyzer, Agilent Technologies) to assess the size range efficiently recovered.

When eluting from silica slurries or columns, the volume recovered is typically in the range 50–100 µl. Depending on preference or the kit used, this may be either 1 × TE buffer or a molecular biology grade of water. Either is suitable for PCR of aDNA templates. Ideally, initial screening for the target of interest is carried out without delay, but if not convenient, samples may be stored at –20°C until assayed. To avoid multiple freeze–thaw insults, it is recommended to divide the eluate immediately so that two or more aliquots are retained for later testing.

## PCR

An early recommendation for PCR of aDNA material was that the amplicon size should be kept below 200 bp (Cano *et al.*, 1996). Experience has since shown that many positive cases would have been missed if the amplicon targeted was >100 bp. Generally, 'the smaller the better' is the rule for aDNA work. There are lower limits for amplicon size, as specificity and ease of authentication are important considerations. Amplicon sizes in the range 80–100 bp remain practical as they permit conventional sequencing to confirm specificity and similarly, allow singe nucleotide polymorphisms (SNPs) to be determined without the need for cloning. Repetitive elements in genomes make good targets for aDNA PCR as the chances of detecting the pathogen template increases accordingly. The occurrence of deletions in a pathogen can be exploited to design extremely specific PCR methods with primers flanking the deletion and a reporter probe hybridizing over the deletion breakpoint. PCR primers can be designed with the aid of sites such as http://biotools.umassmed.edu/bioapps/primer3_www.cgi. If selecting the oligonucleotides manually, it is worthwhile checking the properties of putative primers with one of the many bioinformatics tool available online, such as the oligonucleotide properties calculator (available at http://www.basic.northwestern.edu/biotools/oligocalc.html). This calculator provides basic data on melting temperature and self-complementarity and provides a link to NCBI/BLAST for specificity searches.

Core commercial PCR kits are useful for PCR optimization, which is particularly important for aDNA work. A kit employing a hot start Taq polymerase is preferred for specificity when amplifying targets from a complex milieu, which will invariably contain environmental bacteria. There is rarely any need for nested PCR or second rounds of amplification if the initial PCR method is fully optimized. Use of core kits means that key components can be varied and the effects on PCR efficiency determined. The two most important considerations are the free magnesium ion concentration and measures to overcome the presence any PCR inhibitors in the reaction. Inhibitors are frequently co-purified during the extraction step. As a general rule, final concentration of $MgCl_2$ of 2 mM is appropriate for most PCR applications. Experience has shown that this should be raised to 3 mM when using dual-labelled hydrolysis probes in qPCR. Higher concentrations of magnesium can be deleterious to the underlying PCR, so kits with higher or unspecified concentrations should be avoided. Non-acetylated BSA is extremely useful for counteracting PCR inhibitors and can be included in the master mix without effecting PCR efficiency. Fig. 8.1 shows the benefits of BSA addition to the qPCR mix in MTB complex positive samples containing environmental inhibitors. An alternative way of overcoming inhibition in problematic samples is to augment the PCR master mix with additional Taq polymerase. For example, the typical recommendation of 0.5 U of Taq polymerase for qPCR can be increased to 2 or 3 U for problematic aDNA templates. The typical small aDNA PCR products may be analysed on 3% or 4% gel electrophoresis with appropriate size markers.

Preferably, if a qPCR platform is available, this offers a number of advantages over conventional

a) No BSA

b) With BSA

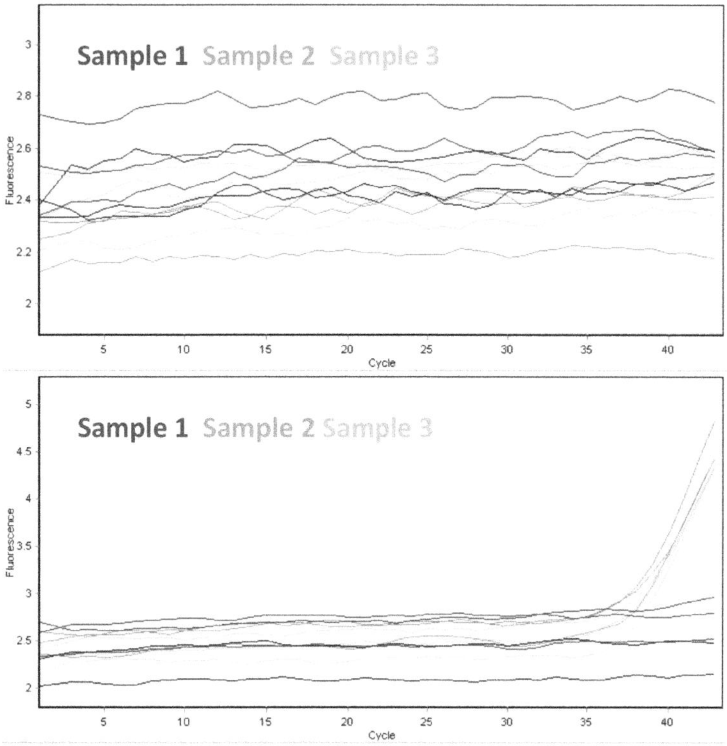


**Figure 8.1** Real-time PCR for a marker of MTB complex DNA run on a Corbett RotorGene™ 3000 platform. The extracts contain co-purified PCR inhibitors of soil origin. Panel (a) shows amplification profiles of extracts run using a Qiagen hot start kit without added BSA. Panel (b) shows the same extracts with the inclusion of non-acetylated BSA at a final concentration of 1 mg/ml. Four samples previously negative are now positive.

PCR for aDNA studies, as it does for other molecular biological applications. Some of these advantages are presented below.

- Product formation can be monitored easily in real time with intercalating dyes such as SYBR®Green or EVA®Green (http://www.gene-quantification.de/hrm-dyes.html). These are easily added to any PCR and are cheap to use. To some extent, specificity of products can be checked with dissociation curves (melt temperature analysis) at the end of the run and compared with a reference standard. This is not infallible, as non-specific products may have similar G + C content and melting profiles. Therefore, specificity should always be confirmed with gel electrophoresis and sequencing. Specific fluorimetric probes can be designed to the amplicon core region to provide extra confidence of product identity during the qPCR run, but these products should also be confirmed by sequencing.
- Multiple no template control samples can be included in the reaction and monitored throughout the run, to check for any evidence of cross-contamination. When using an intercalating dye, it should be noted that if primer-dimer formation occurs, this will result in a positive amplification curve. Melt analysis will reveal the lower melting temperature ($T_m$) of the primer-dimers compared to the target PCR product. Alternatively, the machine software can be set to acquire data above the $T_m$ of non-specific products formed during the course of the run, so that these smaller molecules do not contribute to the real-time amplification profile. As with conventional PCR, primer-dimer formation can be a useful indicator that inhibition was not a significant factor in a negative result.

- To rule out false negative results due to PCR inhibition, an internal control (IC) assay can be multiplexed with the locus of interest. An example of an inhibition monitoring IC assay is the competitive IC model using an artificial template. This template contains the same primer annealing regions as the PCR target but a different sequence between the primer binding sites. Amplification of this construct is monitored with a second probe designed to the central region but labelled with a different fluorophore e.g. FAM cf. HEX. Spiking of samples and water blanks with a constant amount of artificial template can help identify inhibition due to an increase in the quantification cycle (Cq) in affected samples, compared to the control Cq determined from water blanks. The Cq is defined as the cycle number at which fluorescence first increases above a defined threshold level. An alternative way of assessing levels of PCR inhibition in DNA extracts is to add replicates of them to positive controls and compare the Cqs to the effect of adding water. As this method generates the same amplicon as the aDNA target region, this method of inhibition monitoring should be performed separately and after monitoring of samples for pathogen aDNA. A universal qPCR has been developed for assessing inhibition called the 'SPUD' assay (Nolan *et al.*, 2006). Primers are used to amplify a unique sequence from the potato genome, the *Solanum tuberosum phyB* gene. The primers chosen lack homology with any other known sequence which means this IC assay can be used for all samples with the exception of the potato. A single inhibition assay can be used by all groups for monitoring PCR efficiency regardless of organism under investigation.
- Genuine aDNA positives tend to have high Cq values and product is typically seen after more than 30 cycles of amplification. Profiles with low Cq values may be an indicator of contamination, although strong positives are sometimes found. For example, Fig. 8.2 shows an RT-PCR run on the MxPro 3005P platform (Stratagene) in which four cases of LL, diagnosed on osteological grounds, were screened for evidence of *M. leprae* DNA using primers targeting the RLEP repetitive element. This reveals a range of Cq values, which indicates the pathogen survival and different mycobacterial burdens in the various burials and skeletal elements. Whilst some extracts have Cqs between 30 and 40, two replicates exhibit Cqs in the range 23–31 and are therefore stronger positives. In qPCR reactions that are 100% efficient, a variation in Cq of approximately 3.32 cycles equates to a 10-fold difference in template concentration. qPCR is a rapid means for identifying which cases may be good candidates for genotyping studies, which invariably require amplification of single copy loci.

## Validation of PCR findings

The chances of finding positive samples will vary with local site conditions and samples gathered, even varying between different elements from the same burials. The most likely outcome will be a negative one, even when skeletal lesions are unequivocal. Over 16 years of testing hundreds of tuberculosis cases, about 20% have proven positive to some extent and around a third of the positive cases are sufficiently robust to allow genotyping through amplification of single copy loci. DNA preservation at some sites, such as Vác in Hungary, where conditions favoured a natural mummification, has resulted in a much higher yield of MTB complex PCR positives (69%) samples (Fletcher *et al.*, 2003a).

Having obtained positive results from archaeological or archival material, it is then incumbent upon the researchers to prove that their data is authentic. Cooper and Poinar (2000) proposed a number of tests and measures that should be followed for aDNA validation. The majority of these common sense measures have been adopted by workers in the field. They should not be seen as a complete list or proscriptive as different sub areas of aDNA study may require different approaches. Similarly, advances in technologies such as next generation sequencing and other emerging disciplines will likely provide further opportunities for validation. The core measures for authentication are:

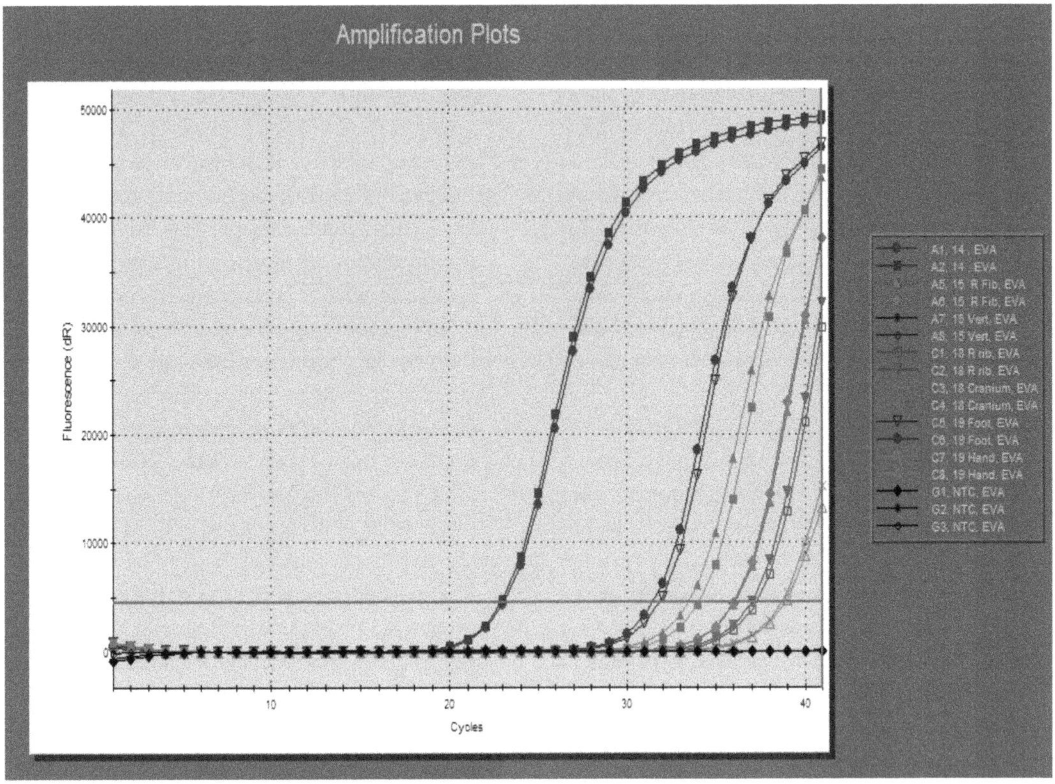


**Figure 8.2** Real-time PCR for RLEP, a marker of *M. leprae* DNA amplified from six extracts prepared from four individual cases of lepromatous leprosy. Extracts were assayed in duplicate. The range of Ct values reflects both the mycobacterial burden preserved in individual cases and skeletal component sampled. The most strongly positive extract (lowest Ct value) was taken from the rhinomaxillary region of the affected skeleton.

- *Reproducibility.* Positive results should be reproducible. Extracts should be assayed on more than one occasion and replicates of each extract included to establish a consensus of the data. PCR reactions rarely require more than 1 or 2 µl of template, so replication need not deplete precious resources or limit further genotyping. Robust positives should be reproducible in experiments separated in time and when run by different operators. Repeated freezing and thawing may result in some DNA degradation, apparent as right-shifts/increases in Cq values on qPCR platforms. It is therefore advantageous to split DNA extracts into two or more aliquots if qPCR is to be undertaken at a later date.
- *Independent replication.* Most importantly, key data (at least 10%) must be replicated at an independent centre that is competent at working with low-copy PCR. If only one or two cases are under study, then it is feasible to replicate all data. For this undertaking, separate samples must be prepared independently in each laboratory, using the same protocols and measures to prevent cross-contamination. Matching PCR platforms in different centres is expensive and rarely achievable. In practice, we have found that use of different PCR kits, PCR platforms and operators does not hinder replication of robust positives between centres and is only an issue when dealing with extremely degraded material.
- *Control amplifications.* It is essential to prepare extraction controls whenever new samples are prepared. These are dummy reactions, lacking sample, taken through the extraction process. They are used to show that extraction kits and other reagents such as solvents and elution

buffers are free of contamination. Template blanks are included in all PCR experiments. In these, water or DNA elution buffer replaces the sample extract. These not only monitor the PCR reagents but also any sporadic tube-to-tube contamination that may arise during the course of the amplification. This route for contamination certainly does occur, probably as a result of PCR tube lid fatigue after a large number of thermal cycles. For this reason, it is good practice to include a number of template blanks in each run.

- *Confirmation of PCR specificity*. Sequencing of the PCR product(s) should be undertaken to confirm their identity. Gel electrophoresis may indicate that a product of the correct size has been produced but this alone cannot be relied upon. Direct sequencing of products, using both forward and reverse primers is necessary to authenticate the result. It is a good idea to confirm each new region amplified at least once for each case studied. If studying human DNA as part of the validation process, it will also be necessary to clone the PCR products and sequence a representative number of the colonies (e.g. 20–30) to look for evidence of exogenous sequences and damage in the archival DNA templates. Occasionally, unrelated fragments will become incorporated into the product as a result of 'jumping PCR' (Pääbo *et al.,* 1990). Cloning of amplicons from microbial genomes will also be required if the products are too small for conventional sequencing.

We have in the past compared direct sequencing with sequencing of cloned products from loci used in VNTR genotyping of *M. leprae* (Taylor *et al.,* 2006). For the ML0058c (21–3) locus of *M. leprae,* 32 clones (from three samples) were sequenced and all showed two copies of the minisatellite. For the ($GTA^9$) locus, 15 clones showed nine copies and one clone showed eight copies. One further clone showed a GTA to ATA mutation attributed to Taq polymerase error. For the TTC ($AGA^{20}$) locus, a total of 16 clones showed 12 copies and four contained 11 copies. There was one clone with a TTC to ATC mutation due to Taq error. The few clones that where discrepancies of the VNTR numbers found probably reflect enzyme 'slippage' rather than a mixed population of isolates (Lovett, 2004).

- *Appropriate molecular behaviour*. DNA templates extracted from archival or archaeological remains will invariably be fragmented, with sizes and actual damage dependent on the conditions and period of time since deposition. It should be possible to demonstrate to what extent templates recovered from such contexts are degraded. This can be achieved with PCR using a series of forward primers with one reverse primer spanning a range of amplicon sizes, e.g. from 80 to 300 bp. A point will be reached when PCR fails due to lack of any template of that length. The amounts of any particular template size present may be determined using qPCR platforms using either relative or 'absolute' qPCR software, the latter requiring comparison against a dilution series of DNA standards. This is an area yet to be tackled systematically in aDNA pathogen studies, although many reports mention success at one target size and failure of a longer one. Fletcher and colleagues (2003a) looked at survival of different template sizes of MTB complex DNA in Hungarian mummy samples and showed that in nested IS*6110* positive samples, it was possible to demonstrate an inverse relationship between percentage of positive cases and amplicon target size using data from different gene targets. In an earlier study of LL, we were able to show different quantities of *M. leprae* DNA in samples taken from multiple sites on the skeleton (Taylor *et al.,* 2006). Sites with lesions characteristic of leprosy showed greater mycobacterial burdens of *M. leprae* DNA compared with macroscopically uninvolved bones. We have also often noted that LL cases often have a greater mycobacterial burden around the rhinomaxillary area compared to other affected sites like the bones of the feet, legs and hands.
- *Genotyping*. This is pivotal in authentication of aDNA findings relating to pathogens based on the detection of a single genetic locus. Amplification of even two or three additional loci is extremely useful in validation. The targets chosen may be single nucleotide

polymorphisms (SNPs), deletions or variable nucleotide tandem repeats (VNTR). The former two provide informative phylogenetic information that will help place an organism within evolutionary models and schemes (Brosch *et al.*, 2002). VNTR typing, though less helpful in evolutionary terms, can be helpful in further typing isolates with similar or identical genotypic markers. For example, VNTR typing of leprosy isolates can differentiate strains with identical SNP type and subtype. This is illustrated in Fig. 8.3, which shows sequencing of the ($GTA^9$) variable locus from two cases of LL with a 3K genotype and two others of 3M genotype. All display different copy numbers of the GTA repeat microsatellite. In fact, all of the cases of skeletal leprosy that we have studied to date have shown distinct genetic profiles when VNTR typed with just three loci (Taylor and Donoghue, 2011).

In the case of organisms of the MTB complex, phylogenetic schemes prepared with either informative SNPs (Baker *et al.*, 2004) or deletions (Gagneux *et al.*, 2006) show overall

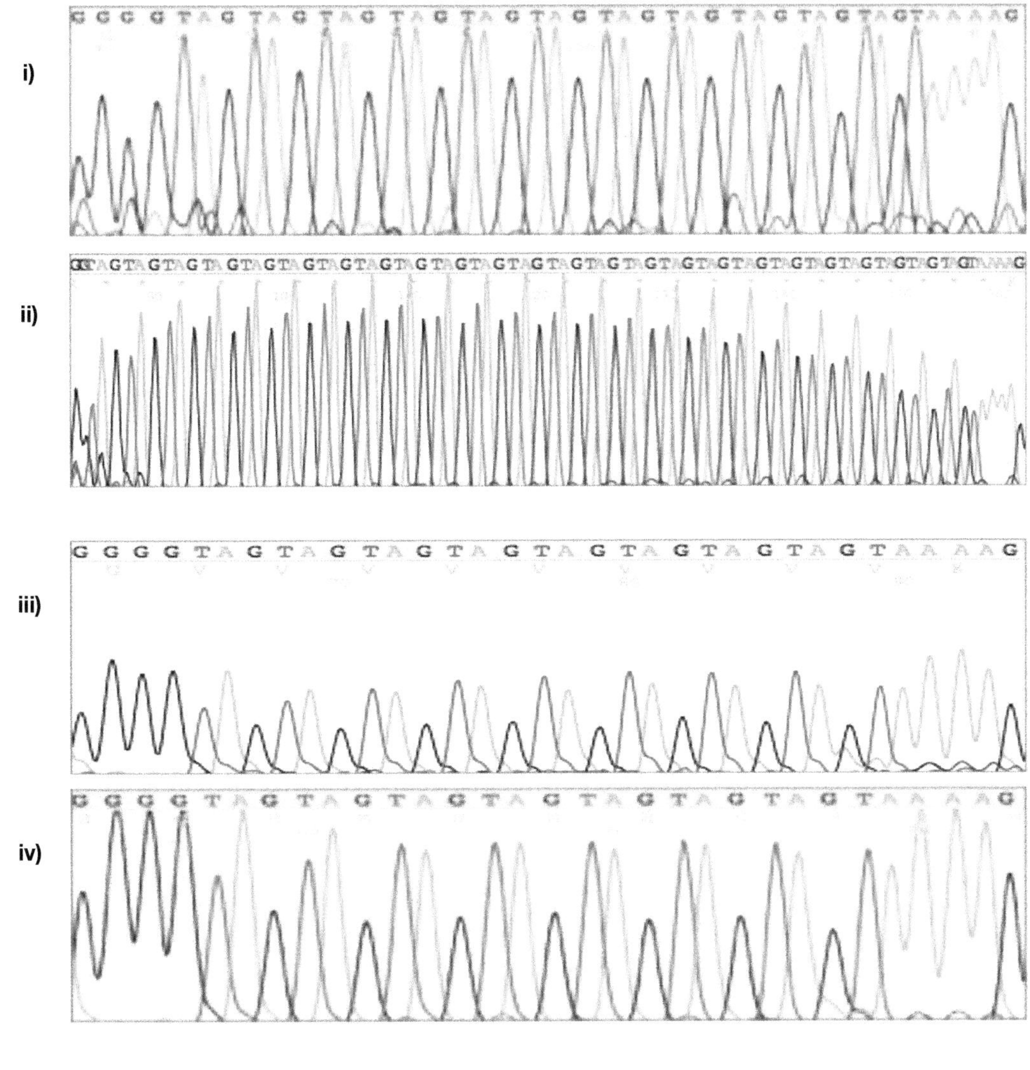


**Figure 8.3** (i and ii) VNTR typing of two cases of LL of genotype 3K using the polymorphic (GTA)9 locus of *M. leprae*. Burial KK02 (i) shows 11 copies and burial KD271 (ii) 24 copies. (iii and iv) VNTR typing of two cases of LL of genotype 3M using the polymorphic (GTA)9 locus of *M. leprae*. Burial 503 (iii) shows eight copies and burial 188 (iv) seven copies.

agreement. This can be used to identify the species/strain and place an ancient isolate. For example, SNP analysis can be used to find the major genotypic group to which an isolate of *M. tuberculosis* belongs (Sreevatsan *et al.*, 1997). These investigators identified three major genotypic groups of MTB complex organisms based on polymorphisms at the *KatG*$^{263}$ and *GyrA95* codons. Group 1 organisms may or may not have undergone loss of the TbD1 deletion. TbD1 is a 2,153 bp fragment containing the *mmpS6* and part of the *mmpL6 gene*. TbD1 is intact in Indo-Oceanic and both African lineages, but lost from East Asian, East-African-Indian and Euro-American strains. Groups 2 and 3 (Euro-American lineage), being derived from Group 1 organisms that have lost TbD1, should also exhibit loss of this region. In turn, this is correlated with a 7 bp loss in the *pks 15/1* locus of *M. tuberculosis*. Similarly, mycobacteria on the *M. bovis* lineage will have lost region of difference 9 (RD9), as well as RD7, RD8 and RD10. Classic *M. bovis* is associated with loss of RD4 (Gordon *et al.*, 1999) and this correlates with a *pncA*$^{57}$ CAC to GAC mutation and an *oxyR*$^{285}$ G to A mutation. These and other loci can be studied to ensure that the sequences retrieved from archival materials make good phylogenetic sense. The field of aDNA is set to benefit from the far larger amounts of data to be gained from microarray enrichment and next generation sequencing techniques which have recently come to be applied to appropriate archaeological cases (Bouwman *et al.*, 2012).

- *Analysis of associated remains.* It has been suggested that when examining human remains, animal or other associated remains from the same sites should be studied at the same time to assist with validation. Moreover, soil or other environmental samples can be used to validate findings in either human or animal remains. This can be undertaken whether studying human aDNA or aDNA from pathogens that may infect either man or other animal species. The demonstration of faunal remains with amplifiable aDNA will show that conditions have favoured survival of DNA *per se*. There is less scope for contamination with modern animal DNA, particularly in laboratories where they are infrequently studied. Evidence of DNA from zoonotic pathogen species demonstrated in human burials should be sought in relevant animal species, if available. As mentioned above, faunal remains do pose several challenges in their own right because of the disarticulated and often scavenged nature of the material. However, we have recently been able to demonstrate MTB complex DNA in several species with osteological evidence of probable tuberculosis infection including horse and pig (Taylor, Wooding and Knüsel, unpublished observations). This required redesign of several MTB complex PCR methods to focus on fragment sizes <100 bp in length and use of qPCR hydrolysis probes for increased specificity.
- *Isolation of other biomarkers of disease.* It may be possible to detect independent evidence of microbial infection through detection of stable pathogen biomarkers other than aDNA. For example, mycolic acid lipid biomarkers may be found in human and animal remains that can provide evidence of mycobacterial diseases like tuberculosis and leprosy. Extracted long chain fatty acids are derivatized to pyrenebutyric acid (PBA) and pentafluorobenzyl (PFB) mycolates and detected using either normal or reverse phase fluorescence HPLC and compared with standards. These traces reveal characteristic patterns for the individual components of alpha, methoxy and keto-mycolates as well as the total mycolates in the material. Another class of mycobacterial lipids, the mycocerosic acids, has also been investigated as biomarkers for tuberculosis diagnosis in archaeological remains. Mycocerosic acids are long-chain, multimethyl-, branched-chain fatty acids that are present in *M. tuberculosis, M. bovis, M. gastri, M. haemophilum, M. kansasii, M. leprae, M. marinum* and *M. ulcerans* (Redman *et al.*, 2009).

The advantage of mycolate biomarker analysis is that there is no need for amplification, which reduces the potential for cross contamination. Moreover, mycolates can be extracted from the same sample as aDNA in an integrated protocol. Mycolic acid analyses has been used to provide a second line of evidence for tuberculosis from a number of archaeological contexts with human remains identified as MTB complex positive by PCR (Gernaey *et al.*, 2001; Donoghue *et al.*, 2004, 2010; Hershkovitz *et al.*, 2008; Redman *et al.*, 2009). The same techniques have recently been applied to a case of LL from Central Asia (Taylor *et al.*, 2009). Here, only the α-mycolates and keto-mycolates were detected, consistent with *M. leprae* which lacks the methoxy-mycolates present in *M. tuberculosis*. Mycolic acid analysis is a technique that could be applied to other mycobacterial species and also to *Nocardia, Rhodococcus, Corynebacterium* and related bacteria (Minnikin *et al.*, 1980; Sutcliffe, 2010). The potential also exists for detection of mycobacterial proteins, peptides and sugars.

- Another example of the use of alternative/complementary biomarkers is in the study of *Yersinia pestis* and the aetiology of black death performed by Haensch and colleagues (2010). They used the plague dipstick assay to supplement aDNA studies and to gather information from remains where DNA had not survived. This rapid diagnostic test for plague is an immunochromatographic assay that detects the F1 envelope glycoprotein specific to *Y. pestis* (Chanteau *et al.*, 2003; Bianucci *et al.*, 2008).

The above is a list of eight suggestions for validation. It will not always be possible to tick every box due to sample availability, cost or finding appropriate collaborators in other specialties. Moreover, it is not necessarily a complete list, as different areas of aDNA research may require other approaches or emphasis.

## Conclusions

The field of aDNA pathogen research has come a long way since the early days less than two decades ago. Like other areas of molecular biology, it has benefited from advances in molecular biological techniques, refinements of PCR platforms and the many pathogen whole-genome sequencing projects. This is set to continue with the emerging techniques of whole genome amplification and high-throughput sequencing, which will take the amount of recovered data to the next level. The scepticism that greeted some early aDNA findings has helped strengthen the discipline by providing improved working practices and means of validation. The field has witnessed developments in related biomolecular areas, which provide additional means for studying pathogens in human and faunal remains. Archaeology has always benefited from diverse specialties and aDNA analysis is surely now established as one of those sciences contributing both to archaeology as well as to our understanding of past diseases and plagues.

## References

Afghani, B., and Stutman, H.R. (1996). Polymerase chain reaction for diagnosis of *M. tuberculosis:* comparison of simple boiling and a conventional method for DNA extraction. Biochem. Mol. Med. *57*, 14–18.

Aldous, W.K., Pounder, J.I., Cloud JL, and Woods GL. (2005). Comparison of six methods of extracting *Mycobacterium tuberculosis* DNA from processed sputum for testing by quantitative real-time PCR. J. Clin. Microbiol. *43*, 2471–2473.

Baker, L., Brown, T., Maiden, M.C., and Drobniewski, F. (2004). Silent nucleotide polymorphisms and a phylogeny for *Mycobacterium tuberculosis*. Emerg. Infect. Dis. *10*, 1568–1577.

Bathurst, R.R., and Barta, J.L. (2004). Molecular evidence of tuberculosis induced hypertrophic osteopathy in a 16th-century Iroquoian dog. J. Archaeol. Sci. *31*, 917–925.

Bianucci, R., Rahalison, L., Massa, E.R., Peluso, A., Ferroglio, E., and Signoli, M. (2008). Technical note: a rapid diagnostic test detects plague in ancient human remains: an example of the interaction between archaeological and biological approaches (southeastern France, 16th-18th centuries). Am. J. Phys. Anthropol. *136*, 361–367.

Boom, R., Sol, C.J.A., Salimans, M.M.M., Jansen, C.L., Wertheim-van Dillen, P.M.E., and van der Noordaa, J. (1990). Rapid and simple method for purification of nucleic acids. J. Clin. Microbiol. *28*, 495–503.

Bouwman, A.S., and Brown, T.A. (2002). Comparison between silica-based methods for the extraction of DNA from human bones from 18th–19th century London. Anc. Biomol. *4*, 173–178.

Bouwman, A.S., and Brown, T.A. (2005). The limits of biomolecular palaeopathology: ancient DNA cannot be used to study venereal syphilis. J. Archaeol. Sci. 32, 703–713.

Bouwman, A.S., Kennedy, S.L., Müller, R., Stephens, R.H., Holst, M., Caffell, A.C., Roberts, C.A., and Brown, T.A. (2012). Genotype of a historic strain of Mycobacterium tuberculosis. Proc. Natl. Acad. Sci. U.S.A. *109*, 18511–18516.

Braun, M., and Cook, D.C. (1998). DNA from *Mycobacterium tuberculosis* complex identified in North

American, pre-Columbian human skeletal remains. J. Archaeol. Sci. *25*, 271–277.

Brennan, P.J., and Nikaido, H. (1995). The envelope of mycobacteria. Ann. Rev. Biochem. *64*, 29–63.

Brosch, R., Gordon, S.V., Marmiesse, M., Brodin, P., Buchrieser, C., Eiglmeier, K., Garnier, T., Gutierrez, C., Hewinson, G., Kremer, K., *et al.* (2002). Proc. Natl. Acad. Sci. U.S.A. *99*, 3684–3689.

Cano, R.J. (1996). Analysing ancient DNA. (Review). Endeavour *20*, 162–167.

Chanteau, S., Rahalison, L., Ralafiarisoa, L., Foulon, J., Ratsitorahina, M., Ratsifasoamanana, L., Carniel, E., and Nato, F. (2003). Development and testing of a rapid diagnostic test for bubonic and pneumonic plague. Lancet *361*, 211–216.

Christopher-Hennings, J., Dammen, M.A., Weeks, S.R., Epperson, W.B., Singh, S.N., Steinlicht, G.L., Fang, Y., Skaare, J.L., Larsen, J.L., Payeur, J.B., *et al.* (2003). Comparison of two DNA extractions and nested PCR, real-time PCR, a new commercial PCR assay, and bacterial culture for detection of *Mycobacterium avium subsp. Paratuberculosis* in bovine feces. J. Vet. Diagn. Invest. *15*, 87–93.

Cooper, A., and Poinar, H. (2000). Ancient DNA: do it right or not at all. Science *289*, 1139.

David, H.L., Levy-Frebault, V., Dauguet, C., and Grimont, F. (1984). DNA from *Mycobacterium leprae*. Acta Leprol. *2*, 129–136.

Dieffenbach, C.W., and Dveksler, G.S. (1993). Setting up a PCR laboratory. Genome Res. *3*, S2–S7.

Dissing, J., Kristinsdottir, M.A., and Friis, C. (2008). On the elimination of extraneous DNA in fossil human teeth with hypochlorite. J. Archaeol. Sci. *35*, 1445–1452.

Donoghue, H.D. (2008). Molecular palaeopathology of human infectious disease. In Advances in Human Palaeopathology, R. Pinhasi and S. Mays, eds. (John Wiley and Sons Ltd., Chichester, UK), pp. 147–176.

Donoghue HD. (2011). Insights gained from palaeomicrobiology into ancient and modern tuberculosis. (Review). Clin. Microbiol. Infect. *17*, 821–829.

Donoghue, H.D., Spigelman, M., Zias, J., Gernaey-Child, A.M., and Minnikin, D.E. (1998). *Mycobacterium tuberculosis* complex DNA in calcified pleura from remains 1400 years old. Lett. Appl. Microbiol. *27*, 265–269.

Donoghue, H.D., Holton, J., and Spigelman, M. (2001). PCR primers that can detect low levels of *Mycobacterium leprae* DNA. J. Med. Microbiol. *50*, 177–182.

Donoghue, H.D., Gladkowska-Rzeczycka, J., Marcsik, A., Holton, J., and Spigelman, M. (2002). *Mycobacterium leprae* in archaeological specimens. In The Past and Present of Leprosy, Archaeological, Historical, Palaeopathological and Clinical Approaches, BAR International Series 1054, C.A. Roberts, M.E. Lewis, and K. Manchester, eds. (Archaeopress, Oxford, UK), pp. 271–285.

Donoghue, H.D., Spigelman, M., Greenblatt, C.L., Lev-Maor, G., Ka, G., Matheson, C., Vernon, K., Nerlich, A.G., and Zink, AR. (2004). Tuberculosis: from prehistory to Robert Koch, as revealed by ancient DNA. Lancet Infect. Dis. *4*, 584–592.

Donoghue, H.D., Marcsik, A., Matheson, C., Vernon, K., Nuorala, E., Molto, J.E., Greenblatt, C.E., and Spigelman, M. (2005). Co-infection of *Mycobacterium tuberculosis* and *Mycobacterium leprae* in human archaeological samples: a possible explanation for the historical decline of leprosy. Proc. Biol. Sci. *272*, 389–394.

Donoghue, H.D., Lee, OY-C., Minnikin, D.E., Besra, G.S., Taylor, J.H., and Spigelman, M. (2010). Tuberculosis in Dr Granville's mummy: a molecular re-examination of the earliest known Egyptian mummy to be scientifically examined and given a medical diagnosis. Proc. Biol. Sci. *277*, 51–56.

Drancourt, M., and Raoult, D. (2005). Palaeomicrobiology: current issues and perspectives. Nat. Rev. Microbiol. *3*, 23–35.

Ferreira, L.F., Britto, C., Cardoso, M.A., Fernandes, O., Reinhard, K., and Araújo, A. (2000). Paleoparasitology of Chagas disease revaled by infected tissues from Chilean mummies. Acta Tropica *75*, 79–84.

Fletcher, H., Donoghue, H.D., Holton, J., Pap, I., and Spigelman, M. (2003a). Widespread occurrence of *Mycobacterium tuberculosis*-DNA from 18th–19th Century Hungarians. Am. J. Phys. Anthropol. *120*, 144–152.

Fletcher, H.A., Donoghue, H.D., Taylor, G.M., van der Zanden, A.G.M., and Spigelman, M. (2003b). Molecular analysis of *Mycobacterium* tuberculosis DNA from a family of 18th century Hungarians. Microbiology *149*, 143–151.

Gagneux, S., DeRiemer, K., Van, T., Kato-Maeda, M., de Jong, B.C., Narayanan, S., Nicol, M., Niemann, S., Kremer, K., Gutierrez, M.C., *et al.* (2006). Variable host–pathogen compatibility in *Mycobacterium tuberculosis*. Proc. Natl. Acad. Sci. U.S.A. *103*, 2869–2873.

Gernaey, A.M., Minnikin, D.E., Copley, M.S., Dixon, R.A., Middleton, J.C., and Roberts, C.A. (2001). Mycolic acids and ancient DNA confirm an osteological diagnosis of tuberculosis. Tuberculosis *281*, 259–265.

Gilbert, M.T.P., Hansen, A.J., Willerslev, E., Barnes, I., Rudbeck, L., Barnes, I., Lynnerup, N., and Cooper A. (2003). Characterisation of genetic caused by postmortem damage. Am. J. Hum. Genet. *72*, 48–61.

Gilbert, M.T., Tomsho, L.P., Rendulic, S., Packard, M., Drautz, D.I., Sher, A., Tikhonov, A., Dalén, L., Kuznetsova, T., Kosintsev, P., *et al.* (2007). Whole-genome shotgun sequencing of mitochondria from ancient hair shafts. Science *317*, 1927–1930.

Gordon, S.V., Brosch, R., Billault, A., Garnier, T., Eiglmeier, K., and Cole, S.T. (1999). Identification of variable regions in the genomes of tubercle bacilli using bacterial artificial chromosome arrays. Mol. Microbiol. *32*, 643–655.

Guhl, F., Jaramillo, C., Yockteng, R., Vallejo, G.A., and Cárdenas-Arroyo, F. (1997). *Trypanosoma cruzi* DNA in human mummies. Lancet *349*, 1370.

Haas, C.J., Zink, A., Palfi, G., Szeimies, U., and Nerlich, AG. (2000). Detection of leprosy in ancient human skeletal remains by molecular identification of *Mycobacterium leprae*. Am. J. Clin. Pathol. *114*, 428–436.

Haensch, S., Bianucci, R., Signoli, M., Rajerison, M., Schultz, M., Kacki, S., Vermunt, M., Weston, D.A.,

Hurst, D., Achtman, M., *et al.* (2010). Distinct clones of *Yersinia pestis* caused the Black Death. PLoS Pathog. *6*, e1001134.

Hansen, A., Willerslev, E., Wiuf, C., Mourier, T., and Arctander, P. (2001). Statistical evidence for miscoding lesions in ancient DNA templates. Mol. Biol. Evol. *18*, 262–265.

Hansen, A.J., Mitchell, D.L., Wiuf, C., Paniker, L., Brand, T.B., Binladen, J., Gilichinsky, D.A., Ronn, R., and Willerslev, E. (2006). Crosslinks rather than strand breaks determine access to ancient DNA sequences from frozen sediments. Genetics *173*, 1175–1179.

Heginbotham, M.L., Magee, J.T., and Flanagan, P.G. (2003). Evaluation of the Idaho Technology Light-Cycler PCR for the direct detection of *Mycobacterium tuberculosis* in respiratory specimens. Int. J. Tuberc. Lung. Dis. *7*, 78–83.

Hershkovitz, I., Donoghue, H.D., Minnikin, D.E., Besra, G.S., Lee, O Y-C., Gernaey, A.M., Galili, E., Eshed, V., Greenblatt, C.L., Lemma, E., *et al.* (2008). Detection and molecular characterization of 9000-Year old *Mycobacterium tuberculosis* from a Neolithic settlement in the eastern Mediterranean. PLoS One *3*, e3426.

Heyn, P., Stenzel, U., Briggs, A.W., Kircher, M., Hofreiter, M., and Meyer, M. (2010). Roadblocks on paleogenomes-polymerase extension profiling reveals the frequency of blocking lesions in ancient DNA. Nucleic Acids Res. *38*, e161.

Jaeger, L.H., Leles, D., Lima, V.D., Silva, L.D., and Iñiguez, A.M. (2011). *Mycobacterium tuberculosis* complex detection in human remains: tuberculosis spread since the 17th century in Rio de Janeiro, Brazil. Infect. Genet. Evol. *12*, 642–648.

Kelley, M.A., and Micozzi, M.S. (1984). Rib lesions in chronic pulmonary tuberculosis. Am. J. Phys. Anthropol. *65*, 381–386.

Knapp, M., Clarke, A.C., Horsburgh, K.A., and Matisoo-Smith, E.A. (2011). Setting the stage – building and working in an ancient DNA laboratory. Ann. Anat. *194*, 3–6.

Likovský, J., Urbanová, M., Hájek, M., Černý, V., and Čech, P. (2006). Two cases of leprosy from Žatec (Bohemia), dated to the turn of the 12th century and confirmed by DNA analysis for *Mycobacterium leprae*. J. Archaeol. Sci. 33, 1276–1283.

Lindahl, T. (1993). Instability and decay of the primary structure of DNA. Nature *362*, 709–715.

Lindahl, T., and Nyberg, B. (1972). Rate of depurination of native deoxyribonucleic acid. Biochemistry *11*, 3610–3618.

Locke, E.A. (1915). Secondary hypertrophic osteoarthropathy and its relation to simple club fingers. Arch. Intern. Med. *15*, 659–713.

Loreille, O., Roumat, E., Verneau, O., Bouchet, F., and Hänni, C. (2001). Ancient DNA from Ascaris: extraction amplification and sequences from eggs collected in coprolites. Int. J. Parasitol. *31*, 1101–1106.

Lovett, S.T. (2004). Encoded errors: mutations and rearrangements mediated by misalignment at repetitive DNA sequences. Mol. Microbiol. *52*, 1243–1253.

Manchester, K. (1991). Tuberculosis and leprosy: evidence for interaction of disease. In Human Paleopathology: Current Syntheses and Future Options, D.C. Ortner and A.C. Aufderheide eds. (Smithsonian Institution Press, Washington, DC), pp. 23–35.

Mangiapan, G., Vokura, M., Schouls, L., Cadranel, J., Lecossier, D., Van Embden, J., and Hance, A.J. (1996). Sequence capture-PCR improves detection of Mycobacterial DNA in clinical specimens. J. Clin. Microbiol. *34*, 1209–1215.

Mateiu, L.M., and Rannala, B.H. (2008). Bayesian inference of errors in ancient DNA caused by *postmortem* degradation. Mol. Biol. Evol. 25, 1503–1511.

Mays, S.A., and Taylor, G.M. (2002). Osteological and biomolecular study of two possible cases of hypertrophic osteoarthropathy from Mediaeval England. J. Archaeol. Sci. *29*, 1267–1276.

Mays, S.A., and Taylor, G.M. (2003). A first prehistoric case of tuberculosis from Britain. Int. J. Osteoarchaeol. *13*, 189–196.

Mays, S.A., Taylor, G.M., Legge, A.J., Young, D.B., and Turner-Walker, G. (2001). A palaepathological and biomolecular study of tuberculosis in a medieval skeletal collection from England. Am. J. Phys. Anthropol. *114*, 298–311.

Mays, S., Fysh, E., and Taylor, G.M. (2002). Investigation of the link between visceral surface rib lesions and tuberculosis in a medieval skeletal series from England using ancient DNA. Am. J. Phys. Anthropol. *119*, 27–36.

Minnikin, D.E., Hutchinson, I.G., Caldicot, A.B., and Goodfellow, M. (1980). Thin-layer chromatography of methanolysates of mycolic acid-containing bacteria. J. Chromatog. A *188*, 221–233.

Monot, M., Honoré, N., Garnier, T., Araoz, R., Coppée, J.Y., Lacroix, C., Sow, S., Spencer, J.S., Truman, R.W., Williams, D.L., *et al.* (2005). On the origin of leprosy. Science *308*, 1040–1042.

Monot, M., Honoré, N., Garnier, T., Zidane, N., Sherafi, D., Paniz-Mondolfi, A., Matsuoka, M., Taylor, G.M., Donoghue, H.D., Bouwman, A., *et al.* (2009). Phylogeography of leprosy. Nat. Genet. *41*, 1282–128.

Montiel, R., Garcia, C., Canadas, M.P., Isidro, A., Guijo, J.M., and Malgosa, A. (2003). DNA sequences of *Mycobacterium leprae* recovered from ancient bones. FEMS Microbiol. Lett. *226*, 413–414.

Murphy, E.M., Chistov, Y.K., Hopkins, R., Rutland, P., and Taylor, G.M. (2009). Tuberculosis among Iron Age individuals from Tyva, South Siberia: palaeopathological and biomolecular findings. J. Archaeol. Sci. *36*, 2029–2038.

Mutolo, M.J., Jenny, L.L., Buszek, A.R., Fenton, T.W., and Foran, D.R. (2011). Osteological and molecular identification of brucellosis in ancient Butrint, Albania. Am. J. Phys. Anthropol. *147*, 254–263.

Nerlich, A.G., Schraut, B., Dittrich, S., Jelinek, T., and Zink, A.R. (2008). *Plasmodium falciparum* in Ancient Egypt. Emerg. Infect. Dis. *14*, 1317–1318.

Niederhauser, C., Höfelein, C., Wegmüller, B., Lüthy, J., and Candrian, U. (1994). Reliability of PCR decontamination systems. Genome Res. *4*, 117–123.

Nolan, T., Hands, R.E., Ogunkolade, W., and Bustin, S.A. (2006). SPUD: a quantitative PCR assay for the detection of inhibitors in nucleic acid preparations. Anal. Biochem. *351*, 308–310.

Nuorola, E. (1999). Tuberculosis on the 17th century man-of-war Kronan. Int. J. Osteoarchaeol. *9*, 344–348.

Ortner, D.J., and Putschar, W.G.J. (1985). Identification of Pathological Conditions in Human Skeletal Remains (Smithsonian Institution Press, Washington, DC).

Pääbo, S., Irwin, D.M., and Wilson, A.C. (1990). DNA Damage promotes jumping between templates during enzymatic amplification. J. Biol. Chem. *265*, 4718–4721.

Poinar, H.N., Hofreiter, M., Spaulding, W.G., Martin, P.S., Stankiewicz, B.A., Bland, H., Evershed, R.P., Possnert, G., and Pääbo, S. (1998). Molecular coproscopy: dung and diet of the extinct ground sloth *Nothrotheriops shastensis*. Science *281*, 402–406.

Rafi, A., Spigelman, M., Stanford, J., Lemma, E., Donoghue, H.D., and Zias, J. (1994). DNA of *Mycobacterium leprae* detected by PCR in ancient bone. Int. J. Osteoarchaeol. *4*, 287–290.

Raoult, D., Aboudharam, G., Crubézy, E., Larrouy, G., Ludes, B., and Drancourt, M. (2000). Molecular identification by 'suicide PCR' of *Yersinia pestis* as the agent of medieval Black Death. Proc. Natl. Acad. Sci. U.S.A. *97*, 12800–12803.

Raoult, D., Dutour, O., Houhamdi, L., Jankauskas, R., Fournier, P.E., Ardagna, Y., Drancourt, M., Signoli, M., La, V.D., Macia, Y., *et al.* (2006). Evidence for louse-transmitted diseases in soldiers of Napoleon's Grand Army in Vilnius. J. Infect. Dis. 193, 112–120.

Redman, J.E., Shaw, M.J., Mallet, A.I., Santos, A.L., Roberts, C.A., Gernaey, A.M., and Minnikin, D.E. (2009). Mycocerosic acid biomarkers for the diagnosis of tuberculosis in the Coimbra Skeletal Collection. Tuberculosis (Edinb) *89*, 267–277.

Resnick, D., and Niwayama, G. (1981). Diagnosis of Bone and Joint Disorders (WB Saunders, London).

Rohland, N., and Hofreiter, M. (2007). Ancient DNA extraction from bones and teeth. Nat. Protoc. *2*, 1756–1762.

Roring, S., Hughes, M.S., Skuce, R.A., and Neill, S.D. (2000). Simultaneous detection and strain differentiation of *Mycobacterium bovis* directly from bovine tissue specimens by spoligotyping. Vet. Microbiol. *74*, 227–236.

Rothschild, B.M., Martin, L.D., Lev, G., Bercovier, H., Bar-Gal, G.K., Greenblatt, C., Donoghue, H., Spigelman, M., and Brittain, D. (2001). Mycobacterium tuberculosis complex DNA from an extinct bison dated 17,000 years before the present. Clin. Infect. Dis. 33, 305–311.

Sallares, R., and Gomzi, S. (2001). Biomolecular archaeology of malaria. Anc. Biomol. 3, 195–213.

Salo, W., Aufderheide, A.C., Buikstra, J.E., and Holcomb, T.A. 1994. Identification of *Mycobacterium tuberculosis* DNA in a pre-Columbian Peruvian mummy. Proc. Natl. Acad. Sci. U.S.A. *91*, 2091–2094.

Schwarz, C., Debruyne, R., Kuch, M., McNally, E., Schwarcz, H., Aubrey, A.D., Bada, J., and Poinar, H. (2009). New insights from old bones: DNA preservation and degradation in permafrost preserved mammoth remains. Nucleic Acids Res. 37, 3215–3229.

Smith, C.I., Chamberlain, A.T., Riley, M.S., Stringer, C., and Collins, M.J. (2003). The thermal history of human fossils and the likelihood of successful DNA amplification. J. Hum. Evol. 45, 203–217.

Spigelman, M., and Donoghue, H.D. (2001). Brief communication: unusual pathological condition in the lower extremities of a skeleton from ancient Israel. Am. J. Phys. Anthropol. *114*, 92–93.

Spigelman, M., and Lemma, E. (1993). The use of the polymerase chain reaction to detect *Mycobacterium tuberculosis* in ancient skeletons. Int. J. Osteoarchaeol. *3*, 137–143.

Spigelman, M., Matheson, C., Lev, G., Greenblatt, C., and Donoghue, H.D. (2002). Confirmation of the presence of *Mycobacterium tuberculosis* complex-specific DNA in three archaeological specimens. Int. J. Osteoarchaeol. *12*, 393–401.

Sreevatsan, S., Pan, X., Stockbauer, K.E., Connell, N.D., Kreiswirth, B.N., Whittam, T.S., and Musser, J.M. (1997). Restricted structural gene polymorphism in the Mycobacterium tuberculosis complex indicates evolutionary recent global dissemination. Proc. Natl. Acad. Sci. U.S.A. *94*, 9869–9874.

Stiller, M., Green, R.E., Ronan, M., Simons, J.F., Du, L., He, W., Egholm, M., Rothberg, J.M., Keates, S.G., Ovodov, N.D., *et al.* (2006). Patterns of nucleotide misincorporations during enzymatic amplification and direct large-scale sequencing of ancient DNA. Proc. Natl. Acad. Sci. U.S.A. *103*, 13578–13584.

Sutcliffe, I.C. (2010). A phylum level perspective on bacterial cell envelope architecture. Trends Microbiol. *18*, 464–470.

Suzuki, K., Takigawa, W., Tanigawa, K., Nakamura, K., Ishido, Y., Kawashima, A., Wu, H., Akama, T., Sue, M., Yoshihara, A., *et al.* (2010). Detection of *Mycobacterium leprae* DNA from archaeological skeletal remains in Japan using whole genome amplification and polymerase chain reaction. PLoS One 5, e12422.

Takade, A., Umeda, A., Matsuoka, M., Yoshida, S-I., Nakamura, M., and Amako, K. (2003). Comparative studies on the cell structures of *Mycobacterium leprae* and *M. tuberculosis* using the electron microscopy freeze-substitution technique. Microbiol. Immunol. *47*, 265–270.

Taylor, G.M., and Donoghue, H.D. (2011). Multiple loci variable number tandem repeat (VNTR) analysis (MLVA) of *Mycobacterium leprae* isolates amplified from European archaeological human remains with lepromatous leprosy. Microbes Infect. *13*, 923–929.

Taylor, G.M., Crossey, M., Saldanha, J.A., and Waldron, T. (1996). Detection of *Mycobacterium tuberculosis* bacterial DNA in medieval human skeletal remains using polymerase chain reaction. J. Archaeol. Sci. *23*, 789–798.

Taylor, G.M., Goyal, M., Legge, A.J., Shaw, R.J., and Young, D. (1999). Genotypic analysis of *Mycobacterium tuberculosis* from medieval human remains. Microbiology *145*, 899–904.

Taylor, G.M., Widdison, S., Brown, I.N., Young, D.B., and Molleson, T. (2000). A case of lepromatous leprosy from 13th century Orkney. J. Archaeol. Sci. *27*, 1133–1138.

Taylor, G.M., Mays, S., Legge, A.J., Ho, T., and Young, D.B. (2001). Genotypic analysis of *Mycobacterium tuberculosis* from human remains. Anc. Biomol. *3*, 267–280.

Taylor, G.M., Young, D.B., and Mays, S.A. (2005). Genotypic analysis of the earliest known prehistoric case of tuberculosis in Britain. J. Clin. Microbiol. *43*, 2236–2240.

Taylor, G.M., Watson, C.L., Lockwood, D.N.J., and Mays, S.A. (2006). Variable nucleotide tandem repeat (VNTR) typing of two cases of lepromatous leprosy from the archaeological record. J. Archaeol. Sci. *33*, 1569–1579.

Taylor, G.M., Murphy, E., Hopkins, R., Rutland, P.C., and Chistov, Y. (2007). First report of *Mycobacterium bovis* DNA in archaeological human remains. Microbiology *153*, 1243–1249.

Taylor, G.M., Blau, S., Mays, S.A., Monot, M., Lee, OY-C., Minnikin, D.E., Besra, G.S., Cole, S.T., and Rutland, P.C. (2009). *Mycobacterium leprae* genotype amplified from an archaeological case of lepromatous leprosy in Central Asia. J. Archaeol. Sci. *36*, 2408–2414.

Taylor, G.M., Mays, S.A., and Huggett, J.F. (2010). Ancient DNA (aDNA) studies of man and microbes: general similarities, specific differences. Int. J. Osteoarchaeol. *20*, 747–751.

Tell, L.A., Foley, J., Needham, M.L., and Walker, R.L. (2003). Comparison of four rapid DNA extraction techniques for conventional polymerase chain reaction testing of three Mycobacterium spp. that affect birds. Avian Dis. *47*, 1486–1490.

Watson, C.L., and Lockwood, D.N.J. (2009). Single nucleotide polymorphism analysis of European archaeological *M. leprae* DNA. PLoS One *4*, e7547.

Winters, M., Barta, J.L., Monroe, C., and Kemp, B.M. (2011). To clone or not to clone: method analysis for retrieving consensus sequences in ancient DNA samples. PLoS One *6*, e21247.

Zhang, Z.Q., and Ishaque, M. (1997). Evaluation of methods for isolation of DNA from slowly and rapidly growing mycobacteria. Int. J. Lepr. other Mycobact. Dis. *65*, 469–476.

Zink, A.R., Sola, C., Reischl, U., Grabner, W., Rastogi, N., Wolf, H., and Nerlich, A.G. (2003). Characterization of *Mycobacterium tuberculosis* complex DNAs from Egyptian mummies by spoligotyping. J. Clin. Microbiol. *41*, 359–367.

Zink, A.R., Spigelman, M., Schraut, B., Greenblatt, C.L., Nerlich, A.G., and Donoghue, H.D. (2006). Leishmaniasis in Ancient Egypt and Upper Nubia. Emerg. Infect. Dis. *12*, 1616–1617.

# Part III

# From Bench to Bedside

# Point-of-care Nucleic Acid Testing: User Requirements, Regulatory Affairs and Quality Assurance

9

Angelika Niemz, Tanya M. Ferguson and David S. Boyle

**Abstract**

Nucleic acid testing (NAT) for *in vitro* diagnostic (IVD) use has moved beyond high-complexity reference laboratories to moderate complexity centralized laboratories, and to certain professional point of care (POC) settings in developed and developing countries. To develop, launch, and sustainably implement a POC NAT system requires compliance with regulatory affairs and quality assurance requirements, and long term demonstration of the system's clinical utility. This report discusses the pros and cons of POC versus centralized laboratory testing, general clinical utility considerations, use environment, and user requirements in developed versus developing countries. The report further provides an overview of the applicable regulatory requirements in different regions around the globe, and describes quality assurance considerations during test execution by the end user, which can be especially challenging for POC testing.

## Nucleic acid testing – key applications

Nucleic acid testing (NAT) for medical applications, also called molecular diagnostics, is a rapidly growing segment of the *in vitro* diagnostics (IVD) industry. Infectious disease-related NAT currently accounts for approximately two thirds of the global molecular diagnostics market (Rosen, 2011), and includes disease diagnosis, viral load monitoring, pathogen typing, drug susceptibility testing, and blood bank screening (Tang *et al.*, 1997; Vernet, 2004; Yang and Rothman, 2004; Espy *et al.*, 2006). This chapter focuses on infectious disease related IVD NAT for indications with a clear need to perform testing at the point of care (POC) (Holland and Kiechle, 2005; Yager *et al.*, 2008). In the developed world, POC NAT has focused on the rapid diagnosis and containment of hospital acquired (nosocomial) infections, such as methicillin-resistant *Staphylococcus aureus* (MRSA) (Wolk *et al.*, 2009) and *Clostridium difficile* (Kuijper *et al.*, 2006; Bauer *et al.*, 2009). Another application for POC NAT is the antepartum screening to minimize mother-to-child transmission of pathogens associated with high infant morbidity and mortality, such as Group B *Streptococci* (GBS) (De Tejada *et al.*, 2011). Recent disease outbreaks such as the swine flu pandemic (influenza H1N1) and SARS (coronavirus) have prompted the development of POC systems for the rapid diagnosis of respiratory pathogens (Letant *et al.*, 2007; Abe *et al.*, 2011; Loeffelholz *et al.*, 2011; Poritz *et al.*, 2011). In developing countries, applications for NAT focus on endemic infectious diseases (Yager *et al.*, 2008). Currently infant HIV-1 diagnosis and HIV-1 viral load testing are the two most common NATs performed in low resource settings (Stevens *et al.*, 2008; Stevens *et al.*, 2010). Significant investments have further been made in NAT for TB diagnosis and drug susceptibility testing (Balasingham *et al.*, 2009; McNerney and Daley, 2011; Niemz and Boyle, 2012).

Beyond infectious diseases, molecular diagnostics includes testing for inherited diseases and disease predisposition, pharmacogenetic tests to tailor personalized therapy, oncology-related diagnostics, prenatal diagnostics, and tissue typing for transplantation (Rosen, 2011). Applications beyond infectious diseases tend to focus on genotyping, which is also required for pathogen typing, surveillance, and drug susceptibility testing. Overall, the same fundamental principles apply, and similar technical challenges have to be overcome. Tests for inherited coagulation disorders and related to anticoagulant therapy are already offered on existing NAT platforms that can be implemented in certain POC settings. Other indications will likely follow suit.

Non-clinical applications of NAT have an easier path to market entry due to lower regulatory hurdles, and often serve as a stepping stone towards clinical IVD applications. For example, the Cepheid GeneXpert system was initially developed and launched for Anthrax testing in postal facilities (Ulrich *et al.*, 2006), then received FDA approval for a range of IVD applications (Boehme *et al.*, 2010; Miller *et al.*, 2010; Marner *et al.*, 2011; Spencer *et al.*, 2011). Idaho technologies developed the Razor for multiplexed detection of biothreat agents, then adopted the technology into the now FDA approved FilmArray platform for multiplexed detection of respiratory pathogens (Poritz *et al.*, 2011). A variety of portable ruggedized NAT platforms have been developed for the rapid molecular detection of biothreat agents by first responders (Higgins *et al.*, 2003; Perdue, 2003; Lim *et al.*, 2005). More recently this field has expanded into agricultural, veterinary, and food-safety applications, with systems under development or commercially available by Axxin, BioHelix, Eiken, Epistem, Ionian Technologies, Lumora, OptiGene, Qiagen, TwistDx Ltd, Ustar Biotechnologies, and Spartan. Many of these portable NAT platforms include the same inherent characteristics required for POC clinical diagnostics, and key technical advances often appear first in areas outside of IVD applications.

## Testing sites and system requirements – centralized laboratories

To date, NAT is almost exclusively performed in centralized laboratories using high-end instrumentation and skilled personnel (Espy *et al.*, 2006). Even in developed countries with well-established networks of clinical laboratories, molecular diagnostics testing is performed by a relatively small percentage thereof, typically high-complexity hospital and reference laboratories. For example, out of the ~35,000 clinical laboratories in the US performing moderate or high complexity testing (CLIA database https://www.cms.gov/CLIA/), only ~800 (~2%) perform molecular diagnostic testing.

Nucleic acid amplification testing is susceptible to amplicon carry-over. Therefore, according to Title 42 of the US Code of Federal Regulations (42 CFR 493.1101a3), molecular amplification procedures that are not contained in closed systems require unidirectional workflow and separate rooms for pre- and post-amplification steps, with appropriate air flow control and other restrictions. Therefore, unless NAT is performed in a closed system, molecular diagnostics operations require facilities separate from the routine clinical laboratory. Most nucleic acid amplification methods, such as PCR, require purified nucleic acids. Sample preparation for NAT therefore significantly increases the overall turnaround time, testing complexity, and the potential for contamination. Laboratory-developed tests (LDTs), which constitute a significant portion of currently available molecular diagnostics, generally involve manually performed or only partially automated assays, with complex assay procedures that require highly trained operators. Broader adoption of NAT in clinical laboratories requires easy to use, integrated, automated, closed system, contamination controlled NAT platforms with appropriate regulatory approval, which can be operated by medical technicians with moderate skill levels in routine centralized clinical laboratories. Interestingly, integrated closed system NAT platforms developed for POC applications that fulfil these criteria may in the long term disseminate molecular diagnostics more broadly into clinical laboratories. For example, Cepheid originally developed the GeneXpert for field-based biothreat detection, and then expanded to near patient infectious disease diagnosis. However, Cepheid also markets the GeneXpert system to moderate complexity centralized clinical laboratories, especially the 16 and 48 module platforms that can accommodate higher throughput.

Centralized testing has the benefit of a controlled environment, economy of scale, suitable quality assurance and oversight over skilled and properly trained personnel. However, centralized testing introduces additional time delays and logistical challenges. For centralized NAT in developed countries, tests can be executed by the laboratory typically in less than 5 hours (Merel, 2005), especially for routine infectious disease related applications. The total turn-around time from sample draw to clinically actionable result however is on the order of several days, especially since most testing is sent out to reference laboratories. Such delays are not optimal when a rapid answer is required to initiate appropriate treatment or prevent disease transmission (Holland and Kiechle, 2005; Clerc and Greub, 2010), or when patients might be lost to follow-up (Khamadi *et al.*, 2008). POC testing significantly reduces these logistical challenges and time delays caused by pre- and post-analytical steps, including sample shipment and return of results. The overall turnaround time for NAT also is prolonged because most molecular diagnostic platforms operate in batch mode (College of American Pathologists, 2011). Specimens received by the laboratory may sit 'on hold' until a sufficient number of samples have been received to warrant a run, or until the previous run is completed. NAT is performed in batch mode in part because standard PCR thermocyclers can only run one thermal profile at a time, which typically takes more than one hour and has to be completed before new samples can be processed. In contrast, most IVD clinical chemistry systems and immunoanalysers run in random access mode, meaning that samples are processed individually, and that urgent samples (STAT samples, acronym derived from statim (latin), which means urgent) can jump into the queue for priority handling, which reduces their turnaround time. POC NAT systems tend to be random access analysers, or analysers that process small batches of samples, designed for lower throughput but rapid turnaround time. Modular random access systems, such as the Cepheid GeneXpert or the Nanosphere Verigene (College of American Pathologists, 2011), can accommodate higher throughput by increasing the number of modules per platform.

Centralized laboratory testing poses a challenge for developing countries with very limited financial resources. Despite pressing needs, developing countries are often unable to implement a network of appropriately equipped and staffed centralized laboratories to offer the needed services effectively to all patients, especially in rural areas (Puren *et al.*, 2010; Stevens and Marshall, 2010). In many developing countries, NAT is only performed at the national reference laboratory level. Following the WHO endorsement of the Cepheid GeneXpert system, NAT for TB is now performed in referral laboratories and high-end microscopy centres (VanRie *et al.*, 2010).

However, the majority of patients in developing countries are served exclusively by peripheral primary health clinics (WHO *et al.*, 2006), which only offer simple non-instrumented diagnostics such as lateral flow immunoassay rapid diagnostic tests (RDTs) that can be used to diagnose HIV and malaria, but not nucleic acid testing, which is required for infant HIV diagnosis (Khamadi *et al.*, 2008), HIV viral load monitoring (Stevens *et al.*, 2010), and for diagnosis of active TB (WHO, 2011b). Countries such as South Africa are aiming to expand the number of centralized laboratories to address this unmet need (Stevens and Marshall, 2010). However, these laboratories are frequently over-burdened, and the infrastructure for sample transport and result dissemination is insufficient. One study (Khamadi *et al.*, 2008) found that laboratories in Kenya required on average 13.5 days for HIV infant diagnosis via NAT, from sample receipt at the laboratory, to dispatching of results. The overall turn-around times from blood draw to result receipt by the client however ranged from one to three months. With such long delays, many patients are lost to follow up, and infants with deteriorating health left untreated. This illustrates that the infrastructure required for effective shipping of samples and mailing/reporting of results is as critical as the laboratory capacity. POC testing can significantly reduce the turnaround time, enabling testing, result review, and treatment initiation in one patient visit.

## Testing sites and system requirements – POC settings

POC testing occurs near the patient in a variety of primary, secondary and tertiary care settings (Table 9.1), and is performed by a range of operators with varying skill levels. Within the foreseeable future, POC NAT in developed countries will likely penetrate only into more advanced secondary and tertiary care POC environments, especially considering that NAT is not yet established even in the majority of centralized laboratories.

Testing to control nosocomial infections such as MRSA and *C. difficile* may be performed in hospital admission units, wards, and intensive care units (Gilligan *et al.*, 2010; Noren *et al.*, 2011). Testing for perinatal infections such as GBS may be performed in maternity wards (Jordan *et al.*, 2010; Young *et al.*, 2011). At some point, testing may advance into primary care settings such as physician's office labs, especially in the context of managing a potential pandemic disease outbreak (e.g. influenza). In developed countries, these POC settings typically have uninterrupted electrical line power, refrigerated reagent storage capabilities, supply management with the appropriate cold chain, and a clean, temperature controlled environment. However, the available space is limited, and any equipment needs to have a small footprint. Hospital POC testing often occurs in a hectic, high pressure environment, performed by nurses or other health care workers who have a high workload and only minimal training in how to perform IVD tests. POC systems therefore have to be easy to use, provide clearly interpretable, clinically actionable results with rapid turnaround, and have to incorporate safeguards to minimize operator errors. In developed countries, POC systems especially for professional applications frequently consist of disposable fully integrated cartridges combined with a small bench-top or handheld instrument. POC testing ideally should have clinical and analytical performance characteristics similar to laboratory-based tests (Nichols, 2003; Ehrmeyer and Laessig, 2007), although this is not always the case. POC tests are therefore often used as initial screening tools, with required laboratory follow up confirmation (Clerc and Greub, 2010). The per-test cost of POC testing is usually higher than for equivalent laboratory-based methods, although better patient management and appropriate disease containment can reduce overall healthcare costs while improving patient outcomes (Gutierres and Welty, 2004).

**Table 9.1** General point of care testing environments

| Primary care | Secondary and tertiary care |
|---|---|
| Home | Emergency room |
| Managed care facilities | Intensive care unit |
| Community pharmacy | Operating room |
| Physician's office | Hospital ward |
| Workplace clinic | Maternity ward |
| STD clinic | Hospital admissions unit |
| Ambulance/first responder | Outpatient clinic |

Although developed and developing countries share the core requirements for POC testing listed above, additional more stringent product requirements apply in the developing world (WHO *et al.*, 2006; Urdea *et al.*, 2006; Yager *et al.*, 2008; Weigl *et al.*, 2009; Peeling and Mabey, 2010), due to the lack of supporting infrastructure. Developing countries have very limited financial resources, therefore costs for reagents, disposables and instruments need to be kept to a minimum for any type of diagnostic. In addition, donor-driven procurement strategies promote single purchases of large amounts of reagents, which create logistical concerns regarding the appropriate long term storage of products prior to use. The available infrastructure, number of trained health care workers, and menu of tests offered is significantly lower in developing compared to developed countries (Table 9.2) and within developing countries decreases further as the settings become more remote (WHO *et al.*, 2006). Most patients only have access to remote primary care facilities. To implement POC testing in developing countries, reagents and supplies have to be provided to thousands of remote testing sites, and an equally large number of unskilled operators have to be recruited, trained, and retained to execute a much smaller number of tests per operator than performed in maybe a dozen centralized laboratories in the country. POC testing faces significant

**Table 9.2** Health care factors illustrating disparities between the developed and developing world (adapted from Peters *et al.*, 2008)

| | Services available per 1000 population | | |
|---|---|---|---|
| WHO region | Hospital beds | Doctors | Nurses |
| Europe | 64 | 3.2 | 7.4 |
| Americas | 25 | 1.94 | 4.88 |
| Southeast Asia | 9 | 0.52 | 0.81 |
| Africa | <1 | 0.21 | 0.93 |

challenges related to logistics, economy of scale, and quality assurance.

Only a small number of hospitals and urban health clinics in developing countries have what has been categorized as advanced to moderate laboratory infrastructure (Urdea *et al.*, 2006), which includes a dedicated laboratory space with access to electricity (not necessarily uninterrupted) and clean water, suitable supply management, with cold chain and cold storage capabilities for sensitive reagents. In these settings, testing is performed by trained personnel with physician oversight, invasive specimens such as venepuncture blood can be obtained and processed, and biosafety containment is available for high risk applications such as TB diagnosis via culture (Fig. 9.1A and B). Biosafety precautions rapidly decline as settings become more remote, ranging from TB sample handling in simple dead air boxes (Fig. 9.1C), to open sample processing at the bench (Fig. 9.1D).

Most of the population in developing countries only has access to peripheral health clinics at the level of a microscopy centre (Fig. 9.2), or to community-based primary care facilities, which have minimal to no laboratory infrastructure. POC NAT in these settings is challenging, but can impact significantly more patients than centralized testing. Remote microscopy centres and primary care facilities either have unreliable or no access to electricity, which means that systems have to operate independent of line power, using e.g. batteries or solar power. Ambient temperatures are often

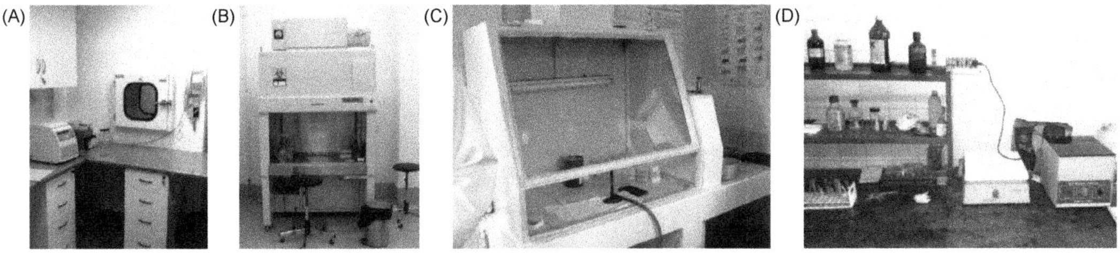

**Figure 9.1** Decreasing levels of laboratory infrastructure and biosafety precautions in the developing world: (A) and (B) reference laboratory with BSL3 level TB culture facility; (C) simple dead air box for TB testing in a district hospital; (D) open sputum slide processing area at an urban TB clinic. (Photographs courtesy of Gerard Cangelosi, Tanya Ferguson, and PATH).

**Figure 9.2** Peripheral laboratory settings of developing countries. (A) Crowded bench area with microscope and centrifuge; (B) sample processing area; (C) sink area, where slides and tubes are recycled. Infrastructure is very limited, there is no complex and expensive instrumentation, power and clean water is sporadic or absent. (Photographs courtesy of Tanya Ferguson and PATH).

high and uncontrolled, cold storage capabilities and a cold chain for reagent supply are generally not available. Therefore reagents need to remain stable for long periods at extreme temperatures, ideally 24 months storage at 40°C, 70% humidity, plus stability for 48 hours at 50°C, and daily thermal cycling of 25°C to account for transport stress (Niemz and Boyle, 2012).

Supply chain and inventory management is a major challenge (Fig. 9.3A and B); therefore tests with fewer independent components are preferable, and test kits have to be small in size due to limitations during shipping and storage. Since access to clean water is lacking, all liquid reagents have to be included in the assay kit, and liquid reagent volumes have to be kept to a minimum. In developing countries, POC testing systems often have to be transported via unpaved roads to remote sites, are therefore subjected to dust, vibration, and impact shock. These POC testing systems therefore have to be rugged, compact, and portable. Since waste management can be a serious issue (Fig. 9.3C), the disposables volume should be kept to a minimum.

Remote locations in developing countries typically operate without physician oversight, and tests are performed by minimally trained or untrained community healthcare workers, or volunteers. Therefore, tests have to be extremely simple to use, require minimal training, and must be resilient towards operator errors. Quality assurance and quality control are significant challenges especially in remote settings, and therefore have to be considered during IVD system development. To reduced lost to follow up testing and treatment needs to be performed in the same visit, therefore tests need to have rapid turnaround times. Most of these settings cannot perform invasive specimen acquisition, such as venepuncture blood draw, therefore testing is restricted to non-invasive specimens such as swabs, urine, saliva, sputum, and capillary blood obtained via finger-prick. Although microscopy centres routinely handle sputum, blood, and other infectious specimens, biosafety precautions are minimal to absent, therefore testing devices should suitably contain infectious agents, with appropriate disposal of biohazardous materials. For nucleic acid amplification testing, contamination of the test site with amplicons is an additional concern, since amplicon carry over contamination causes false positives and unreliable results. The process therefore needs to incorporate provisions related to the correct handling and disposal of amplified positive reactions, which should remain enclosed within the system, and should not be handled manually.

It is not possible to acquire and maintain expensive instrumentation in these settings, due to cost, complexity, and logistical constraints. Additional required hardware, such as an external laptop computer, may break down or disappear, rendering the system non-functional. Therefore, POC testing in developing countries needs to be either minimally instrumented and maintenance free, such as the sensor strips and handheld readers used for self-blood-glucose-monitoring, or non-instrumented, such as lateral flow RDTs. Existing equipment available in remote locations of developing countries, if any, does not go beyond a microscope and possibly a centrifuge. Therefore POC tests ideally should not rely on additional

**Figure 9.3** Logistical challenges of clinical laboratories in low resource settings, related to (A) supply constraints, (B) storage of laboratory supplies, and (C) bio-hazardous waste disposal (Photographs courtesy of PATH and Tanya Ferguson).

instruments such as water baths, precision pipettors, or even timers. The acronym ASSURED (Urdea *et al.*, 2006), introduced in 2003 by the WHO, summarizes the ideal characteristics of an IVD test that can be used in all healthcare levels of low resource developing countries. ASSURED stands for Affordable, Sensitive, Specific, User-friendly (simple to perform in a few steps with minimal training), Robust (e.g. can be stored at elevated temperature) and Rapid (results available in <30 minutes), Equipment-free (or minimally instrumented, cannot require line power), Deliverable to those in need (supply chain limitations).

On the other hand, certain requirements in developing countries are less stringent than in the developed world. For example, in most developing countries, infectious disease diagnosis can be made based exclusively on POC testing, while laboratory confirmation is required in the developed world (example: HIV diagnosis, (Wesolowski *et al.*, 2011)). Decision thresholds are often higher (e.g. HIV viral load monitoring (Smith *et al.*, 2009)), especially in cases with a clear clinical need and no suitable diagnostic capabilities, which leads to LOD and analytical sensitivity requirements that are less stringent than in developed countries.

## Regulatory requirements for IVD manufacturers

In most countries around the globe, IVD manufacturers have to meet country- or region-specific regulatory requirements to legally market an IVD.

In the United States, IVDs are regulated by the Food and Drug Administration (FDA) (FDA, 2012c), in most cases through the Center for Devices and Radiological Health (CDRH), although some IVD products, such as HIV testing, are regulated through the Center for Biologics Evaluation and Research (CBER). According to Title 21 of the US Code of Federal Regulations (21 CFR), all IVD products are subject to general controls, which include registration of the manufacturer and listing of the product with the FDA (21 CFR 807), labelling requirements (21 CFR 809), which includes the package inserts with claims and directions for use, cGMP manufacturing as described in the Quality Systems Regulation framework (QSR, 21 CFR 820), and medical device reporting (MDR, 21 CFR 803), which includes post-market surveillance and adverse event reporting to the FDA. In all cases, the manufacturer is legally responsible for appropriate performance of the IVD in the marketplace, and is subject to inspections by the FDA.

The FDA classifies IVD products as class I (low risk), class II (moderate risk), or class III (high risk) devices (21 CFR 862, 864, 866). IVD NAT kits or systems generally are class II or III, and their development is subject to Design Controls (QSR, 21 CFR 820 Subpart C) (FDA, 2012b), to ensure that specific design requirements have been met spanning from user needs assessment to final launch of a medical device (Fig. 9.4). Documentation related to these key steps becomes part of the FDA submission package.

To be considered Class II, a device cannot be life sustaining and has to be substantially equivalent to a legally marketed device. Class II devices usually are approved via the Premarket Notification (510(k)) route (21 CFR 807 Subpart E), which mainly requires non-clinical laboratory studies to demonstrate analytical performance compared to a predicate device, such as evaluation of accuracy and precision, analytical specificity and sensitivity. Filing fees for standard a 510k submission are around $5000, and the applicant can request status information after 90 days. In contrast, Class III devices are considered to be life sustaining, meaning of high importance in

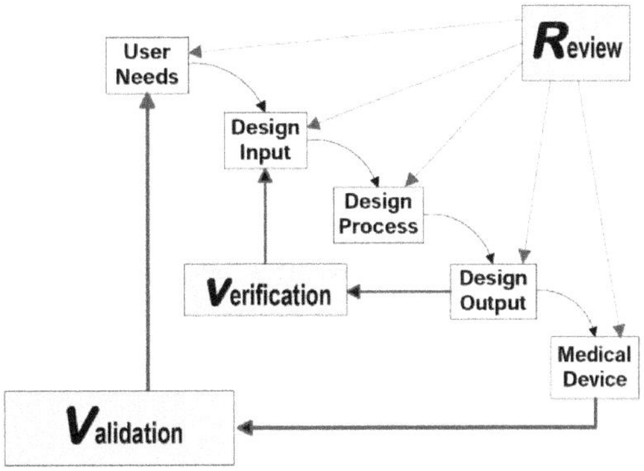


**Figure 9.4** Design Controls required as part of the Quality Systems Regulation (QSR) framework for IVD product development in the US (FDA, 2012b).

preventing impairment of human health, may present a potential unreasonable risk of illness or injury, or may lack a predicate device, and required Pre-Market Approval (PMA, 21 CFR 814). A PMA submission requires 'scientific evidence providing a reasonable assurance that the device is safe and effective for its intended use', which in practice means demonstration of appropriate clinical sensitivity and specificity through clinical studies that are often lengthy and expensive. PMA submissions are reviewed by the FDA within 180 days (or longer in the case of additional inquiries), and the filing fees alone for a standard PMA submission are ~$260,000. The total cost and time required to develop, evaluate, and obtain approval for a Class III device is therefore substantially higher than for a class II device. Nucleic acid amplification tests for HIV are automatically classified as class III devices and require a PMA. Until recently, the same has been said for TB, which has deterred many companies from developing IVD NATs for the diagnosis of TB which has low incidence and limited market potential in the United States. Successful dissemination of the Cepheid GeneXpert MTB/MR if assay for TB diagnosis in low resource settings of developing countries has facilitated the recent FDA clearance of this system in the US as a class II device.

In the US, additional regulatory requirements are posed on IVD tests under the Clinical Laboratory Improvement Amendments (CLIA, 42 CFR 493). Although CLIA primarily regulates testing sites, as discussed in the next section, the imposed restrictions impact the development of IVDs marketed to these settings. According to CLIA, IVDs are categorized into high complexity, moderate complexity (42 CFR 493.17), or waived tests (42 CFR 493.15), based on difficulty of test execution, including required user training and skills, system troubleshooting, equipment maintenance, and the degree of judgement required to interpret results (Gutierres and Welty, 2004; Ehrmeyer and Laessig, 2007). IVD testing sites have to be certified, or have to obtain a certificate of waiver, which dictates the types of tests that may be performed at the site. Centralized laboratories generally are certified as either high or moderate complexity. High complexity sites tend to be large hospital laboratories or reference laboratories. Hospital POC operations that oversee testing at the secondary and tertiary testing sites listed in Table 9.1 may have moderate complexity certification, as do a small number of high-end Physician Office laboratories. The vast majority of POC testing sites however can only perform waived tests.

Although CLIA is administered through the Centre for Medicare and Medicaid Services (CMS), the FDA is responsible for CLIA categorization of IVDs. Along with the 510(k) and PMA application, IVD manufactures submit supporting data for device/test categorization to the FDA, which then determines the complexity category, and notifies the manufacturers and CMS. The CLIA categorization of all IVDs marketed in the US is also available in a searchable database on the FDA website (FDA, 2012a). The vast majority of IVD NATs are categorized as high complexity tests, therefore can only be executed by high complexity hospital laboratories or reference laboratories. In 2006, Cepheid was the first company to obtain moderate complexity designation for an IVD NAT, the GeneXpert GBS test. Currently, more than a dozen moderate complexity IVD NATs are on the market in the US (selected examples listed in Table 9.3), however each year their number is rapidly increasing. Most of these systems are clearly intended for use in moderate complexity centralized laboratories. The Cepheid GeneXpert however has been implemented in hospital POC settings (Brenwald *et al.*, 2010; De Tejada *et al.*, 2011), and other platforms likely will follow suite, such as the Iquum LIAT (Tanriverdi *et al.*, 2010), Biocartis Apollo, Enigma ML, and the Alere NAT Analyzer and iNAT, which have been designed for use in POC environments.

Although there currently are no CLIA waived IVD NAT systems on the market, it is anticipated that such systems will become available in the future. To obtain a CLIA waiver, an IVD test has to be 'so simple and accurate as to render the likelihood of erroneous results by the user negligible' or 'determined to pose no unreasonable harm to the patient if performed incorrectly' (Ehrmeyer and Laessig, 2007). Since incorrectly performed tests in general can harm a patient, waived tests have to be simple, accurate, and virtually fool proof. This has important implications in developing IVD NATs for POC use. Tests that require

**Table 9.3** FDA cleared IVD NATs with moderate complexity designation as of December 2012 (FDA, 2012a)

| Company | Platform | Test name | Targeted analyte(s) | Specimen | NA amplification | Detection | Integration level/ disposables | Effective date |
|---|---|---|---|---|---|---|---|---|
| Cepheid | GeneXpert | GBS | Group B streptococci | Vaginal/rectal swab | PCR | Real-time fluorescence | Integrated S/A/D (one cartridge) | 2006 |
| Cepheid | GeneXpert | MRSA | MRSA (mecA gene) | Nasal swab | PCR | Real-time fluorescence | Integrated S/A/D (one cartridge) | 2007 |
| Cepheid | GeneXpert | MRSA/SA | MRSA + *Staphylococcus aureus* | Blood culture, skin/soft tissue | PCR | Real-time fluorescence | Integrated S/A/D (one cartridge)) | 2008 |
| Cepheid | GeneXpert | *Clostridium difficile*/Epi | *C. difficile* (tcdB gene) + epidemic 027 strain | Unformed stool | PCR | Real-time fluorescence | Integrated S/A/D (one cartridge) | 2009 |
| Cepheid | GeneXpert | vanA | Vancomycin-resistant *Enterococci* (VRE) | Swab | PCR | Real-time fluorescence | Integrated S/A/D (one cartridge) | 2010 |
| Cepheid | GeneXpert | Flu | Influenza A, Influenza B, Influenza A H1N1 | Nasopharyngeal wash | PCR | Real-time fluorescence | Integrated S/A/D (one cartridge) | 2011 |
| Nanosphere | Verigene | Respiratory viruses | Influenza A/B, respiratory syncytial virus (RSV) | Nasopharyngeal swab | PCR | Array scanner (scattered light) | Two instruments, one cartridge | 2010[1] |
| Nanosphere | Verigene | Gram-positive blood culture | Panel of staphylococci, *Streptococcus*, *Enterococcus*, *Micrococcus* and *Listeria* spp., mecA, vanA and vanB resistance markers | Gram-positive blood culture | PCR | Array scanner (scattered light) | Two instruments, one cartridge | 2012 |
| Nanosphere | Verigene | *C. difficile* | *C. difficile* (tcdA and tcdB gene) + epidemic 027 strain | Unformed stool | PCR | Array scanner (scattered light) | Two instruments, one cartridge | 2012[2] |
| Becton Dickinson | BD MAX | GBS | Group B streptococci | Lim broth culture ($^{3}$18 h) of swabs | PCR | Real-time fluorescence | Integrated S/A/D (many disposables) | 2010 |

**Table 9.3** Continued

| Company | Platform | Test name | Targeted analyte(s) | Specimen | NA amplification | Detection | Integration level/ disposables | Effective date |
|---|---|---|---|---|---|---|---|---|
| Becton Dickinson | BD MAX | MRSA | MRSA (mecA gene) | Nasal swab | PCR | Real-time fluorescence | Integrated S/A/D (many disposables) | 2012 |
| Meridian Bioscience | Illumipro | Illumigene *C. difficile* | *C. difficile* (tcdA gene) | Unformed stool | LAMP | Turbidity | Manual S, automated A/D, (many disposables) | 2010 |
| Iquum | LIAT | Influenza A/B | Influenza A, Influenza B | Nasopharyngeal swab | PCR | Real-time fluorescence | Integrated S/A/D (one cartridge) | 2011 |
| BioHelix | BESt cassette | IsoAmp HSV | HSV1, HSV2 (type common) | Genital/oral swabs | HDA | NALF | Manual S, A, D, (many disposables) | 2011 |
| Idaho technology | FilmArray | Respiratory pathogen panel | Adenovirus, coronavirus (HKU1, NL63, OC43, 229E), Meta-pneumovirus, influenza A (H1, H3, 2009 H1), influenza B, parainfluenza virus (1–4), rhinovirus, enterovirus, respiratory syncytial virus, *Bordetella pertussis*, *Chlamydophila pneumoniae*, *Mycoplasma pneumoniae* | Nasopharyngeal swab (in viral transport medium) | PCR | Real-time fluorescence | Integrated S/A/D (many disposables) | 2011/2012 |

[1]FDA cleared 2009 as high complexity test, re-classified 2010 as moderate complexity test. [2]CLIA classification not yet posted at the time of this review, but presumably moderate complexity device. S, sample preparation; A, amplification; D, detection.

precision pipetting, reagent reconstitution, multiple disposables, additional instrumentation such as centrifuges and water baths, and many operator steps that require precise timing likely will not receive waived status. Waived tests in general are 'sample in-answer out' tests, which are either instrument free or consist of a compact instrument in conjunction with a disposable that has all reagents on board. Examples for waived tests include self-blood-glucose-monitoring devices, simple lateral flow RDTs, or systems such as the Abbott iStat and the BioSite Triage used for clinical chemistry or immunoassays performed in POC hospital settings. Passing the CLIA waived hurdle will be critical in order for IVD NATs to become truly suitable for POC applications in the US.

In Europe, prior to 1998, most of the member states had varying and in some cases minimal regulatory requirements for IVDs. In 1998, the European medical device directives were passed, and have been adopted by the member countries in the subsequent years, including the IVD specific directive 98/79/EC (European Parliament, 1998; Williams, 2011). Since the end of 2003, all IVD products introduced to the market in member states of the European Union (EU) and the European Economic Area (EEA) have to comply with these directives. IVDs sold in Europe have to display the Conformité Europeène (CE) mark, which indicates that the manufacturer declares full compliance with EU Directives and has completed an appropriate conformity assessment. Each country designates a 'competent authority' (CA), sometimes also called 'notified authority' (NA) that acts on behalf of its government in a function similar to the FDA in the US. The competent authority in the UK is the Medical and Health Products Regulatory Authority (MHRA), in Germany it is the Zentralstelle der Länder für Gesundheitsschutz (ZLG) and the Paul-Ehrlich-Institut, and in France it is the Agence Française de Sécurité Sanitaire des Produits de Santé (AFSSAPS). The CA is responsible for ensuring that all medical devices sold in that country meet the essential requirements of the EU medical device directives, and do not compromise the health and safety of patients and users. The actual tasks related to dossier review, site examination, etc. are not executed by the CA, but by one of several independent entities within that country, referred to as Notified Bodies (NB). For example, SGS is one of the NBs in the UK that evaluate IVDs, and the TÜV is one of the IVD related NBs in Germany. Four-digit identification numbers are assigned to these NBs. For all conformity assessments carried out through a NB, this number has to be listed on the device below the CE mark. A list of CAs and NBs for all EU and EEA countries is provided through the EC NANDO database (European Commission, 2012b).

All EU device directives, including 98/79/EC, consist of articles that lay out the general framework of requirements and definitions, supplemented by Annexes that contain more detailed information concerning the involved processes. According to 98/79/EC (European Parliament, 1998), IVDs are classified into one of four categories: High risk products listed in Annex II Lists A are IVDs related to testing for HIV, HTLV, and hepatitis. Products considered to be of medium risk are detailed in Annex II lists B, and include infectious disease related IVDs testing for rubella, toxoplasmosis, cytomegalovirus, and chlamydia. List B also includes self-blood-glucose-monitoring devices. A third category includes all other POC IVDs for self-testing. The equivalent US category would be CLIA waived products approved for home use. All other IVDs are in a fourth, general category, and are considered to be of low risk and the required EC declaration of conformity process is described in Annex III; the IVD manufacturer can self-certify these products, without external review by a NB. Self-certification requires the IVD manufacturer to compile the required technical documentation, and conduct appropriate in house conformity assessment. For self-certified products, no NB number is displayed under the CE mark. For the third category of self-testing devices, IVD manufacturers have to submit the dossier to a NB, which then examines the evidence and may require clarification and additional testing. These products also have to undergo usability testing with lay persons. Additional requirements apply for IVDs listed in Annex II Lists A or B, which range in scope depending on risk category (European Parliament, 1998; Williams, 2011). For all Annex II devices, the manufacturer has to go through a NB. If a quality management system is required, it is often established according to the international harmonized standard ISO 13485, which in many respects is similar to the QSR framework in the US. For all IVDs, EU regulations require registration of IVD manufactures and products with the competent authority, appropriate labelling, post market surveillance, and adverse event reporting.

When compared to the US regulatory system, the EU regulatory framework requires less clinical evidence that an IVD is safe and effective (Stynen, 2011). Overall, the EU IVD regulatory requirements are less stringent than in the US, therefore IVDs often are first launched in Europe, in many cases years before entering the US market, especially if a lengthy and expensive PMA approval process is required in the US. The discrepancies between US and European device categorization and regulatory requirements are sometimes striking. For example, TB related IVD NATs until recently were considered class III high risk in the US, requiring a PMA, while the same products fall in the general low risk category in the EU, which can be self-certified. The Cepheid GeneXpert MTB/RIF Assay, which diagnoses TB and identifies rifampin resistance mutations, is CE marked, thus can be used as IVD in Europe. Following WHO endorsement in 2010, the system has undergone large-scale global roll-out in most high TB burden countries. However, until receiving FDA clearance in July 2013 as the first class II TB diagnosis NAT, this assay in the USA was listed as Research Use Only (RUO), and could not be used for TB clinical diagnosis.

The EU IVD directive 98/79/EC is undergoing revisions. Following public consultation and feedback from stakeholders (European Commission, 2011), the European Commission published a proposal for a revised IVD directive in September of 2012 (European Commission, 2012c). To become binding law, the proposed changes have to be adopted through the ordinary legislative procedure by the European Parliament and Council – a process not yet completed at the time this review was written. Although it is likely that the proposed changes will be adopted, full implementation will involve a transition period of several years.

**Table 9.4** Risk-based classification of IVDs according to Global Harmonization Task Force guidance

| Class | Individual risk | | Public health risk | Example |
|---|---|---|---|---|
| A | Low | and | Low | Specimen receptacles |
| B | Moderate | and | Low | Pregnancy self-testing |
| C | High | and/or | Moderate | Blood glucose self-testing |
| D | High | and | High | HIV tests |

According to the proposal, IVD classification will change to a risk-based general system, rather than a stagnate listing of specific moderate and high risk tests. The proposed new risk-based classification system is aligned with the Global Harmonization Task Force (GHTF) guidance document GHTF/SG1/N045:2008, which classifies IVDs into four classes based on individual and public health risk (Table 9.4). This new risk-based classification further changes the requirements for conformity assessment, aligned with GHTF principles on conformity assessment, described in document GHTF/SG1/N046:2008.

Class D is analogous to the current high risk Annex IIa list of IVDs, Class C includes Annex IIb and many self-testing products such as blood glucose monitoring. For these IVDs, the conformity assessment process will not change significantly. However, of the IVDs currently in the general low risk category, it is anticipated that ~20,000 will be re-classified as Class B, and ~9800 will be re-classified as Class C devices (European Commission, 2012a). For these devices, requirements regarding conformity assessment, quality management, and involvement of notified bodies will become more stringent. For IVDs re-classified as Class B or C, a notified body has to assess the manufacturer's quality system. For devices re-classified as Class C, a notified body also has to review the design documentation and conformity assessment dossier.

The proposed revisions further clarify what clinical data and evidence is required to obtain regulatory approval, but specifically state that demonstrating clinical utility is not a precondition for marketing a new IVD. This contrasts with the US regulatory system, but was viewed as important to not stifle innovation. The proposed revisions also clarify requirements for POC IVDs, specifically related to more stringent risk analysis and clear instructions for use, to ensure high quality of test execution by end users who are not clinical laboratory professionals.

Countries such as Canada, Japan, Australia, India and China have their own regulatory bodies and approval processes for IVDs (WHO *et al.*, 2006; Williams, 2011). However, out of the 193 WHO Member States, less than one-third have a regulatory system for IVDs (Vercauteren, 2011). In general, governmental institutions in these countries, and international aid organizations that support test rollout, still have to endorse a test. These organizations often consider regulatory approval in the US or Europe as part of their decision, and may independently evaluate the test performance prior to endorsement (WHO *et al.*, 2006). Recently, the WHO has increased its role in the regulation of IVDs (Vercauteren, 2011; WHO, 2011a). This is particularly relevant since IVDs are increasingly manufactured in the high burden low resource countries where they are most needed, which saves cost, but also decreases the regulatory oversight over the development and manufacturing process. WHO now requires prequalification of diagnostics (PQ Dx) for IVDs related to HIV/AIDS, malaria, and hepatitis B/C. The scope likely will be expanded in the future. The PQ Dx requirement applies to new products without prior regulatory approval, and to products already established in low resource settings, that have been approved in the US or Europe, although existing and approved products can utilize a fast track option.

The standard PQ Dx process is depicted in Fig. 9.5. The dossier submitted by the IVD manufacturer to the WHO includes information on how the device was designed, developed, and manufactured, the implemented quality management system (in accordance with ISO 13485), risk management, data related to verification and validation of analytical and clinical performance

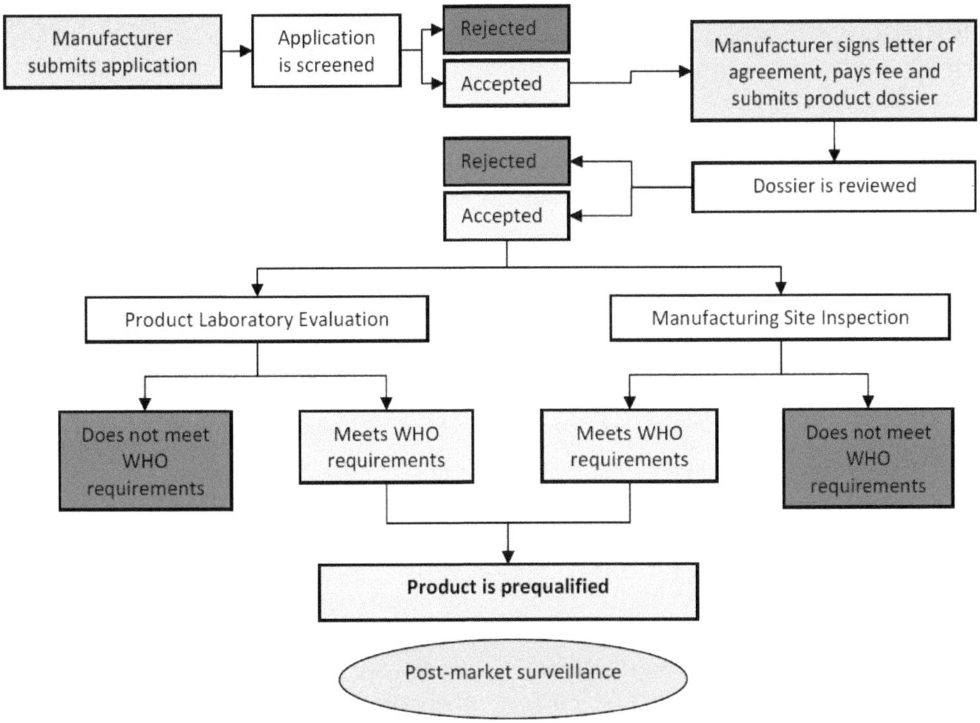


**Figure 9.5** Overview of the WHO Prequalification for Diagnostics (PQ Dx) process (WHO, 2011a).

specifications, stability studies, appropriate labelling and directions for use. The WHO reviews the dossier, inspects the manufacturing sites, and one or multiple WHO collaborating centre(s) perform an independent evaluation of test performance under the intended operating conditions. The first applications were received by the WHO in April of 2010, the first PQ Dx was issued in December of 2010 (Vercauteren, 2011). By December 2012, eleven different IVDs, plus variants thereof, have obtained PQ Dx status (WHO, 2012). Four of these prequalified IVDs are central laboratory-based NATs for HIV viral load monitoring. Only IVDs that have passed the PQ Dx process are included in the WHO list of prequalified diagnostics eligible to be invited into the procurement processes of the WHO and other UN agencies. Other governmental and aid organizations may also use the WHO list of prequalified products to guide their procurement decisions. A significant number of IVD NAT systems are currently under development or close to market release for near patient HIV viral load monitoring and infant HIV diagnosis. The WHO PQ Dx process will become an essential requirement in the development of these and other IVD NATs designed for near patient testing in international markets.

The U.S. Agency for International Development (USAID) also has implemented an approval process for HIV related IVDs. According to the US Acquisition and Assistance Policy Directive (AAPD 07-05) (USAID, 2007), only HIV tests on the USAID waiver list can be used in projects funded through USAID or the President's Emergency Plan for AIDS Relief (PEPFAR). To be included on the waiver list, products either require approval through the FDA or other 'Stringent Regulatory Agencies' (e.g. Canada, EU, Japan), or have to be USAID evaluated (USAID, 2010). This approval process again requires a dossier with technical specifications, product information, including shelf life, manufacturing sites, and certification of appropriate quality management. All test kits are then evaluated by the US Centers for Disease Control and Prevention (CDC), and have to meet a set of minimum requirements. Although the USAID waiver list includes only lateral flow RDT immunoassays for HIV diagnosis as of 2012, IVD NATs may be included in the future as part of the broader roll-out of IVD NAT for HIV viral load monitoring and infant diagnosis.

## Quality assurance for IVD test execution

Laboratories and many POC sites performing IVD tests are required to implement quality systems to ensure appropriate test execution, accurate results, and therefore proper patient care. For POC settings, ensuring the quality of test execution is a challenge, and a key reason cited for keeping a test in the centralized laboratory (Ehrmeyer and Laessig, 2007). Quality issues are even more significant for POC testing in high burden low resource settings of developing countries (Peeling and Mabey, 2010). In designing and developing POC NAT systems, these constraints have to be kept in mind.

In the US, oversight over IVD testing according to CLIA regulations (42 CFR 493) applies to any facility that examines specimens derived from the human body to diagnose, prevent, or facilitate treatment of human diseases, or to assess the health of patients (Lewin Group, 2008). CLIA regulations also apply to most of the POC testing performed in professional health care settings as listed in Table 9.1. Self-testing at home does not fall under CLIA regulations, but all tests used outside of professional healthcare settings have to be CLIA waived and approved for home use. Professional IVD testing sites must be certified to receive payments from Medicare and Medicaid (administered by CMS) or from private insurance companies. IVD testing sites that hold a certificate of waiver (42 CFR 493.15) can only perform waived tests. Testing sites with a certificate of waiver still are required to follow the manufacturers' instructions for performing these test, including all applicable quality control protocols, have to be registered with CMS, have to pay biennial fees, and can be inspected on an ad hoc basis, but are not subject to routine inspections, and are not required to implement an extensive quality management system as required for labs performing non-waived tests. All testing sites performing non-waived high or moderate complexity tests have to either hold a certificate of compliance or a certificate of accreditation, are inspected on a biennial basis, and are subject to more stringent requirements. Accreditation and inspection is typically performed by independent third parties, such as the Joint Commission on Accreditation of Healthcare Organizations (JCAHO), the College of American Pathologists (CAP), or the Commission on Office Laboratory Accreditation (COLA).

High and moderate complexity sites, including POC testing sites, have to implement a quality management system, which includes quality assurance (QA), quality control (QC), and proficiency testing (PT). QA focuses on 'doing things right', which includes implementing proper managerial oversight, policies, standard operating protocols, personnel training, and record maintenance. QC and PT focus on 'getting the right answer'. QC is internally administered by the testing site, and involves daily monitoring of instruments and test execution using controls that are often included in or sold along with the IVD test kit. During the design and development of an IVD, it is therefore imperative to include the appropriate internal and external controls, and appropriate QC protocols. PT is an external quality assurance (EQA) mechanism to ensure that testing sites and their personnel can obtain accurate answers for the tests performed at that site. Four times a year, an outside entity, such as CAP, sends to the testing site a set of standardized samples containing the analytes of interest, which are then tested at that site under regular operating conditions by personnel approved to perform the tests. Results are reported back to the external entity, and are required to fall within accepted ranges (Ehrmeyer *et al.*, 1990). If an operator fails PT, the operator or lab may be barred from executing a given test. In extreme cases, the lab may lose its CLIA certification.

Other countries have implemented similar regulatory frameworks for IVD testing sites, but the requirements can vary widely. Efforts are under way to facilitate global harmonization in this area. For example, the European proficiency testing information system (EPTIS) is a database system aimed at documenting and coordinating proficiency testing around the globe (EPTIS, 2012). EPTIS was initially founded in 1998 by 16 European countries, and now has expanded to ~40 partner organizations from all continents. The internationally recognized standard ISO 15189, first published in 2003, aims to harmonize best practices regarding quality management for clinical laboratories. Many countries and accrediting organizations have since adopted this standard. As a supplement, ISO 22870 provides additional specific requirements for professional point-of-care testing (i.e. testing performed by a healthcare organization, as opposed to patient self-testing). This standard was first published in 2006, but within the next decade it will likely form the basis for quality management of IVD testing in POC settings globally. In developing countries without existing programs, the WHO and CDC have implemented efforts to facilitate laboratory accreditation and quality management (Gershy-Damet *et al.*, 2010). For example, South Africa is the only country in sub-Saharan Africa with an appropriate accreditation and quality assurance system for clinical laboratories. The WHO-AFRO initiative (Gershy-Damet *et al.*, 2010) aims to build capacity in other sub-Saharan countries through a step-wise laboratory accreditation approach, including provisions for appropriate QA, QC and PT, that eventually will lead to ISO 15189 compliance. Similar efforts are under way to facilitate accreditation of laboratories performing TB diagnosis (Global Laboratory Initiative, 2012). In high burden low resource settings, ensuring quality of POC IVD testing is extremely challenging. For example, a study of HIV testing using lateral flow RDTs at 38 primary health clinics or community health centres in South Africa (Begg *et al.*, 2010) found that only 3.4% of the 265 tests observed were fully compliant with the required testing procedure. One frequently observed deficiency was inaccurate incubation timing: the average observed incubation time (5 minutes) was significantly shorter than the required incubation time (15 minutes), which can lead to false negative results. Causes for this discrepancy included intentional staff error, lack of staff training, plus in many cases, no timer was available at the testing site. Another frequent deficiency was the lack of a second test performed to confirm an initial positive result. This was often related to stock shortage of test strips. QC protocols and QC frequency were also highly variable. Such shortcomings can compromise proper patient care, undermine the medical value of POC testing, and eventually jeopardize support from local governments and outside aid organizations. IVDs developed for these settings therefore need to minimize the number and complexity of user steps, and the potential of error during test execution. Ideally, a POC testing system should not rely on additional instrumentation, even as simple as a timer, should have QC built into the test execution process, and should have some provisions to facilitate external oversight.

In recent years, nucleic acid testing is increasingly used for diagnosis of active tuberculosis, following endorsement of the Cepheid GeneXpert MTB/Rif test by the WHO. Most countries with high TB burdens have implemented EQA programs for sputum smear microscopy (Global Laboratory Initiative, 2012), but currently no well-established or standardized EQA programs exists for TB nucleic acid testing. However, in a pilot study in South Africa (Scott *et al.*, 2011), dried culture spots of inactivated *M. tuberculosis* at various concentrations were sent out to peripheral sites, which were instructed to re-constitute and analyse the samples using the GeneXpert MTB/RIF, then report back the results. Dried culture spots were found to be safe, easy to transport, and suitable for verification purposes. However, this type of proficiency testing sample does not take sputum sample matrix effects and variability related to sample preparation into account.

Most current EQA efforts for POC IVD testing perform 'spot checks', and may not adequately capture problems or lapses that occur on an ongoing basis. Many existing central laboratory-based clinical chemistry and immunoanalyzers in developed countries feature internet-based remote analyser monitoring (Heierman *et al.*, 2007). These systems recognize instrument malfunctions and automatically notify internal and external stakeholders, such as quality assurance programs or the IVD manufacturer. If a problem is recognized, technical support can be deployed to correct the malfunction before serious consequences are encountered. Remote connectivity further facilitates ongoing external review of quality control data. Monitoring the number of tests run on the device can inform the timing of inventory replenishment and preventative maintenance. Software upgrades can be easily uploaded onto the analyser. Remote connectivity also enables IVDs to interface with laboratory information systems (LIS) or other electronic medical systems, which facilitates result recording and reimbursement. Some NAT systems suitable for near patient use already incorporate such features. For example, the Cepheid GeneXpert software enables remote machine health monitoring and LIS interfacing in accordance with HL7 and ATSM standards. Internet connection is established through an external computer terminal. The Nanosphere Verigene reader enables LIS connectivity without an external PC. The MolBio TrueLab system (Niemz and Boyle, 2012), designed to enable NAT for TB diagnosis in low resource settings, also enables remote connectivity through embedded electronics with an Android operating system. Overall, IVD manufacturers and end users benefit from such added features through longer instrument lifetime, better quality data, and improved workflow.

## Conclusions

NAT is rapidly advancing from high-end reference laboratories to moderate complexity centralized laboratories, with many new systems and assays that have recently entered the market (Table 9.3) or are in late-stage development. However, NAT for near patient testing is still in the early stages, and has been applied thus far only in a limited number of professional POC settings (Gilligan *et al.*, 2010; Jordan *et al.*, 2010; Noren *et al.*, 2011; Young *et al.*, 2011). Deciding if a test is appropriate for POC use requires a clear understanding of its clinical utility. Moving an IVD test out of a centralized laboratory to near patient settings only makes sense if this leads to better patient management, overall reduced health care cost, reduced disease transmission, and therefore improved overall public health. In so justified cases, technology development has to be guided by a clear understanding of the use environment and user requirements. In addition, medical diagnostics is a heavily regulated industry, which poses a challenge especially for smaller companies entering the market with new technologies initially developed for less stringently regulated applications, such as biodefence, agriculture, and food safety. The situation is complicated by the highly variable and rapidly emerging regulatory frameworks around the globe. Beyond initial approval, sustainable implementation of POC NAT systems has to consider quality assurance and quality control at the end user side. IVD developers therefore should early on incorporate into their systems design provisions that can facilitate QA/QC in the intended use setting.

## References

Abe, T., Segawa, Y., Watanabe, H., Yotoriyama, T., Kai, S., Yasuda, A., Shimizu, N., and Tojo, N. (2011). Point-of-care testing system enabling 30 min detection of influenza genes. Lab Chip *11*, 1166–1167.

Balasingham, S.V., Davidsen, T., Szpinda, I., Frye, S.A., and Tonjuin, T. (2009). Molecular diagnostics in tuberculosis basis and implications for therapy. Mol. Diagn. Ther. *13*, 137–151.

Bauer, M.P., van Dissel, J.T., and Kuijper, E.J. (2009). *Clostridium difficile*: controversies and approaches to management. Curr. Opin. Infect. Dis. *22*, 517–524.

Begg, K., Tucker, T., Manyike, P., Ramabulane, F., Smith, Y., Miller, R., Young, T., Eksteen, D., Galloway, M., Tichapondwa, S., *et al.* (2010). Analysis of POCT/VCT performed at South African primary health care clinics. Available at: http://www.sead.co.za/downloads/POCT-clinics-2011.pdf (accessed 10 September 2013).

Boehme, C.C., Nabeta, P., Hillemann, D., Nicol, M.P., Shenai, S., Krapp, F., Allen, J., Tahirli, R., Blakemore, R., Rustomjee, R., *et al.* (2010). Rapid molecular detection of tuberculosis and rifampin resistance. N. Engl. J. Med. *363*, 1005–1015.

Brenwald, N.P., Baker, N., and Oppenheim, B. (2010). Feasibility study of a real-time PCR test for meticillin-resistant *Staphylococcus aureus* in a point of care setting. J. Hosp. Infect. *74*, 245–249.

Clerc, O., and Greub, G. (2010). Routine use of point-of-care tests: usefulness and application in clinical microbiology. Clin. Microbiol. Infect. *16*, 1054–1061.

College of American Pathologists (2011). CAP survey: automated molecular platforms. Available at: http://www.cap.org/apps/docs/cap_today/0811/0811_CAPTODAY_AutomatedMolecularPlatformsGuide.pdf (accessed 24 February 2012).

Ehrmeyer, S.S., and Laessig, R.H. (2007). Point-of-care testing, medical error, and patient safety: a 2007 assessment. Clin. Chem. Lab. Med. *45*, 766–773.

Ehrmeyer, S.S., Laessig, R.H., Leinweber, J.E., and Oryall, J.J. (1990). 1990 Medicare CLIA final rules for proficiency testing – minimum intralaboratory performance-characteristics (CV and Bias) needed to pass. Clin. Chem. *36*, 1736–1740.

EPTIS (2012). European proficiency testing information system. Available at: www.eptis.bam.de (accessed 25 January 2012).

Espy, M.J., Uhl, J.R., Sloan, L.M., Buckwalter, S.P., Jones, M.F., Vetter, E.A., Yao, J.D.C., Wengenack, N.L., Rosenblatt, J.E., Cockerill, F.R., *et al.* (2006). Real-time PCR in clinical microbiology: applications for a routine laboratory testing. Clin. Microbiol. Rev. *19*, 165–256.

European Commission (2011). Revision of Directive 98/79/EC of the European Parliament and of the Council of 27 October 1998 on *in vitro* diagnostic medical devices. Summary of responses to the Public Consultation. Available at: http://ec.europa.eu/health/medical-devices/files/recast_docs_2008/ivd_pc_outcome_en.pdf ((accessed 25 January 2012).

European Commission (2012a). Impact assessment on the revision of the regulatory framework for medical devices with relevance for IVDs. Available at: http://ec.europa.eu/health/medical-devices/files/revision_docs/revision_ia_part2_annex2_en.pdf (accessed 19 December 2012).

European Commission (2012b). Nando (New Approach Notified and Designated Organisations) Information System. Available at: http://ec.europa.eu/enterprise/newapproach/nando/ (accessed 25 Januray 2012).

European Commission (2012c). Proposal for a regulation of the European parliament and of the council on *in vitro* diagnostic medical devices. Available at: http://ec.europa.eu/health/medical-devices/files/revision_docs/proposal_2012_541_en.pdf (accessed 19 December 2012).

European Parliament (1998). Directive 98/79/EC on *in vitro* diagnostic medical devices. Off. J. Eur. Commun. *41*, L331.

FDA (2012a). CLIA Database. Available at: http://www.accessdata.fda.gov/scripts/cdrh/cfdocs/cfCLIA/search.cfm (accessed 24 January 2012).

FDA (2012b). Design Control Guidance for Medical Device Manufacturers. Available at: http://www.fda.gov/downloads/MedicalDevices/DeviceRegulationandGuidance/GuidanceDocuments/ucm070642.pdf (accessed 24 January 2012).

FDA (2012c). Device Advice: Comprehensive Regulatory Assistance. Available at: http://www.fda.gov/MedicalDevices/DeviceRegulationandGuidance/default.htm (accessed 24 January 2012).

Gershy-Damet, G.M., Rotz, P., Cross, D., Belabbes, E., Cham, F., Ndihokubwayo, J.B., Fine, G., Zeh, C., Njukeng, P.A., Mboup, S., *et al.* (2010). The World Health Organization African region laboratory accreditation process improving the quality of laboratory systems in the African region. Am. J. Clin. Pathol. *134*, 393–400.

Gilligan, P., Quirke, M., Winder, S., and Humphreys, H. (2010). Impact of admission screening for methicillin-resistant Staphylococcus aureus on the length of stay in an emergency department. J. Hosp. Infect. *75*, 99–102.

Global Laboratory Initiative (2012). Stepwise Process towards TB Laboratory Accreditation. Available at: http://www.gliquality.org/ (accessed 19 December 2012).

Gutierres, S.L., and Welty, T.E. (2004). Point-of-care testing: an introduction. Ann. Pharmacother. *38*, 119–125.

Heierman, E, Anderson, D, and Armbruster, D (2007). Remote analyzer monitoring using the internet. Available at: http://www.ivdtechnology.com/article/remote-analyzer-monitoring-using-internet (accessed 19 December 2012).

Higgins, J.A., Cooper, M., Schroeder-Tucker, L., Black, S., Miller, D., Karns, J.S., Manthey, E., Breeze, R., and Perdue, M.L. (2003). A field investigation of Bacillus anthracis contamination of US Department of Agriculture and other Washington, DC, buildings during the anthrax attack of October 2001. Appl. Environ. Microbiol. *69*, 593–599.

Holland, C.A., and Kiechle, F.L. (2005). Point-of-care molecular diagnostic systems – past, present and future. Curr. Opin. Microbiol. *8*, 504–509.

Jordan, J.A., Hall, G., and Davis, T. (2010). Multicenter study evaluating performance of the Smart Group B Streptococcus (GBS) assay using an enrichment protocol for detecting GBS colonization in patients in the antepartum period. J. Clin. Microbiol. *48*, 3193–3197.

Khamadi, S., Okoth, V., Lihana, R., Nabwera, J., Hungu, J., Okoth, F., Lubano, K., and Mwau, M. (2008). Rapid identification of infants for antiretroviral therapy in a resource poor setting: the Kenya experience. J. Trop. Pediatr. *54*, 370–374.

Kuijper, E.J., Coignard, B., and Tull, P. (2006). Emergence of *Clostridium difficile*-associated disease in North America and Europe. Clin. Microbiol. Infect. *12*, 2–18.

Letant, S.E., Ortiz, J.I., tley Tammero, L.F., Birch, J.M., Derlet, R.W., Cohen, S., Manning, D., and McBride, M.T. (2007). Multiplexed reverse transcriptase PCR assay for identification of viral respiratory pathogens at the point of care. J. Clin. Microbiol. *45*, 3498–3505.

Lewin Group (2008). Laboratory medicine: a national status report. Available at: http://www.lewin.com/publications/publication/343/ (accessed 15 November 2013).

Lim, D.V., Simpson, J.M., Kearns, E.A., and Kramer, M.F. (2005). Current and developing technologies for monitoring agents of bioterrorism and biowarfare. Clin. Microbiol. Rev. *18*, 583–607.

Loeffelholz, M.J., Pong, D.L., Pyles, R.B., Xiong, Y., Miller, A.L., Bufton, K.K., and Chonmaitree, T. (2011). Comparison of the FilmArray respiratory panel and prodesse real-time PCR assays for detection of respiratory pathogens. J. Clin. Microbiol. *49*, 4083–4088.

McNerney, R., and Daley, P. (2011). Towards a point-of-care test for active tuberculosis: obstacles and opportunities. Nat. Rev. Microbiol. *9*, 204–213.

Marner, E.S., Wolk, D.M., Carr, J., Hewitt, C., Dominguez, L.L., Kovacs, T., Johnson, D.R., and Hayden, R.T. (2011). Diagnostic accuracy of the Cepheid GeneXpert vanA/vanB assay ver. 1.0 to detect the vanA and vanB vancomycin resistance genes in Enterococcus from perianal specimens. Diagn Microbiol. Infect. Dis. *69*, 382–389.

Merel, P. (2005). Perspectives on molecular diagnostics automation. J. Lab. Autom. *10*, 342–350.

Miller, S., Moayeri, M., Wright, C., Castro, L., and Pandori, M. (2010). Comparison of GeneXpert FluA PCR to direct fluorescent antibody and respiratory viral panel PCR assays for detection of 2009 novel H1N1 influenza virus. J. Clin. Microbiol. *48*, 4684–4685.

Nichols, J.H. (2003). Quality in point-of-care testing. Expert Rev. Mol. Diagn. *3*, 563–572.

Niemz, A., and Boyle, D. (2012). Nucleic acid testing for tuberculosis at the point-of-care in high-burden countries. Expert Rev. Mol. Diagn. *12*, 687–701.

Noren, T., Alriksson, I., Andersson, J., Akerlund, T., and Unemo, M. (2011). Rapid and sensitive loop-mediated isothermal amplification test for Clostridium difficile detection challenges cytotoxin B cell test and culture as gold standard. J. Clin. Microbiol. *49*, 710–711.

Peeling, R.W., and Mabey, D. (2010). Point-of-care tests for diagnosing infections in the developing world. Clin. Microbiol. Infect. *16*, 1062–1069.

Perdue, M.L. (2003). Molecular diagnostics in an insecure world. Avian Dis. *47*, 1063–1068.

Peters, D.H., Garg, A., Bloom, G., Walker, D.G., Brieger, W.R., and Rahman, M.H. (2008). Poverty and access to health care in developing countries. Ann. N.Y. Acad. Sci. *1136*, 161–171.

Poritz, M.A., Blaschke, A.J., Byington, C.L., Meyers, L., Nilsson, K., Jones, D.E., Thatcher, S.A., Robbins, T., Lingenfelter, B., Amiott, E., *et al.* (2011). FilmArray, an automated nested multiplex PCR system for multi-pathogen detection: development and application to respiratory tract infection. PLoS One *6*, e26047.

Puren, A., Gerlach, J.L., Weigl, B.H., Kelso, D.M., and Domingo, G.J. (2010). Laboratory operations, specimen processing, and handling for viral load testing and surveillance. J. Infect. Dis. *201*, S27–S36.

Rosen, S (2011). The World Market for Molecular Diagnostics, 4th edn (Kalorama Information, New York).

Scott, L.E., Gous, N., Cunningham, B.E., Kana, B.D., Perovic, O., Erasmus, L., Coetzee, G.J., Koornhof, H., and Stevens, W. (2011). Dried culture spots for Xpert MTB/RIF external quality assessment: results of a phase 1 pilot study in South Africa. J. Clin. Microbiol. *49*, 4356–4360.

Smith, D.M., May, S.J., Perez-Santiago, J., Strain, M.C., Ignacio, C.C., Haubrich, R.H., Richman, D.D., Benson, C.A., and Little, S.J. (2009). The use of pooled viral load testing to identify antiretroviral treatment failure. AIDS 23, 2151–2158.

Spencer, D.H., Sellenriek, P., and Burnham, C.A. (2011). Validation and implementation of the GeneXpert MRSA/SA blood culture assay in a pediatric setting. Am. J. Clin. Pathol. *136*, 690–694.

Stevens, W.S., and Marshall, T.M. (2010). Challenges in implementing HIV load testing in South Africa. J. Infect. Dis. *201*, S78–S84.

Stevens, W., Sherman, G., Downing, R., Parsons, L.M., Ou, C.Y., Crowley, S., Gershy-Damet, G.M., Fransen, K., Bulterys, M., Lu, L., *et al.* (2008). Role of the laboratory in ensuring global access to ARV treatment for HIV-infected children: consensus statement on the performance of laboratory assays for early infant diagnosis. Open AIDS J. *2*, 17–25.

Stevens, W.S., Scott, L.E., and Crowe, S.M. (2010). Quantifying HIV for monitoring antiretroviral therapy in resource-poor settings. J. Infect. Dis. *201*, S16–S26.

Stynen, D. (2011). Revision of Europe's IVD Directive 98/79/EC. IVD Technol. *14*, 38.

Tang, Y.W., Procop, G.W., and Persing, D.H. (1997). Molecular diagnostics of infectious diseases. Clin. Chem. *43*, 2021–2038.

Tanriverdi, S., Chen, L.J., and Chen, S.Q. (2010). A rapid and automated sample-to-result HIV load test for near-patient application. J. Infect. Dis. *201*, S52-S58.

de Tejada, B.M., Pfister, R.E., Renzi, G., Francois, P., Irion, O., Boulvain, M., and Schrenzel, J. (2011). Intrapartum group B Streptococcus detection by rapid polymerase chain reaction assay for the prevention of neonatal sepsis. Clin. Microbiol. Infect. *17*, 1786–1791.

Ulrich, M.P., Christensen, D.R., Coyne, S.R., Craw, P.D., Henchal, E.A., Sakai, S.H., Swenson, D., Tholath, J., Tsai, J., Weir, A.F., *et al.* (2006). Evaluation of the Cepheid GeneXpert system for detecting Bacillus anthracis. J. Appl. Microbiol. *100*, 1011–1016.

Urdea, M., Penny, L.A., Olmsted, S.S., Giovanni, M.Y., Kaspar, P., Shepherd, A., Wilson, P., Dahl, C.A., Buchsbaum, S., Moeller, G., *et al.* (2006). Requirements for high impact diagnostics in the developing world. Nature *444* (Suppl. 1), 73–79.

USAID (2007). AAPD 07 – 05: USAID List of Approved HIV/AIDS Test Kits. Available at: http://www.usaid.gov/business/business_opportunities/cib/pdf/aapd07_05.pdf (accessed 25 January 2012).

USAID (2010). HIV/AIDS Rapid Test Kits: Process for USAID Approval and Technical Guidance. Available at: http://www.usaid.gov/our_work/global_health/aids/TechAreas/treatment/testkit_explain.pdf (accessed 25 January 2012).

VanRie, A., Page-Shipp, L., Scott, L., Sanne, I., and Stevens, W. (2010). Xpert((R)) MTB/RIF for point-of-care diagnosis of TB in high-HIV burden, resource-limited countries: hype or hope? Expert Rev. Mol. Diagn. *10*, 937–946.

Vercauteren, G. (2011). Programme update: WHO Prequalification of Diagnostics. Available at: http://www.who.int/diagnostics_laboratory/110404_pqdx_stakeholders_final.pdf (accessed 25 January 2012).

Vernet, G. (2004). Molecular diagnostics in virology. J. Clin. Virol. *31*, 239–247.

Weigl, B.H., Boyle, D.S., de los Santos, T., Peck, R.B., and Steele, M.S. (2009). Simplicity of use: a critical feature for widespread adoption of diagnostic technologies in low-resource settings. Expert Rev. Med. Devices *6*, 461–464.

Wesolowski, L.G., Delaney, K.P., Hart, C., Dawson, C., Owen, S.M., Candal, D., Meyer, W.A. 3rd, Ethridge, S.F., and Branson, B.M. (2011). Performance of an alternative laboratory-based algorithm for diagnosis of HIV infection utilizing a third generation immunoassay, a rapid HIV-1/HIV-2 differentiation test and a DNA or RNA-based nucleic acid amplification test in persons with established HIV-1 infection and blood donors. J. Clin. Virol. *52* (Suppl. 1), S45–S49.

WHO (2011a). Overview of the Prequalification of Diagnostics Assessment Process. Available at: http://www.who.int/diagnostics_laboratory/evaluations/110322_pqdx_007_pq_overview_document_v4.pdf (accessed 25 January 2012).

WHO (2011b). WHO warns against the use of inaccurate blood tests for active tuberculosis. Available at: http://www.who.int/mediacentre/news/releases/2011/tb_20110720/en/ (accessed 23 January 2012).

WHO (2012). WHO list of prequalified diagnostic products. Available at: http://www.who.int/diagnostics_laboratory/evaluations/PQ_list/en/index.html (accessed 25 January 2012).

WHO, TDR, and FIND (2006). Diagnostics for tuberculosis: global demand and market potential. Available at: http://www.who.int/tdr/publications/documents/tbdi.pdf (accessed 15 November 2013).

Williams, S (2011). Understanding the EC Directive 99/79/EC on In Vitro Diagnostic Medical Devices. Available at: http://www.cn.sgs.com/sgs-ssc-white-paper-on-ec-directive-9879ec-in-vitro-diagnostic-medical-devices-en-10.pdf (accessed 25 January 2012).

Wolk, D.M., Picton, E., Johnson, D., Davis, T., Pancholi, P., Ginocchio, C.C., Finegold, S., Welch, D.F., de Boer, M., Fuller, D., *et al.* (2009). Multicenter evaluation of the Cepheid Xpert methicillin-resistant *Staphylococcus aureus* (MRSA) test as a rapid screening method for detection of MRSA in nares. J. Clin. Microbiol. *47*, 758–764.

Yager, P., Domingo, G.J., and Gerdes, J. (2008). Point-of-care diagnostics for global health. Annu. Rev. Biomed. Eng. *10*, 107–144.

Yang, S., and Rothman, R.E. (2004). PCR-based diagnostics for infectious diseases: uses, limitations, and future applications in acute-care settings. Lancet Infect. Dis. *4*, 337–348.

Young, B.C., Dodge, L.E., Gupta, M., Rhee, J.S., and Hacker, M.R. (2011). Evaluation of a rapid, real-time intrapartum group B streptococcus assay. Am. J. Obstet. Gynecol. *205*, 372–376.

# Point-of-care Nucleic Acid Testing: Clinical Applications and Current Technologies

# 10

Angelika Niemz, David S. Boyle and Tanya M. Ferguson


## Abstract

Nucleic acid testing (NAT) has greatly enhanced the diagnosis of infectious diseases, but typically requires a central laboratory with highly skilled operators and complex equipment. This paradigm is not ideal for indications that require a rapid turn-around or for settings where a complex laboratory infrastructure is not available. In the past ten years, novel assays and hardware solutions have been developed that have advanced NAT into peripheral laboratories. We anticipate that NAT in the near future will move to near patient settings in hospitals and physician's offices in developed countries, and to low resource primary care settings in developing countries. In this review we discuss selected clinical applications that can benefit from disseminated NAT, and highlight emerging diagnostic technologies for sample preparation, nucleic acid amplification and detection. We further describe systems that integrate several or all of these components to enable highly sensitive, rapid, and user friendly NAT for infectious disease diagnosis outside of the traditional laboratory.


## Nucleic acid testing – key applications

In 2011, the global molecular diagnostics, or nucleic acid testing (NAT) market was approximated at 3.8 billion USD (Rosen, 2012), of which 60% (US$2.28 billion) were attributed to infectious disease testing, and 23% (US$890 million) to blood bank screening for infectious pathogens. All other applications not related to infectious diseases accounted for 17% (US$630 million). Although many of these other applications, especially those related to pharmacogenetics and oncology, are rapidly gaining in importance, this chapter will focus on infectious disease related NAT. To date in the US, NAT is almost exclusively performed in centralized laboratories designated as high complexity sites under the US Clinical Laboratory Improvement Amendments (CLIA), using sophisticated instrumentation and skilled personnel (Espy *et al.*, 2006). However, more affordable, simplified, and miniaturized systems are being developed and commercialized. Many new NAT platforms and tests have entered the market in the last five years that can be used in laboratories with CLIA moderate complexity designation (see Chapter 9), and it is anticipated that within the next few years CLIA waived NAT systems will emerge, that can effectively penetrate into professional point of care testing sites such as hospital wards and physician office labs. In developed countries, NAT applications that likely will move towards near patient testing include the rapid diagnosis and containment of hospital acquired infections, influenza and other respiratory pathogens, and maternal antepartum screening to minimize mother-to-child disease transmission. In developing countries, NAT applications focus on endemic infectious diseases such as HIV and tuberculosis. Disseminated NAT for these applications can improve patient care, reduce disease transmission rates, and decrease overall healthcare costs.

Each of these applications is briefly described herein, with selected examples of commercially available products. These examples illustrate the range of available technologies, including assays based on the polymerase chain reaction (PCR) plus various isothermal methods, the different levels of integration, and the gradual shift of NAT from high towards moderate complexity central laboratories. Several existing platforms have potential for eventual implementation in POC settings. The clinical applications and associated NAT systems listed herein are just a selection exemplifying the rapid developments in this field.

## Nosocomial or hospital-acquired infections (HAIs)

HAIs are transmitted to a patient within a healthcare setting. The two most common causative agents for HAIs are *Clostridium difficile* (Kuijper *et al.*, 2006; Bauer *et al.*, 2009) and methicillin-resistant *Staphylococcus aureus* (MRSA) (Chambers and Deleo, 2009; Otto, 2012).

*C. difficile* has recently surpassed MRSA as the leading HAI (Rupnik *et al.*, 2009). *C. difficile*-associated disease (CDAD) manifests as severe and recurrent antibiotic-associated diarrhoea, which can lead to pseudomembranous colitis. CDAD is caused by *C. difficile* strains that express two protein toxins, Tox A and Tox B, encoded by the tcdA and tcdB genes, respectively, which damage the epithelial layer of the colon. In recent years, a hypervirulent *C. difficile* strain (BI/NAP1/027) has emerged, with higher sporulation rates and increased levels of toxin production due to mutations in the regulatory gene tcdC (Carter *et al.*, 2011). Current NATs (Table 10.1) are used to detect toxigenic *C. difficile* in unformed stool samples of symptomatic patients.

Hospital acquired MRSA can give rise to surgical site infection, catheter-related infections, sepsis, and pneumonia. NAT has been used to identify asymptomatic MRSA carriers admitted to the hospital by testing nasal swabs (Wolk *et al.*, 2009; Gilligan *et al.*, 2010). NATs are also used to detect MRSA in symptomatic patients, e.g. by testing a tissue biopsy or blood cultures (Spencer *et al.*, 2011). Hospital-acquired infections can also be caused by enterococci found in the normal intestinal flora, which acquire resistance to many antibiotics, including vancomycin. NAT has been used to identify patients who carry (Marner *et al.*, 2011) or have been infected by vancomycin-resistant enterococci (VRE) (Barken *et al.*, 2007). Multiplexed NAT can enable appropriate diagnosis and treatment of sepsis, which can be caused by many different pathogens, including MRSA and VRE. Commercially available NATs for MRSA and VRE are listed in Table 10.2.

## Influenza and respiratory infections

The Influenza virus causes the seasonal flu, but more virulent forms can cause major global pandemics. The influenza H1N1 (swine flu) and H5N1 (avian flu) outbreaks in 2009 and 2006/2007, respectively (Peiris *et al.*, 2007; Neumann *et al.*, 2009), have led to a recent increased demand for improved-NAT-based diagnostic and surveillance tools (Abe *et al.*, 2011; Loeffelholz *et al.*, 2011). Many NAT systems for influenza diagnosis have recently entered the market (Table 10.3), with others in late stage development. Cepheid anticipates that the GeneXpert Flu assay will eventually attain CLIA waived status, thereby enabling implementation in physician office labs. Another fully integrated PCR-based platform, the LIAT influenza test by Iquum, also has potential for implementation in physician office labs. The LIAT is more compact and less expensive than the Cepheid GeneXpert. Alere is developing a compact inexpensive NAT system based on isothermal amplification, called the iNAT, with a clinical trial for an influenza assay ongoing in 2012 (Alere, 2012), and commercial launch anticipated in 2013. The iNAT platform has been designed specifically for POC use in a physician office lab.

Likewise, members of the coronavirus family can cause the common cold, but a more virulent strain caused the 2003 outbreak of the severe acute respiratory syndrome (SARS), resulting in approximately 8000 cases, with 750 deaths. The rapid differential diagnosis of such severe respiratory infections is complicated by the large number of viral and bacterial pathogens that can cause similar clinical symptoms. Therefore, new platforms are emerging that can detect a panel of pathogens in a highly multiplexed manner (Letant *et al.*, 2007; Poritz *et al.*, 2011).

## Perinatal disease transmission

Mother-to-child transmission of group B streptococci (GBS) at birth (intrapartum) causes neonatal disease in 0.53 out of 1000 live births

**Table 10.1** NAT systems for diagnosis of *Clostridium difficile*

| Test (manufacturer) | Targeted analyte(s) | Specimen | Amplification/ detection method | Integration level | Targeted use site | Approval (year) |
|---|---|---|---|---|---|---|
| GeneOhm *C. difficile* (Becton Dickinson) | *C. difficile* (tcdB gene) | Unformed stool | PCR/real-time fluorescence | Manual S, automated A/D | HC central lab | CE-IVD/510k (2008) |
| ProGastro CD assay (Gen-Probe/ Prodesse) | *C. difficile* (tcdB gene) | Unformed stool | PCR/real-time fluorescence | Automated S, A/D – two instruments | HC central lab | CE-IVD/ 510k (2009) |
| GeneXpert *C. difficile* (Cepheid) | *C. difficile* (tcdB gene) | Unformed stool | PCR/real-time fluorescence | Automated integrated S/A/D | MC central lab/POC potential | CE-IVD/510k (2009) |
| Illumigene *C. difficile* (Meridian Bioscience) | *C. difficile* (tcdA gene) | Unformed stool | LAMP/real-time turbidity | Manual S, automated A/D | MC central lab | CE-IVD/ 510k (2010) |
| GeneXpert *C. difficile*/Epi (Cepheid) | *C. difficile* (tcdB gene) + epidemic 027 strain | Unformed stool | PCR/real-time fluorescence | Automated integrated S/A/D | MC central lab/POC potential | CE-IVD/510k (2011) |
| Simplexa *C. difficile* (Focus Diagnostics) | *C. difficile* (tcdB gene) | Unformed stool | PCR/real-time fluorescence | Manual S, automated A/D | HC central lab | CE-IVD/ 510k (2012) |
| Portrait Toxigenic *C. difficile* assay (Great Basin) | *C. difficile* (tcdB gene) | Unformed stool | HDA/end-point hybridization | Automated integrated S/A/D | MC central lab/POC potential | CE-IVD/ 510k (2012) |
| Verigene *C. difficile* (Nanosphere) | *C. difficile* (tcdA and tcdB gene) + epidemic 027 strain | Unformed stool | PCR/end-point array detection | Automated S/A, D – two instruments | MC central lab | 510k (2012) |
| AmpliVue *C. difficile* (Quidel) | *C. difficile* (tcdA gene) | Unformed stool (swab) | HDA/end-point NALF | Manual S, A, D (NALF cassette) | MC central lab | CE-IVD/ 510k (2012) |
| IMDx m2000 *C. difficile* (Abbott) | *C. difficile* (tcdA and tcdB gene) + epidemic 027 strain | Unformed stool | PCR/real-time fluorescence | Automated S, A/D – two instruments | HC central lab | CE-IVD |

Information based on FDA medical device databases, package inserts, and company websites. PCR: Polymerase Chain Reaction, LAMP: Loop Mediated Amplification. HDA: Helicase Dependent Amplification. NALF: Nucleic Acid Lateral Flow. S: sample preparation; A: amplification; D: detection. CE-IVD: compliant with the European IVD directive 98/79/EC. 510k: US FDA cleared, with approval year in parentheses. The approval year for CE-IVD certification cannot readily be looked up, thus is omitted. HC and MC denote US CLIA High Complexity, and Moderate Complexity designation. For products approved only outside the US, the complexity designation is based on the author's judgement, and CLIA designation for other FDA cleared tests performed on the same platform.

**Table 10.2** NAT systems for diagnosis of *Staphylococcus aureus*, MRSA, and other HAIs

| Test (manufacturer) | Targeted analyte(s) | Specimen | Amplification/ detection method | Integration level | Targeted use site | Approval (year) |
|---|---|---|---|---|---|---|
| GeneXpert MRSA (Cepheid) | MRSA (mecA gene) | Nasal swab | PCR/real-time fluorescence | Automated integrated S/A/D | MC central lab/POC potential | CE-IVD/510k (2007) |
| GeneOhm Staph SR (Becton Dickinson) | *S. aureus* + MRSA | Positive blood culture | PCR/real-time fluorescence | Manual S, automated A/D | HC central lab | 510k (2007) |
| GeneXpert MRSA/SA (Cepheid) | MRSA + *S. aureus* | Blood culture, skin/soft tissue | PCR/real-time fluorescence | Automated integrated S/A/D | MC central lab | CE-IVD/510k (2008) |
| GeneOhm MRSA ACP (Becton Dickinson) | MRSA (mecA + OrfX gene) | Nasal swab | PCR/real-time fluorescence | Manual S, automated A/D | HC central lab | CE-IVD/510k (2009) |
| GeneXpert vanA (Cepheid) | Vancomycin resistant *Enterococci* (VRE) | Swab | PCR/real-time fluorescence | Automated integrated S/A/D | MC central lab/POC potential | 510k (2009) |
| GeneXpert vanA/vanB (Cepheid) | Vancomycin resistant *Enterococci* (VRE) | Swab | PCR/real-time fluorescence | Automated integrated S/A/D | MC central lab/POC potential | CE-IVD |
| IMDx m2000 VanR (Abbott) | Vancomycin resistant *Enterococci* (VRE) | Rectal swab, stool sample | PCR/real-time fluorescence | Automated S, A/D – two instruments | HC central lab | CE-IVD |
| Light Cycler MRSA (Roche) | MRSA | Nasal swab | PCR/real-time fluorescence | Manual S, automated A/D | HC central lab | CE-IVD/510k (2009) |
| NucliSense EasyQ MRSA (bioMérieux) | MRSA | Nasal swab | NASBA/real-time fluorescence | Manual S, automated A/D | HC central lab | CE-IVD/510k (2011) |
| GeneOhm VANR (Becton Dickinson) | Vancomycin resistant enterococci (vanA and vanB genes) | Perianal or rectal swab | PCR/real-time fluorescence | Manual S, automated A/D | HC central lab | CE-IVD/510k (2011) |
| Verigene Gram-positive Blood Culture (Nanosphere) | Panel of *Staphylococci*, *Streptococcus*, *Enterococcus*, *Micrococcus* and *Listeria* spp., mecA, vanA and vanB resistance markers | Gram-positive blood culture | PCR/end-point array detection | Automated S/A, D – two instruments | MC central lab | CE-IVD/510k (2012) |
| BD MAX MRSA (Becton Dickinson) | MRSA (mecA gene) | Nasal swab | PCR/real-time fluorescence | Automated integrated S/A/D | MC central lab | CE-IVD/510k (2012) |

Information based on FDA medical device databases, package inserts, and company websites. NASBA: Nucleic Acid Sequence Based Amplification. S: sample preparation; A: amplification; D: detection. CE-IVD: compliant with the European IVD directive 98/79/EC. 510k: US FDA cleared, with approval year in parentheses. The approval year for CE-IVD certification cannot readily be looked up, thus is omitted. HC and MC denote US CLIA High Complexity, and Moderate Complexity designation. For products approved only outside the US, the complexity designation is based on the author's judgement, and CLIA designation for other FDA cleared tests performed on the same platform.

**Table 10.3** NAT systems for diagnosis of influenza and respiratory diseases

| Test (manufacturer) | Targeted analyte(s) | Specimen | Amplification/ detection method | Integration level | Targeted use site | Approval (year) |
|---|---|---|---|---|---|---|
| xTAG Respiratory Viral Panel (Luminex/ Abbott[1]) | Influenza A (H1, H3, H5[1]), Influenza B, Respiratory Syncytial Virus (RSV A, RSV B), Coronavirus[1] (5 strains), Parainfluenza virus 1–3, 4[1], Metapneumovirus, Rhinovirus, Adenovirus, and Enterovirus[1] | Nasopharyngeal swabs | PCR/bead-based flow cytometry | Manual S, A, automated D | HC central lab | CE-IVD[1]/510k (2008/2011) |
| Verigene Respiratory Virus Plus (Nanosphere) | Influenza A (H1, H3)/B, RSVA/B | Nasopharyngeal swab | PCR/end-point array detection | Automated S/A, D – two instruments | MC central lab | CE-IVD/510k (2009/2011) |
| GeneXpert Flu (Cepheid) | Influenza A, Influenza B, Influenza A H1N1 | Nasopharyngeal wash or swab | PCR/real-time fluorescence | Automated integrated S/A/D | MC central lab/POC potential | CE-IVD/510k (2011) |
| FilmArray Respiratory Pathogen Panel (BioFire Diagnostics, formerly Idaho Technology) | Adenovirus, Coronavirus (HKU1, NL63, OC43, 229E), Meta-pneumovirus, Influenza A (H1, H3, 2009 H1), Influenza B, Parainfluenza Virus (1–4), Rhinovirus, Enterovirus, Respiratory Syncytial Virus, *Bordetella pertussis*, *Chlamydophila pneumoniae*, *Mycoplasma pneumoniae* | Nasopharyngeal swab (in viral transport medium) | PCR/real-time fluorescence detection in array format | Automated integrated S/A/D | MC central lab/POC potential | CE-IVD/510k (2011/2012) |
| ProFlu + assay (Gen-Probe/ Prodesse) | Influenza A, Influenza B, Respiratory Syncytial Virus (RSV) | Nasopharyngeal swab (in viral transport medium) | PCR/real-time fluorescence | Automated S, A/D – two instruments | HC central lab | CE-IVD/510k (2008) |
| LIAT Influenza A/B (Iquum) | Influenza A/B | Nasopharyngeal swab | Real-time PCR | Automated integrated S/A/D | MC central lab/POC potential | CE-IVD/510k (2011) |
| Molecular Influenza A + B Assay (Quidel) | Influenza A/B | Nasal swab/ nasopharyngeal swab | PCR/real-time fluorescence | Automated S, A/D – two instruments | HC central lab | CE-IVD/510k (2011/2012) |
| Molecular RSV + hMPV (Quidel) | RSV, Metapneumovirus (hMPV) | Nasal swab/ nasopharyngeal swab | PCR/real-time fluorescence | Automated S, A/D – two instruments | HC central lab | CE-IVD/510k (2013) |
| Artus Influenza A/B RG (Qiagen) | Influenza A/B | Nasopharyngeal swab | PCR/real-time fluorescence | Automated S, A/D – two instruments | HC central lab | CE-IVD/510k (2012) |

**Table 10.3** Continued

| Test (manufacturer) | Targeted analyte(s) | Specimen | Amplification/ detection method | Integration level | Targeted use site | Approval (year) |
|---|---|---|---|---|---|---|
| Simplexa Flu A/B RSV Direct (Focus Diagnostics) | Influenza A, Influenza B, Respiratory Syncytial Virus (RSV) | Nasopharyngeal swab | PCR/real-time fluorescence | Automated A/D (no S required) | MC central lab | CE-IVD/510k (2012) |
| eSensor Respiratory Viral Panel (GenMark) | Influenza A (H1, H3, H1N1), Influenza B, Respiratory Syncytial Virus (RSV A, RSV B), Parainfluenza Virus 1–3, Metapneumovirus (hMPV), Rhinovirus (HRV), Adenovirus B/E and C | Nasopharyngeal swab | PCR/ multiplexed electrochemical detection | Automated S, semi-automated A, automated D – three instruments | HC central lab | 510k (2012) |
| Seegene Anyplex II RV16 | Adenovirus; Influenza A/B, Parainfluenza viruses 1/2/3/4; Rhinovirus A/B/C; Respiratory syncytial virus A/B; Bocavirus 1/2/3/4; Coronavirus 229E, NL63 OC43; Metapneumovirus; Enterovirus | Nasopharyngeal swab/aspirates, Bronchoalveolar lavage | Multiplexed PCR/real-time fluorescence melt curve analysis | Automated S, A/D – two instruments | MC central lab | CE-IVD |

Information based on FDA medical device databases, package inserts, and company websites. S: sample preparation; A: amplification; D: detection. CE-IVD: compliant with the European IVD directive 98/79/EC. 510k: US FDA cleared, with approval year in parentheses. The approval year for CE-IVD certification cannot readily be looked up, thus is omitted. HC and MC denote US CLIA High Complexity, and Moderate Complexity designation. For products approved only outside the US, the complexity designation is based on the author's judgement, and CLIA designation for other FDA cleared tests performed on the same platform. [1]Abbott markets a CE-IVD marked version of the XTAG respiratory viral panel, with additional analytes not included in the FDA cleared test, indicated by the superscript.

(Melin, 2011), leading to pneumonia, sepsis, or meningitis, with high associated mortality and long-term morbidity. Routine antenatal screening for GBS colonization at 35–37 weeks via culture does not detect all cases, especially for infants delivered pre-term, who are at high risk (Clifford *et al.*, 2012). The Cepheid GeneXpert system has been used to rapidly and accurately detect GBS carriage from a recto-vaginal swab during labour (De Tejada *et al.*, 2011) (Table 10.4).

In developed countries, routine vaccination and other preventative efforts have drastically reduced mother to child transmission rates for many infectious diseases, such as rubella, chickenpox, hepatitis B, toxoplasmosis and syphilis. However, for some pathogens the perinatal disease transmission rate remains high, and other preventative measures are required. For example, the frequency of neonatal HSV infection in the United States ranges from 1 out of 12,500 to 1 out of 1700 live births (Corey and Wald, 2009). Intra partum mother-to-child transmission of HSV causes severe infections of the central nervous system or disseminated infection affecting multiple organs in newborns, with high mortality and long term morbidity (Kimberlin, 2004). A rapid PCR method has been used to detect HSV in genital secretions of the mother within 2 hours during labour (Gardella *et al.*, 2010). Recently, BioHelix commercialized an isothermal NAT with lateral flow readout to detect HSV in moderate complexity laboratories (Table 10.4). Intrapartum

**Table 10.4** NAT systems for GBS and HSV infection

| Test (manufacturer) | Targeted analyte(s) | Specimen | Amplification/ detection method | Integration level | Targeted use site | Approval (year) |
|---|---|---|---|---|---|---|
| GeneXpert GBS (Cepheid) | Group B *Streptococci* | Vaginal/rectal swab | PCR/real-time fluorescence | Automated integrated S/A/D | MC central lab/POC potential | CE-IVD/510k (2006) |
| Smart GBS (Cepheid) | Group B *Streptococci* | Vaginal/rectal swab | PCR/real-time fluorescence | Manual S, Automated A/D | HC central lab | CE-IVD/510k (2006) |
| BD MAX GBS (HandyLab/ Becton Dickinson) | Group B *Streptococci* | Lim broth culture ($^{3}$18 h) of swabs | PCR/real-time fluorescence | Automated integrated S/A/D | MC central lab | CE-IVD/510k (2010/'12) |
| GeneXpert GBS LB (Cepheid) | Group B *Streptococci* | Lim broth culture ($^{3}$18 h) of swabs | PCR/real-time fluorescence | Automated integrated S/A/D | MC central lab | CE-IVD/510k (2012) |
| IMDx m2000 HSV1/2 (Abbott) | HSV1, HSV2 (type specific) | Genital/oral swabs, CSF | PCR/real-time fluorescence | Automated S, A/D – two instruments | HC central lab | CE-IVD |
| Viper ProbeTec HSV (Becton Dickinson) | HSV1, HSV2 (type specific) | Anogenital swabs | SDA/real-time fluorescence | Automated integrated S/A/D | HC central lab | CE-IVD/510k (2011) |
| MultiCode HSV 1 and 2 (EraGen/ Luminex) | HSV1, HSV2 (type specific) | Vaginal lesions | PCR/real-time fluorescence | Automated S, A/D – two instruments | HC central lab | 510k (2011) |
| IsoAmp HSV (BioHelix) | HSV1, HSV2 (type common) | Genital/oral swabs | HDA (end-point NALF) | Manual S, A, D | MC central lab | CE-IVD/510k (2011) |

Information based on FDA medical device databases, package inserts, and company websites. SDA: Strand Displacement Amplification; HDA: Helicase Dependent Amplification; NALF: Nucleic Acid Lateral Flow; S: sample preparation; A: amplification; D: detection. CE-IVD: compliant with the European IVD directive 98/79/EC. 510k: US FDA cleared, with approval year in parentheses. The approval year for CE-IVD certification cannot readily be looked up, thus is omitted. HC and MC denote US CLIA High Complexity, and Moderate Complexity designation. For products approved only outside the US, the complexity designation is based on the author's judgement, and CLIA designation for other FDA cleared tests performed on the same platform.

HSV detection can enable appropriate preventative treatment. Mother-to-child transmission of cytomegalovirus, another member of the human herpes virus family, causes gradual hearing loss and in some cases neurological impairment in approximately 1.26 out of 1000 live births, which has prompted efforts to implement routine PCR-based screening of newborns for CMV infection (de Vries *et al.*, 2011).

## Infant HIV diagnosis

Acquired Immunodeficiency Syndrome (AIDS), caused by the Human Immunodeficiency Virus (HIV), continues to be a major global health priority. Globally, approximately 370,000 infants are newly infected with HIV each year through mother to child transmission *in utero*, during birth, or after birth through breastfeeding (Ciaranello *et al.*, 2011), and approximately 3.4 million children under 15 years of age are living with HIV (Abrams *et al.*, 2012). Rapid diagnosis is essential (Violari *et al.*, 2008), since untreated infant HIV can rapidly progress to AIDS, with over 50% mortality by 2 years of age (Newell *et al.*, 2004). Commonly used rapid serological assays for HIV diagnosis are confounded the persistence of maternal HIV antibodies for up to 18 months (Chantry *et al.*, 1995). The World Health Organization (WHO) and United Nations International Children's Emergency Fund (UNICEF) therefore recommend virological NAT for HIV infant diagnosis, typically based on detecting viral RNA and/or pro-viral DNA from a dried blood spot (Stevens *et al.*, 2008). Testing is performed in centralized facilities, which significantly delays result reporting, with the resulting loss to follow up for treatment proving a substantial challenge. One report has shown that centralized testing resulted in a turnaround of one to three months from the date of sample collection to the date of reporting the result (Khamadi *et al.*, 2008). Decentralized, near patient testing using a suitable POC NAT system may mitigate these logistical challenges and improve treatment rates of infected infants.

## HIV viral load monitoring

Improved access to anti-retroviral therapy (ART) in low resource settings has helped to decrease AIDS-related morbidity and mortality, but increasing rates of treatment failure for first line therapy may undermine the long-term success of ongoing efforts (Gupta *et al.*, 2009; Harries *et al.*, 2010). HIV viral load monitoring, the periodic assessment of the number of HIV virions per ml of plasma of AIDS patients receiving ART, is routinely performed in developed countries to identify treatment failure, prompt adherence counselling and if necessary a change to second-line therapy. Widespread implementation of HIV viral load monitoring in resource limited settings remains challenging (Calmy *et al.*, 2007; Rouet and Rouzioux, 2007). However, the 2010 revisions of the WHO HIV/AIDS treatment guidelines recommend expanding the access to viral load testing to improve the early and accurate diagnosis of treatment failure. NAT is the preferred method for HIV viral load monitoring in developed and developing countries (Fiscus *et al.*, 2006; Rouet and Rouzioux, 2007). Several African countries are utilizing centralized laboratory-based systems for viral load monitoring from dried blood spots sent in from remote areas (Johannessen *et al.*, 2009; Mehta *et al.*, 2009), which introduces logistical challenges and significant time delays. Dry blood spot HIV viral load assays often have reduced sensitivity, and do not always correlate with plasma viral load.

Currently, HIV NAT for viral load testing and infant diagnosis is performed exclusively in high complexity centralized laboratories, using laboratory developed tests, or commercial IVD systems marketed by Roche Molecular Systems, Abbott, Bayer/Siemens, and bioMérieux (Table 10.5) (Stevens *et al.*, 2010). However, this paradigm is likely to change in the near future. Several HIV 1 viral load assays are in development on smaller, fully integrated, and easy to use NAT platforms discussed later on in this chapter, such as the LIAT platform by Iquum (Tanriverdi *et al.*, 2010), and the NAT analyser by Alere (Ullrich *et al.*, 2012). Furthermore, Cepheid, in partnership with the Foundation for Innovative New Diagnostics (FIND), is developing an HIV viral load assay for the GeneXpert. This platform is already disseminated in many countries with high HIV burden, which will facilitate market entry.

## Tuberculosis

Tuberculosis (TB) is another global health threat, with an estimated prevalence of 12 million and incidence of 8.7 million cases in 2011, resulting in 1.4 million deaths (WHO, 2012), many of them amongst HIV/AIDS patients (Perkins and Cunningham, 2007). Approximately 1/3 of the world's population is infected with latent TB (Russell, 2007), which can reactivate if a patient's immune system is weakened. Active TB, most commonly presenting as pulmonary TB (Harries and Dye, 2006), can be diagnosed through smear microscopy or culture-based methods. However, smear microscopy suffers from low sensitivity, while culture methods have very long turn-around times. NAT (Table 10.6) enables sensitive, specific, and comparatively rapid TB diagnosis and detection of drug resistance mutations (Balasingham *et al.*, 2009; Niemz and Boyle, 2012). The MTB/RIF assay performed on the GeneXpert system (Cepheid, USA) (Boehme *et al.*, 2010; Helb *et al.*, 2010; Lawn and Nicol, 2011) received WHO endorsement for TB diagnosis and rifampin resistant testing in district and sub-district level laboratories of developing countries (World Health Organization, 2011), and is undergoing large scale roll-out in several high-TB-burden countries including South Africa. Several smaller battery operated real-time PCR-based

**Table 10.5** NAT systems for HIV viral load monitoring (may also be used for infant diagnosis)

| Test (manufacturer) | Targeted analyte(s) | Specimen | Amplification/ detection method | Integration level | Targeted use site | Approval (year) |
|---|---|---|---|---|---|---|
| Amplicor HIV-1 Monitor (Roche) | HIV-1 (*gag* gene) | Plasma, DBS | PCR/end-point absorbance | Manual or automated S, Automated A/D – one or two instruments | HC central lab | CE-IVD/PMA (1999) |
| NucliSENS HIV-1 QT (bioMérieux) | HIV-1 (*gag* gene) | Plasma, DBS, etc. | NASBA/end-point electro-chemiluminescence | Manual or automated S, Automated A, D – two or three instruments | HC central lab | CE-IVD/PMA (2001) |
| Versant HIV-1 RNA (Bayer) | HIV-1 (*pol* gene) | Plasma | bDNA/end-point chemiluminescence | Manual S, automated A/D | HC central lab | CE-IVD/PMA (2002) |
| NucliSENS EasyQ HIV-1 (bioMérieux) | HIV-1 (*gag* gene) | Plasma, DBS, etc. | NASBA/real-time fluorescence | Automated S, A/D – two instruments | HC central lab | CE-IVD/ PqDX (2011) |
| m2000 RealTime HIV-1 (Abbott) | HIV-1 (*pol Integrase*) | Plasma, DBS | PCR/real-time fluorescence | Automated S, A/D – two instruments | HC central lab | CE-IVD/PMA (2007)/PqDX (2011) |
| COBAS AmpliPrep/ TaqMan HIV-1 (Roche) | HIV-1 (LTR – *gag*) | Plasma, DBS | PCR/real-time fluorescence | Automated S, A/D – two instruments, can be docked | HC central lab | CE-IVD/PMA (2007)/PqDX (2012) |
| Versant HIV-1 RNA kPCR (Bayer/Siemens) | HIV-1 (*pol* gene) | Plasma, DBS, etc. | PCR/real-time fluorescence | Automated S, A/D – two instruments | HC central lab | CE-IVD/ PqDX (2012) |

Information based on FDA medical device databases, package inserts, and company websites. DBS: dried blood spot; NASBA: Nucleic Acid Sequence Based Amplification, bDNA: branched DNA assay; S: sample preparation; A: amplification; D: detection. CE-IVD: compliant with the European IVD directive 98/79/EC. PMA: US FDA Pre-Market Approval, with approval year in parentheses. The approval year for CE-IVD certification cannot readily be looked up, thus is omitted. HC and MC denote US CLIA High Complexity, and Moderate Complexity designation. For products approved only outside the US, the complexity designation is based on the author's judgement, and CLIA designation for other FDA cleared tests performed on the same platform.

systems with decreased instrument cost are in late stage development for TB diagnosis (Niemz and Boyle, 2012), in addition to isothermal nucleic acid amplification reactions coupled to end-point detection, such as loop-mediated amplification (LAMP) with turbidity/fluorescence-based readout (Mitarai *et al.*, 2011), and cross-priming amplification (CPA) with nucleic acid lateral flow readout (Fang *et al.*, 2009). All of these systems are however much less integrated and automated, therefore less user friendly, compared to the Cepheid GeneXpert.

In addition to the diagnosis of TB, there is an emerging need to rapidly assess if the pathogen is drug resistant (Zumla *et al.*, 2012). Multidrug-resistant TB (MDR TB) occurs when *Mycobacterium tuberculosis* acquires resistance to rifampicin and isoniazid, the two most common first line anti-TB drugs. Extensively drug-resistant TB (XDR-TB) is defined as MDR TB with additional resistance to a fluoroquinolone and one of the second line injectable drugs. Patients with MDR and XDR TB are extremely difficult to treat, with high mortality. Recently, Totally Drug Resistant TB (TDR TB) strains have emerged, which have acquired resistance to all first and second line anti-TB drugs. Rapid and accurate TB drug susceptibility testing is crucial to reduce disease transmission and inform treatment decisions. The majority of resistance to rifampin is conferred by mutations within *rpoB*, the gene coding for the RNA polymerase subunit B, the target of rifampin. As ~90% of all rifampicin-resistant TB strains are also isoniazid resistant, several TB NAT systems

**Table 10.6** NAT systems for diagnosis of tuberculosis

| Test (manufacturer) | Targeted analyte(s) | Specimen | Amplification/ detection method | Integration level | Targeted use site | Approval (year) |
|---|---|---|---|---|---|---|
| Amplified MTB Direct (Gen-Probe) | MTB | Sputum, respiratory specimens | TMA/end-point chemiluminescence | Manual S, A, D | HC central lab | PMA (1995) |
| Amplicor MTB (Roche) | MTB | Sputum (smear positive) | PCR/end-point absorbance | Manual S, automated A/D | HC central lab | PMA (1996) discontinued 2010 |
| COBAS TaqMan MTB (Roche) | MTB | Sputum, respiratory specimens | PCR/real-time fluorescence | Manual S, automated A/D | HC central lab | CE-IVD |
| ProbeTec ET MTB complex (Becton Dickinson) | MTB | Sputum, respiratory specimens | SDA/real-time fluorescence | Manual S, automated A/D | HC central lab | CE-IVD |
| GeneXpert MTB/Rif (Cepheid) | MTB + mutations in *rpo B* (RIF resistance) | Sputum | PCR/real-time fluorescence | Automated integrated S/A/D | MC central lab/POC potential | CE-IVD |
| GenoType MTBDRplus (Hain Lifescience) | MTB + mutations in *rpo*, *katG*, *inhA* (RIF + INH resistance) | Sputum (smear positive) or culture | PCR/end-point line probe hybridization | Manual S, A and D | HC central lab | CE-IVD |
| GenoType MTBDRsl (Hain Lifescience) | MTB + mutations in *gyrA*, *embB*, *rrs* (resistance to EMB + FLQ + injectable drugs) | Sputum (smear positive) or culture | PCR/end-point line probe hybridization | Manual S, A and D | HC central lab | CE-IVD |
| Magicplex™ MTB Real-Time Test (Seegene) | MTB | Sputum, stool, blood, pus, tissue, urine, CSF | PCR/real-time fluorescence | Manual S, automated A/D | HC central lab | CE-IVD |
| Anyplex™ II, MTB/MDR/XDR (Seegene) | MTB + mutations in *rpo*, *katG*, *inhA*, *gyrA*, *rrs*, *eis* (resistance to RIF + INH + FLQ + injectable drugs) | Sputum, culture, bronchial wash, tissue | PCR/real-time fluorescence (melt curve analysis) | Manual S, automated A/D | HC central lab | CE-IVD |
| Genedrive TB (Epistem) | MTB + mutations in *rpo B* | Sputum | PCR/real-time fluorescence melt curve analysis | Manual S, automated A/D | MC central lab | CE-IVD |
| Truenat – MTB (MolBio Diagnostics) | MTB | Sputum | PCR/real-time fluorescence | Partially integrated | MC central lab | CE-IVD |
| Loopamp TB detection (Eiken) | MTB | Sputum | LAMP/real-time turbidity or end-point fluorescence | Manual S, automated A/D | MC central lab | CE-IVD |

Information based on FDA medical device databases, package inserts, and company websites. MTB: Mycobacterium tuberculosis complex; MDR: Multi-Drug Resistant; XDR: Extensively Drug Resistant; RIF: rifampicin, INH: isoniazid; EMB: ethambutol; FLQ: fluoroquinolone; TMA: transcription mediated amplification; SDA: Strand Displacement Amplification; LAMP: Loop Mediated Amplification; S: sample preparation; A: amplification; D: detection. CE-IVD: compliant with the European IVD directive 98/79/EC. PMA: US FDA Pre-Market Approval, with approval year in parentheses. The approval year for CE-IVD certification cannot readily be looked up, thus is omitted. HC and MC denote US CLIA High Complexity, and Moderate Complexity designation. For products approved only outside the US, the complexity designation is based on the author's judgement, and CLIA designation for other FDA cleared tests performed on the same platform.

such as the Cepheid GeneXpert and Epistem GeneDrive perform *rpoB* genotyping as a surrogate to identify MDR TB strains. More accurate diagnosis of MDR-TB requires detection of isoniazid resistance, which is typically conferred by mutations in either *katG* or in the *inhA* promoter region. Molecular diagnosis of XDR-TB requires genotyping of many more genetic loci associated with second line anti-TB drug resistance. NAT systems exist for further interrogation of MDR and XDR TB e.g. by Hain Lifescience or Seegene, but these NATs are typically complex, not well integrated or fully automated, preventing effective implementation in high burden low resource settings. Integrated microarray-based systems that enable highly multiplexed *M. tuberculosis* genotyping and drug resistance screening are starting to emerge such as the Akonni TruArray MDR-TB Test, the AutoGenomics INFINITI MDR-TB assays, and the Affymetrix *M. tuberculosis* genome array Genechip.

### Non-clinical applications

Many of the newer technologies described in the remainder of this chapter were initially developed or gained entry to market through products related to non-clinical applications in the areas of agricultural, veterinary sciences, food-safety, and biothreat detection. For example, Cepheid's GeneXpert system was initially launched for Anthrax screening in postal facilities (Ulrich *et al.*, 2006). Further development of the system eventually led to FDA approved and/or CE marked IVD products for a range of infectious disease applications related to TB, HAIs, maternal health, and influenza (Boehme *et al.*, 2010; Miller *et al.*, 2010; Marner *et al.*, 2011; Spencer *et al.*, 2011). NAT systems for non-clinical applications have also been developed and commercialized by other companies, such as Axxin, Epistem, Ionian Technologies, Lumora, Molbio, OptiGene, Qiagen, TwistDx Ltd, and Spartan. These non-clinical applications of NAT have an easier path to market due to less stringent regulation.

## NAT process overview

This section provides the reader with an overview of key steps in and general considerations regarding the NAT process applicable to POC testing. Further details are provided in the case studies discussed in the following section.

### Sample collection, pre-processing, and preservation

Ideally, POC NAT should require only minimally invasive or non-invasive sampling methods that are easy to perform, especially for use in low resource settings (Urdea *et al.*, 2006; Yager *et al.*, 2008). Samples appropriate for POC applications include swabs, sputum, oral fluids, whole blood from a finger or heel prick, urine or faeces. Testing from whole blood often requires a plasma or peripheral blood mononuclear cell (PBMC) or red blood cell (RBC) separation method; urine can vary in pH and electrolyte composition; sputum and stool can be difficult matrices to collect, manipulate, and contain many inhibitory or degradative compounds. In all cases, sample handling protocols have to include appropriate biosafety precautions, such as full containment, user protective methods and if possible, inactivation of pathogens prior to testing, and safeguards to prevent aerosol formation and accidental skin disruption.

Sputum specimens for TB diagnosis via culture or smear microscopy are typically liquefied and decontaminated, using solutions containing, e.g. N-acetyl-l-cysteine and sodium hydroxide. Sputum samples pre-treated in this manner are often used as starting specimens for NAT in centralized laboratories, using laboratory-developed tests (LDTs) (Halse *et al.*, 2011) or commercial IVDs (Dalovisio *et al.*, 1996). Decontamination means organisms other than mycobacteria are killed, which is a pre-requisite for culture-based TB diagnosis. To facilitate appropriate biohazard protection for POC NAT, it is preferable to completely disinfect the sputum, i.e. kill all pathogens including mycobacteria, which requires a different pre-treatment process (Helb *et al.*, 2010).

For applications where samples are collected and then tested at a later time, appropriate sample storage and stability during transport also have to be considered to ensure that the integrity of the target nucleic acid is maintained. This is particularly true for downstream amplification and detection of RNA, which is more susceptible to degradation than DNA. Dry blood spots have been shown to be an effective means of preserving viral RNA and proviral DNA of peripheral blood drawn from heel sticks (infants) or finger sticks (children and adults) (Uttayamakul *et al.*, 2005; Leelawiwat *et al.*, 2008; Ngo-Giang-Huong *et al.*, 2008; Lofgren *et al.*, 2009). However, it is not ideal to perform quantitative HIV viral load monitoring from dry blood spots, since the presence of proviral DNA creates a positive bias at the lower end of the range (Lofgren *et al.*, 2009). Accurate HIV viral load monitoring at the lower end of the range requires separation of plasma from blood. Rapid plasma separation can be accomplished using microfluidic devices (Zhang *et al.*, 2008; Sollier *et al.*, 2009), or simple membranes, such as the Vivid plasma separation membrane from PALL.

## Sample preparation

Nucleic acid sample preparation entails all steps required to isolate and concentrate targeted nucleic acids from a clinical sample in a format that is amenable to downstream amplification and detection. Sample preparation has been recognized as a major hurdle in nucleic acid testing, particularly for POC and near patient applications (Dineva *et al.*, 2007). If not automated, sample preparation traditionally involves cumbersome and lengthy multi-step processes that are prone to error and require ancillary equipment such as pipettes and centrifuges, plus associated disposables. Most infectious disease NAT requires target amplification using polymerases that can be inhibited by compounds found in clinical samples, or introduced during the sample preparation process (such as chaotropic salts and organic solvents). A variety of substances that inhibit PCR have been described (Wilson, 1997; Al-Soud and Radstrom, 2000; Al-Soud and Radstrom, 2001; Radstrom *et al.*, 2004) and if carried over from sample preparation into the downstream reaction, these inhibitors can lead to false negatives. Most commercial IVD NATs incorporate an Internal Amplification Control (IAC), spiked into the sample upstream of sample preparation and then co-amplified with the target, to identify false negatives due to failed sample preparation or polymerase inhibition. Recently, PCR assays tolerant of whole blood have been demonstrated based on the use of PCR enhancer cocktails and mutant Taq polymerases (Kermekchiev *et al.*, 2009; Zhang *et al.*, 2010). Many of the current isothermal technologies, most notably Loop-Mediated Amplification (LAMP), appear to be unaffected by inhibitory compounds, especially those present in whole blood, stool, and urine (Enomoto *et al.*, 2005; Curtis *et al.*, 2008; Francois *et al.*, 2011).

Most sample preparation methods also facilitate nucleic acid concentration, which facilitates sensitive analyte detection. Since the attainable limit of detection (LOD) scales with the sample input volume, the sample preparation process ideally needs to accommodate large sample input volumes, in the order of hundreds of μl to several ml. However, downstream amplification steps typically require small sample input volumes ranging from several μl to tens of μl. Therefore, concentrating the targeted nucleic acids is an important and sometimes overlooked aspect of sample preparation.

The size, complexity, and cost of the overall system are often times dictated by the sample preparation process, not the amplification and detection steps. IVD developers occasionally underestimate these challenges, and focus on systems developed for amplification and detection only. For POC NAT, sample preparation starting from the clinical specimen ideally needs to be integrated and coupled with amplification and detection in an inexpensive, automated, miniaturized, closed system format. Microfluidic devices have been developed to automate individual steps or the entire process of NA sample preparation (Kim *et al.*, 2009). However, these systems are often incapable of processing large sample input volumes, are complex, relatively expensive, and difficult to manufacture. Macrofluidic systems are more appropriate for the sample preparation steps, while downstream amplification and detection involves smaller volumes that can be effectively handled in microfluidic devices.

Sample preparation involves pathogen lysis or disruption to liberate nucleic acids. For most organisms, chemical and/or enzymatic lysis methods are sufficient (Boom *et al.*, 1990; Paule *et al.*, 2004; Niwa *et al.*, 2005; Salazar and Asenjo, 2007). However, difficult to lyse organisms such as *M. tuberculosis, C. difficile, Bacillus anthracis* spores, or *Giardia* cysts are more effectively lysed using mechanical disruption approaches (Amaro *et al.*, 2008; Kaser *et al.*, 2009; de Boer *et al.*, 2010). Once nucleic acids have been released via cell lysis, solid phase extraction (SPE) is most frequently used as a means to isolate and purify the nucleic acids from the sample matrix. For SPE-based methods, nucleic acids present in crude lysate are preferentially bound to a surface such as silica membranes or silica coated magnetic beads (Boom *et al.*, 1990; Berensmeier, 2006; Dauphin *et al.*, 2009), facilitated by lysis/binding buffers that contain chaotropic salts, ethanol, and/or have a low pH. The surface is then washed under conditions that ensure that the nucleic acids remain bound while removing undesirable substances, followed by nucleic acid elution off the surface into a high pH low ionic strength buffer compatible with subsequent amplification. Invitrogen has developed the Charge Switch® method as an alternate to traditional nucleic acid SPE. The Charge Switch approach does not require chaotropic salts and organic solvents, but utilizes pH dependent anion exchange (Price *et al.*, 2009): the bead surface incorporates functional groups that are positively charged at low pH, and neutral at pH 8.5, to bind and elute nucleic acids. This method has been used effectively for HIV RNA extraction from plasma (Dineva *et al.*, 2007).

Filter paper-based extraction tools, such as the Flinders Technology Associates (FTA™) cards commercialized by Whatman (GE healthcare) (Rogers and Burgoyne, 2000; Inoue *et al.*, 2007), can facilitate nucleic acid sample preparation and stabilization, especially from samples collected in remote locations. The paper in these cards is pretreated with chaotropic salts and detergents in a lyophilized format. When the liquid sample is added, cells are lysed upon contact with the treated paper, and liberated nucleic acids are preserved in the dried paper matrix. Pathogen lysis also eliminates biosafety risks. However, the lysis/binding reagents must be removed prior to downstream amplification to prevent polymerase inhibition, e.g. by performing several washes on cut discs removed from the card. During the wash steps, large genomic DNA remains trapped in the paper matrix, but smaller viral RNA genomes may become eluted. Retention of such small genomes within the paper during washing can be promoted by adding high salt concentrations or organic solvents to the wash buffer. These components however have to be removed prior to amplification, which increases the complexity of the process and reduces overall yield. Once sufficiently purified, the FTA paper disc can be directly added to an amplification reaction.

Recently, nitrocellulose strips have been applied for the purification of nucleic acids to detect plant pathogens in agricultural settings (Tomlinson *et al.*, 2010a,b). Specimens are first lysed via chemical and mechanical lysis. A nitrocellulose strip is then placed into the lysis mixture, which flows through the strip via capillary action. A piece of the nitrocellulose strip is then scraped off and placed into the application reaction. This sample preparation method is simple and rapid to perform (<5 minutes) and has been coupled with loop-mediated amplification and real-time PCR. Nitrocellulose membranes have also been applied in a different format to detect plant viruses via PCR-based NAT (Chang *et al.*, 2011). In this case, torn leafs of infected plants were simply blotted onto the nitrocellulose membrane, a sample of the membrane was removed using a hole punch, rinsed in a triton buffer, and then placed into the PCR reaction. This nitrocellulose blotting method demonstrated comparable performance to sample preparation using a commercial FTA card. The inherent simplicity, speed and low cost of nitrocellulose membrane-based nucleic acid sample preparation approached make them worthy of further investigation for use with clinical applications in low resource settings. Similarly, malaria rapid immunologic strips have been demonstrated as a rapid but crude sample preparation method for subsequent PCR analysis (Cnops *et al.*, 2011).

A fundamentally different sample preparation approach involves the concentration and purification of intact microorganisms, typically bacteria, followed by lysis to liberate the genomic material. This approach is frequently applied to isolate mycobacteria from sputum for TB NAT. Mycobacteria in liquefied sputum samples can be enriched through centrifugation, followed by re-suspending the pellet in lysis buffer (Dalovisio *et al.*, 1996; Halse *et al.*, 2011). Not all impurities are removed by this method, which means inhibition can be a concern in subsequent amplification. Mycobacteria can also be separated from the sample matrix through filtration, followed by washing, as implemented in the Cepheid GeneXpert cartridge (Boehme *et al.*, 2010), or can be captured on magnetic beads functionalised with cationic lipophilic polymers (Wilson *et al.*, 2010; Mitarai *et al.*, 2012).

## Amplification

NAT generally involves some mode of amplification to attain the required LOD. This can involve amplification of signal, target, probe, or a combination thereof. Signal amplification such as the bDNA (Urdea, 1993) and hybrid capture (Cox *et al.*, 1995) methods, are often similar to enzyme-linked immunosorbent assays (ELISAs), with a sandwich assay format in conjunction with enzyme-mediated amplification using, for example, horseradish peroxidase or alkaline phosphatase. Enzyme-independent signal amplification methods may involve catalysed silver reduction, as used in the Nanosphere technology (Storhoff *et al.*, 2004), or chemiluminescent reactions based on dye intercalation, such as the hybridization protection assay by Gen-Probe (Nelson, 1998). Polymerase-independent probe amplification methods such as the Invader assay (Kwiatkowski *et al.*, 1999) generate specific detectable DNA sequences through the action of a nuclease, while other methods such as the ligase chain reaction (Wiedmann *et al.*, 1994) are based on oligonucleotide ligation.

Target amplification methods, meaning methods based on the amplification of targeted nucleic acids using polymerases, are most commonly used in NAT for infectious diseases due to their low LOD. The polymerase chain reaction (PCR), the most well-known target amplification method, requires thermocycling to mediate melting of the DNA duplex, primer annealing and extension. Fully integrated devices (Raja *et al.*, 2005; Tanriverdi *et al.*, 2010; Zhou *et al.*, 2010a) and microfluidic lab-on-a-chip systems (Zhang and Ozdemir, 2009; Nagatani *et al.*, 2012) have been developed for miniaturized PCR thermocycling.

Isothermal NA amplification technologies require a single reaction temperature which translates into less complex and expensive instrumentation, a significant advantage for POC applications in low-resource settings. Some isothermal amplification assays may be incubated at close to ambient temperature, depending on the enzyme kinetics involved (Piepenburg *et al.*, 2006). Most isothermal amplification methods however require incubation at temperatures between 50°C–65°C, reflecting the common use of *Bst* DNA polymerase. The optimal incubation temperature is also influenced by the need to destabilize DNA duplexes to facilitate the insertion of the oligonucleotide primers prior to amplification (e.g. LAMP or CPA), other enzymes that may be present in the reaction, and the general reaction design. Isothermal amplification can be performed in water baths or using simple resistive heaters. Isothermal amplification can also be performed in chemically powered reactors for assay incubation. Recently several such devices have been reported (Hatano *et al.*, 2010; Labarre *et al.*, 2010, 2011b; Liu *et al.*, 2011), in the format of simple hand warmers, wherein sodium acetate-based heat generation is manually activated by a kinetic switch (Hatano *et al.*, 2010), or through more sophisticated devices containing phase change materials to modulate a precise incubation temperature (Labarre *et al.*, 2010; Labarre *et al.*, 2011b). These devices are independent from traditional electrical power source or instrument that needs electrical power (e.g. water baths and heat blocks), which permits NAT in either disaster areas or in regions where power is intermittent or completely absent. Unlike batteries, chemically powered heaters can be stored for years without gradual loss of power, making them ideal for use in low resource settings.

Many isothermal amplification methods have been reported, which can be grouped based on the reaction principle (Niemz *et al.*, 2011). Selected examples of each group are discussed herein.

Methods based on RNA transcription (Fig. 10.1) include Transcription Mediated Amplification (TMA, Gen-Probe) (Hofmann *et al.*, 2005) and Nucleic Acid Sequence Based Amplification (NASBA, BioMérieux) (Gracias and McKillip, 2007). Both methods are used in commercially available systems for IVD NAT in centralized laboratories. In NASBA and TMA, target RNA is converted into a double stranded DNA

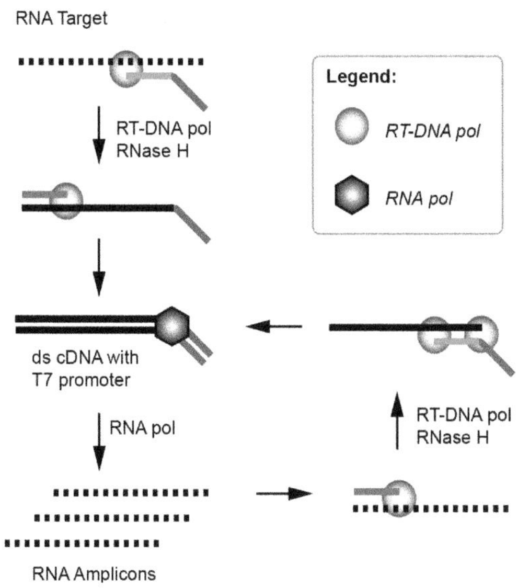


**Figure 10.1** Reaction scheme for Transcription Mediated Amplification (TMA) (Hofmann *et al.*, 2005) and Nucleic Acid Sequence Based Amplification (Gracias and McKillip, 2007). A DNA polymerase with reverse transcriptase activity (RT-DNA pol) converts target RNA to cDNA, RNAse H degrades the RNA strand, priming with a second primer generates a ds-DNA intermediate with T7 promoter region. RNA polymerase (RNA pol) generates RNA amplicons that are converted back into the cDNA intermediate, which leads to exponential amplification.

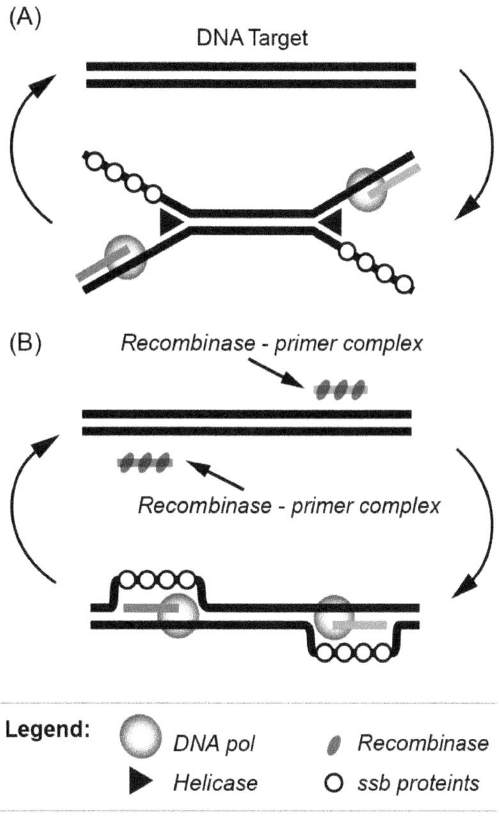


**Figure 10.2** Schematic overview of 'PCR-like' isothermal amplification reactions with enzymatic duplex melting/primer annealing: (A) Helicase Dependent Amplification (HDA): ds target DNA is unwound through helicases, primers anneal and the separated strands are stabilized through single strand binding (ssb) proteins. DNA polymerase (DNA Pol) extends the primers, and the process repeats. (B) Recombinase Polymerase Amplification (RPA): Recombinase – primer complexes scan the ds-DNA target. Once homologous sequences are found, the recombinase facilitates strand exchange and dissociates from the primer. Single strand binding (ssb) proteins attach to the displaced strand to stabilize the primer-target duplex, which is then extended by DNA Pol. The process repeats, generating amplicons in a manner analogous to PCR.

intermediate with promoter region through action of a reverse transcriptase and RNase H, plus one primer with a 5′ overhang that encodes the promoter site. RNA polymerase then generates RNA amplicons from this DNA template in a linear amplification process. The RNA amplicons are again converted back into the double stranded DNA intermediate, which leads to exponential amplification. TMA and NASBA involve the same reaction scheme, but TMA requires only two enzymes (an RT-DNA polymerase with RNase H activity and an RNA polymerase).

Isothermal amplification using DNA polymerase can be achieved through enzymatic duplex melting/primer annealing, as implemented in the Helicase Dependent Amplification (HDA, BioHelix) (Vincent *et al.*, 2004; Jeong *et al.*, 2009) and Recombinase Polymerase Amplification (RPA, TwistDx) (Piepenburg *et al.*, 2006; Lutz *et al.*, 2010). Recently, a hot start HDA format has been described, using blocked primers that cannot be extended by the polymerase until deblocked by a thermostable RNase (Hicke *et al.*, 2012). HDA and RPA have a 'PCR-like' reaction scheme which is relatively simple compared to other isothermal amplification methods. However, HDA and RPA require additional enzymes (DNA helicase and recombinase, respectively), single strand binding proteins, and in the case of RPA an endonuclease and other reagents such as ATP, which increases the complexity and cost of the reaction.

Methods based on strand displacement using polymerases only, with multiple linear primer sets include the Loop-mediated AMPlification (LAMP, Eiken) (Curtis *et al.*, 2009; Mori and Notomi, 2009), Cross-Priming Amplification (CPA, Ustar Biotechnologies) (Fang *et al.*, 2009; Yulong *et al.*, 2010), and the Smart Amplification Process (SMART-AMP) (Mitani *et al.*, 2009). These and other isothermal amplification methods use a set of sacrificial outer bumper primers to initiate the reaction. The bumper primers are no longer required after the initial step. A strand displacing 5′ exo- polymerase, such as large fragment *Bst* polymerase, 'peels off' the DNA strand downstream of the extended primers from the template, which facilitates isothermal amplification. LAMP, CPA and SMART-AMP all create high molecular weight concatenated products rather than amplicons of a uniform discrete size. The reactions are typically performed in the presence of betaine to render the high molecular weight double stranded DNA partially unstable, permitting the insertion of primers for continued amplification. In LAMP and SMART-AMP, the inner primers recognize an additional site within the amplicon, thereby 'fold back' onto the amplicon, which initially creates a dumbbell shaped intermediate and ultimately cauliflower like high MW products. Additional loop or booster primers can be used to speed up isothermal amplification. SMART-AMP further enables SNP genotyping via use of the MutS protein. LAMP is a well-established isothermal amplification method; however designing assays for new targets can be complex.

Cross priming amplification (CPA) (Fang *et al.*, 2009; Yulong *et al.*, 2010; Xu *et al.*, 2012) is a more recent amplification methodology, in some respects similar to LAMP, but with simpler oligonucleotide primer designs. Double crossing CPA involves two crossed inner primers (Fig. 10.3), each with a 5′ sequences identical to the other's priming site. The final product at the end of amplification is a mix of single strands with various secondary structures, branched DNA molecules and double stranded DNA. In single crossing CPA, only one primer is crossed, which reduces the size of the CPA products generated and is thought to better facilitate probe-based detection. CPA has further been demonstrated to amplify RNA targets from viral genomes, provided that the RNA target is first converted to complementary DNA via a reverse transcriptase (RT). The RT component of RT CPA can be performed simultaneously in the same reaction, at the same temperature, and without the need for increased incubation time.

Many isothermal amplification methods use polymerase extension in conjunction with a single-strand cutting event (Fig. 10.4). Strand Displacement Amplification (SDA, Becton Dickinson) (Hellyer and Nadeau, 2004; Mchugh *et al.*, 2004), Nicking Enzyme Amplification Reaction (NEAR, Ionian Technologies) (Maples *et al.*, 2009), and the Nicking Enzyme Mediated Amplification (NEMA, Ustar Biotechnologies) (You *et al.*, 2006) all use similar reaction schemes (Fig. 10.4A), however with important variations. In each case, an intermediate amplicon is generated from the targeted genomic DNA using primers with a 5′ overhang containing the recognition sites for a restriction or nicking endonuclease. This intermediate amplicon contains restriction sites near its 5′ and 3′ ends. SDA and NEMA use bumper primers to facilitate this initial step. NEAR does not use bumper primers, plus it generates shorter amplicons than SDA and NEMA. Next, a single strand cutting event generates nicks in each strand of this intermediate amplicon near its 5′ end. NEAR and NEMA use nicking endonucleases that are inherently single strand cutting. SDA uses a double strand cutting restriction endonuclease in conjunction with phosphorothioate nucleotides that are incorporated into the strand opposite to the primer during polymerization, therefore only the primer strand is cuts by the restriction enzyme. A strand displacing polymerase then extends the nicked strands and 'peels off' the downstream DNA, such that the two halves of the intermediate amplicon separate. The shorter amplicon size used in NEAR also enables thermal dissociation, which facilitates rapid amplification. This nicking and extension process repeats, leading to linear amplification of a shorter DNA sequence which can then be re-primed with the opposite primer

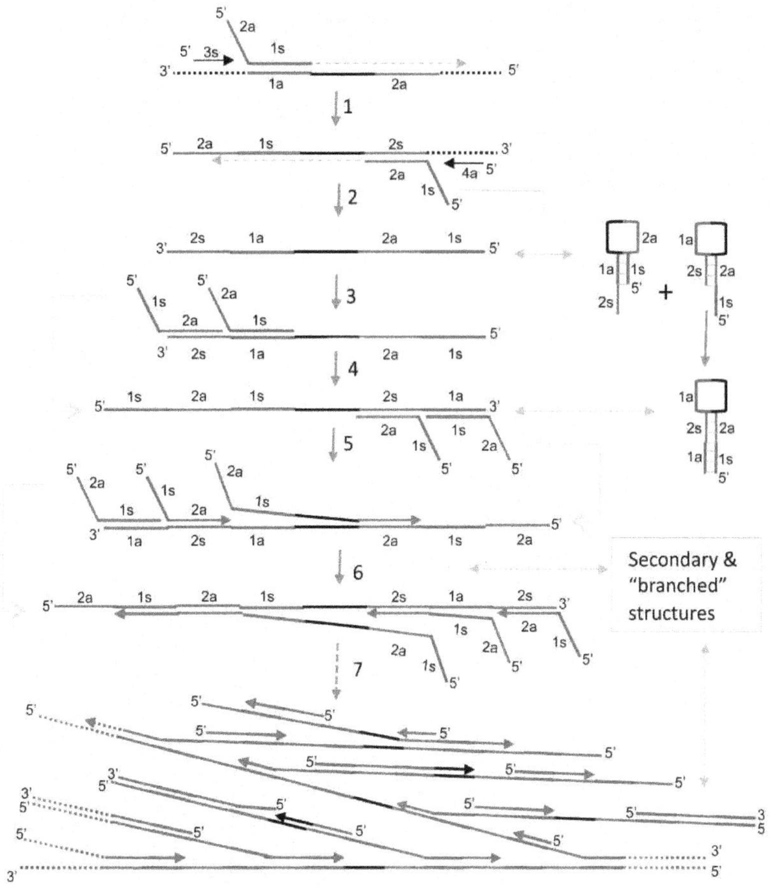


**Figure 10.3** Schematic representation of double crossing CPA: The 3′ end region of the sense cross primer (1s) binds to its complement (its antisense, 1a) on the target DNA and primer extension is facilitated by *Bst* DNA polymerase. The extension from primer 3s displaces the sense cross primer strand (Step 1). In Step2 the same events occur on the antisense strand via the antisense cross primer (2a-1s), its complement 2s and the displacement primer (4a). The products of steps 1 and 2 can exist either as single stranded DNA or as hairpin structures. The 3′ end of the hairpin antisense cross primer product is extended to complement the sense primer region. The initial CPA products made in the first round of double CPA have complementary sequences to both the sense and antisense primers (steps 3 and 4). This results in larger amplicons that contain further complementary sites to both double cross primers. These facilitate yet more cross primer binding to further amplify from the target amplicons in addition to displacing shorter single stranded amplicon products derived from earlier rounds of amplification and these then serve as new templates for yet more amplification (steps 5 and 6). Continuation of the cross priming amplification process results in secondary and branched and other structures of CPA products (steps 6 and 7). Image courtesy of Ustar Biotechnologies.

to generate new intermediate amplicons that feed back into the reaction, thereby leading to crossed exponential amplification. Isothermal Chain Amplification (ICA; RapleGene Inc.) (Jung *et al.*, 2010) uses an analogous reaction scheme (Fig. 10.4B). However, ICA uses DNA-RNA-DNA chimeric primers rather than restriction endonucleases. Following the initial primer extension, an RNase H included in the reaction degrades the RNA sections of the intermediate amplicon, which creates two recessed 3′ hydroxyls that are extended analogous to the nicked strands in the other reaction schemes.

The exponential amplification reaction (EXPAR; Fig. 10.5) amplifies short oligonucleotides at 55°C (Van Ness *et al.*, 2003; Tan *et al.*,

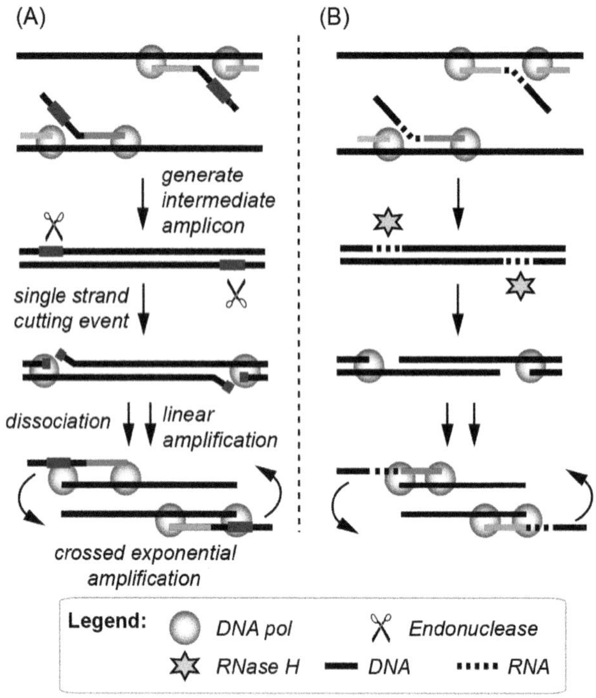


**Figure 10.4** Schematic overview of isothermal amplification methods based on polymerase extension in conjunction with a single-strand cutting event: (A) general reaction scheme employed in Strand Displacement Amplification (SDA), Nicking Enzyme Amplification Reaction (NEAR) and Nicking Enzyme Mediated Amplification (NEMA). (B) general reaction scheme of the Isothermal Chain Amplification (ICA).

2005, 2007). These short trigger oligonucleotides can be generated from adjacent nicking enzyme recognition sites in genomic DNA via the so-called Fingerprinting reaction (Tan *et al.*, 2007). Exponential amplification is facilitated by a template oligonucleotide that contains the reverse complement of a nicking-enzyme-recognition-site plus the reverse complement of the trigger, repeated twice at the 3′ and 5′ end. The trigger oligonucleotide primes the template, and is extended by a polymerase. A linear amplification cycle creates new copies of the trigger oligonucleotide, which activate new templates, thereby enabling exponential amplification.

## Detection

Nucleic acid detection methods fall into two main categories: (1) end-point detection, as with microarrays, chromatographic assays (NALF in antibody dependent or independent formats), or using visual tube inspection, and (2) real-time detection based e.g. on fluorescence or turbidimetry. In all cases, detection methods that can differentiate target from non-target amplicons, and that enable at least a duplex readout, to accommodate detection of target and IAC co-amplified in the same reaction, are preferable.

### End-point detection

Lateral flow (LF) rapid diagnostic tests (RDTs) have long been at the forefront of POC IVD detection schemes, particularly in low resource settings, due to their simplicity, relatively low costs, and non-instrumented format (Posthuma-Trumpie *et al.*, 2009). Simple lateral flow assays are generally qualitative. More sophisticated assay schemes and use of a reader also enable semi-quantitative and quantitative analyte detection, albeit with limited dynamic range. For serological tests that measure host antibodies against a targeted pathogen after seroconversion, e.g. employed for HIV diagnosis (Schito *et al.*, 2010), the sensitivity of simple RDTs is usually sufficient. However, for assays that require antigen detection, the clinical

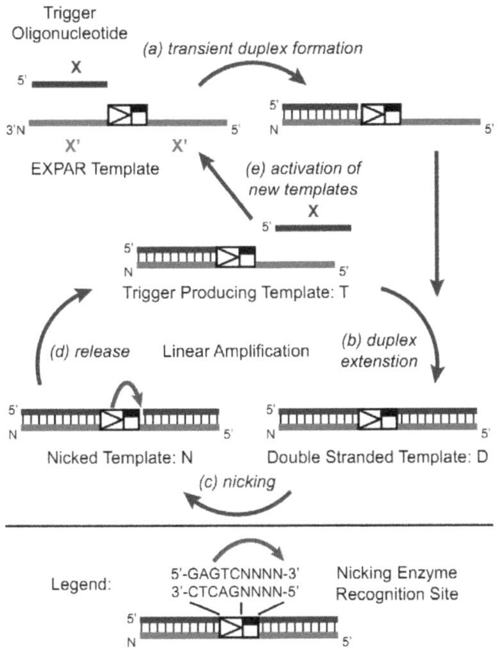


**Figure 10.5** Overview of the exponential amplification reaction EXPAR: (a) Trigger X transiently binds to the complementary recognition sequence at the 3′-end of the amplification template; (b) the trigger sequence is extended by the DNA polymerase, forming the double stranded nicking enzyme recognition site 5′-GAGTCNNNN-3′ on the top strand; (c) the top strand is cleaved through the nicking endonuclease Nt.BstNBI; (d) at the temperature of the reaction (55°C), the newly formed trigger is released from the amplification template. The trigger-producing form of the amplification template re-enters the linear amplification cycle, and new trigger oligonucleotides are generated through duplex extension, nicking, and release; (e) the newly formed trigger oligonucleotides activate additional template sequences, giving rise to exponential amplification of the trigger X.

sensitivity of standard RDTs is often times lacking (Yager *et al.*, 2008). The sensitivity of LF devices can be improved by using fluorophore-, europium chelate-, or up-converting phosphor- containing particles. However, such enhancers require a reader for detection. In nucleic acid lateral flow (NALF), the sensitivity is typically provided through upstream nucleic acid amplification, which readily generates nanomolar concentrations of the labelled target. Therefore, regular coloured conjugates with a visual readout are usually sufficient for NALF. However, NALF using up-converting phosphors and a simple sandwich hybridization format can detect low attomolar concentrations of pathogenic DNA (Corstjens *et al.*, 2003). Pathogens present in clinical samples at relatively high concentrations, such as the human papillomavirus (HPV), can be diagnosed by this up-converting phosphor NALF approach without target amplification (Corstjens *et al.*, 2001).

NALF detection can be coupled to PCR (Dineva *et al.*, 2005; Mens *et al.*, 2008), NASBA (Mugasa *et al.*, 2009), helicase dependent amplification (HDA) (Goldmeyer *et al.*, 2008), RPA (Piepenburg *et al.*, 2006), LAMP (Puthawibool *et al.*, 2009), cross-priming amplification (CPA) (Fang *et al.*, 2009), and other isothermal amplification methods. However, some methods are easier to couple to NALF than others, depending upon the complexity of the amplification scheme and the amplicon produced. NALF requires handling of amplified test reactions, and therefore introduces significant risk of carryover contamination with amplicon DNA. NALF-based devices therefore need to couple the transfer of amplified mastermix to the strip in a fully enclosed device to minimize carryover contamination (Goldmeyer *et al.*, 2008; Fang *et al.*, 2009; Yulong *et al.*, 2010).

Lateral flow tests typically include a control line to ensure that the lateral flow test itself has performed correctly. NALF enables low levels of multiplexing, therefore can be used to detect several pathogens simultaneously (Dineva *et al.*, 2005), include detection of an IAC. A NALF configuration with three lines to detect the target (T), IAC, and the regular NALF hybridization control (HC) is shown in Fig. 10.6.

There are antibody-independent NALF schemes (Fig. 10.6B), which involve the hybridization of amplified products with oligonucleotide-conjugated colorimetric particles, and capture probes immobilized onto a nitrocellulose or nylon membrane. In antibody-dependent NALF schemes (Fig. 10.6A), the incorporation of a hapten such as digoxigenin, biotin, or fluorescein (FAM) into the amplicon is required. Although not an antibody, the streptavidin – biotin interaction can be used in a similar manner. Steric hindrance can be a concern, especially for large amplicons with secondary structures generated e.g. in LAMP and CPA. A specific reaction

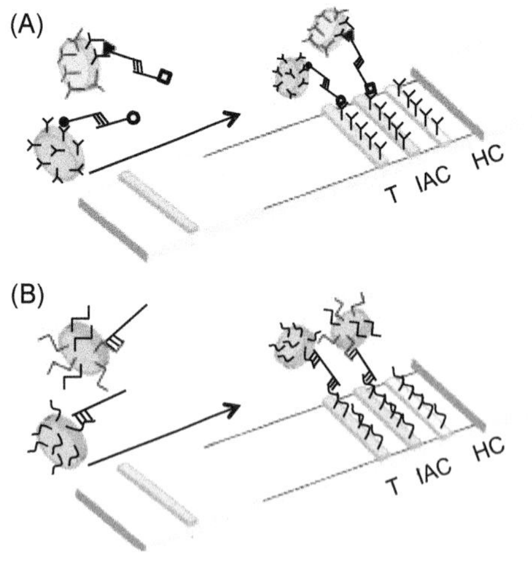


**Figure 10.6** Nucleic Acid Lateral flow with internal amplification control (IAC) and hybridization control (HC) capture lines. (A) Antibody-dependent NALF utilizes antibody-conjugated coloured detection particles and a test strip impregnated with a capture moiety, in conjunction with a dual hapten-labelled target amplicon in order to obtain a positive signal at the test line (T). The hapten labels are introduced via labelled oligonucleotide primers or probes that contribute to amplicon production (B) Antibody-independent NALF involves direct hybridization of unlabelled target amplicon with oligonucleotide-functionalised coloured particles and with oligonucleotides deposited on the LF membrane. The IAC in both scenarios is detected in the same manner; however, a different oligo sequence and/or hapten is detected. The HC detects the antibody or oligo conjugated to the coloured particle to demonstrate sufficient assay performance.

sequence therefore has been incorporated into CPA (Fig. 10.7), which generates small, discrete, dual labelled amplicons for detection.

Several end-point detection formats based on visual tube inspection have been reported for LAMP. These approaches minimize the risk for amplicon carryover, since the reaction tube remains sealed after incubation. The simplest approach relies on changes in turbidity of the reaction solution as nucleotides become incorporated during DNA or RNA polymerization, which occurs as the pyrophosphate ions released into the buffer precipitate in the presence of metal cations such as $Mn^{2+}$ (Mori and Notomi, 2009). When calcein is incorporated into the mastermix, turbidity is enhanced via fluorescence (Boehme *et al.*, 2007). Colorimetric detection, which avoids the need for a UV lamp, can be achieved through the addition of hydroxy naphthol blue (Notomi *et al.*, 2000; Wastling *et al.*, 2010). In a recent review of TB diagnostics, it was noted that the use of non-specific colorimetric indicators with LAMP were not ideal, as only a monoplexed reaction can be performed without a concurrent assay to demonstrate either reagent efficacy, specimen quality or both (McNerney and Daley, 2011). To enable low levels of multiplexing, precipitates of different colours can be obtained through addition of cationic polymers in conjunction with fluorescently labelled primers and probes (Mori *et al.*, 2006). Another approach to LAMP multiplexing has been reported enabling simultaneous detection of four targets in one reaction (Liang *et al.*, 2012). In this approach, the forward inner primers were modified to contain a nicking enzyme recognition site and a short nucleotide sequence encoding each target. The amplified mastermix was digested using a nicking enzyme, and the different bar codes and therefore amplicons were distinguished through pyrosequencing.

Microarrays enable highly multiplexed end-point detection of pathogen panels, or of drug resistance markers. However, conventional microarray detection procedures require complex multi-step procedures that are unsuited for POC applications. Systems have been developed to automate the fluidics and thermal control during hybridization, stringency washing, and the optical readout, but these systems tend to be relatively complex and expensive, and more suited for centralized laboratory settings. Nanosphere Inc. has developed the Verigene platform, a microarray system for multiplexed pathogen detection without polymerase-based target amplification (Storhoff *et al.*, 2004). In the Verigene platform, targeted nucleic acids hybridize to a microarray slide in a sandwich format. The detection probe is conjugated to a gold nanoparticle, which then catalyses silver reduction. The slide is further used as a wave guide: for detection, light is coupled into the slide from the side, and the reader measures evanescent light scattering using relatively simple and robust optics. Several other microarray formats have been developed that may be suitable for NAT outside of a centralized laboratory. Great Basin has coupled isothermal blocked

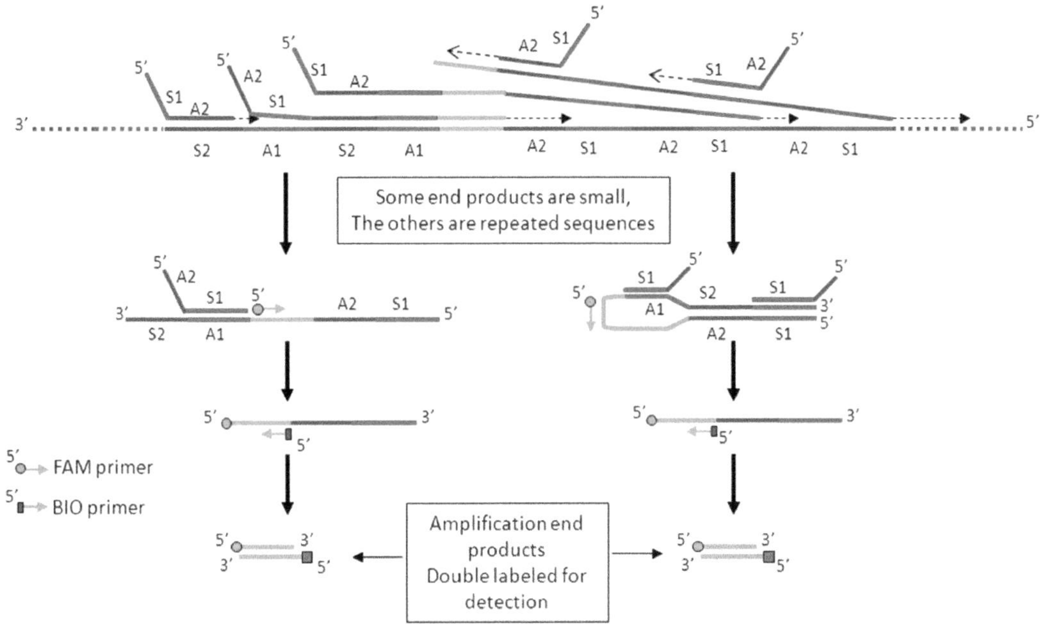


**Figure 10.7** Schematic diagram showing how dual hapten-labelled products are generated during CPA for subsequent NALF detection. The FAM and biotin-labelled primers complement different parts of the targeted DNA sequence on opposing strands, and not the introduced primer sequences, to ensure specificity. The process generates double labelled small amplicons ideal for NALF detection. Image courtesy of Ustar Biotechnologies.

primer HDA with array-based end-point detection. Biotin labelled amplicons are hybridized to a capture probe array, and are then detected via horseradish peroxidase mediated signal amplification (Hicke *et al.*, 2012). ClonDiag (now part of Alere) has developed a miniaturized 'array in a tube' format for multiplexed pathogen identification and drug susceptibility testing (Monecke and Ehricht, 2005; Perreten *et al.*, 2005; Batchelor *et al.*, 2008). Based on technology developed at the Argonne National Laboratory (USA) and Engelhardt Institute (Russia), Akonni has developed a field deployable microarray system, called the TrueArray or TrueDiagnosis platform, which employs a microarray functionalised with a 3D hydrogel matrix. The system has been applied for environmental monitoring of microbial communities in groundwater (Chandler *et al.*, 2010) and, clinical diagnostic applications are under development. On chip isothermal amplification coupled with microarray-based fluorescent readout has been developed for multiplexed pathogen detection (Andresen *et al.*, 2009).

## Real-time detection

Detection of amplicons in real-time, as the reaction proceeds, is superior for quantitative assays that require a wide dynamic range, but typically has involved more expensive and complex equipment than simple end-point methods. For real-time detection methods, the amplification tube typically remains sealed after amplification to prevent amplicon contamination of the test site. The most common real-time detection approach involves real-time PCR, also called qPCR. Low levels of multiplexing can be accomplished via qPCR. For example, the Cepheid GeneXpert enables detection of up to six targets per reaction, through detection probes labelled with different fluorophores that can be resolved through the six-channel optical system (Helb *et al.*, 2010). Microarray formats enable higher levels of multiplexing, which is desirable to detect panels of pathogens or drug resistance mutations. Multiplexed qPCR in a microarray format can be achieved by splitting the amplification mastermix containing the sample into an array of separate microfluidic amplification chambers, each containing primers and probes for a specific target (Morrison *et al.*, 2006; Spurgeon *et al.*, 2008; Poritz *et al.*, 2011). Mastermix splitting and the small volumes present in each chamber can negatively affect the attainable LOD, unless the target is amplified prior to splitting, e.g. through the first round of a nested PCR reaction (Poritz *et al.*, 2011). Recently, ClonDiag/Alere have reported array-based multiplexed qPCR in one reaction chamber without mastermix splitting, through Competitive Reporter Monitored Amplification (Ullrich *et al.*, 2012). In this reaction, array capture probes and amplicons formed in solution compete for binding to target specific fluorescent reporter probes present in the mastermix. As the amplicon concentration increases in the mastermix, the amount of fluorescence measured on the array decreases, enabling quantitative detection of target with a six log dynamic range.

Most isothermal methods can also be monitored in real-time. Real-time fluorescence detection for PCR or isothermal amplification can be mediated by intercalating dyes (Higuchi *et al.*, 1993; Zipper *et al.*, 2004), by oligonucleotide probes that are cleaved during the reaction (Livak *et al.*, 1995), or using probes or primers that change conformation upon amplification of target (Piatek *et al.*, 1998). Real-time detection can also be accomplished based on pyrophosphate ions generated by the reaction, either using turbidity (Noren *et al.*, 2011; Saetiew *et al.*, 2011), fluorescence (Tomita *et al.*, 2008), or a bioluminescent readout, by coupling with luciferase mediated light emission (Gandelman *et al.*, 2010). Recently, a novel scheme for labelling LAMP primers to facilitate multiplexed real-time analysis of up to four multiplexed reactions in a single tube was described (Tanner *et al.*, 2012). The method uses adapted standard LAMP primers that contain a quencher-fluorophore duplex region, which is separated during the target amplification process, creating a fluorescent signal.

## Case studies

The following section presents selected examples of POC NAT systems in development or on the market, with varying levels of integration, ranging from methods that utilize manually performed and separate sample preparation, amplification and detection, to fully integrated systems.

### Stand-alone sample preparation devices

#### OmniLyse/PureLyse (Claremont BioSolutions)

For mechanical lysis of difficult to lyse pathogens, Claremont BioSolutions has developed a miniaturized, battery operated, disposable bead beating system called OmniLyse™ (Doebler *et al.*, 2009). This system utilizes mechanical shear forces to effectively and rapidly lyse hard to lyse microorganisms such as mycobacteria and *Bacillus anthracis* spores (Vandeventer *et al.*, 2011), *Clostridium difficile* or *Giardia* cysts (unpublished data). The PureLyse™ system combines the OmniLyse™ technology with a nucleic acid solid phase capture and extraction protocol that does not introduce inhibitory compounds during the extraction step. These battery-operated, miniaturized, and disposable devices (Fig. 10.8A) are amenable for near patient NAT, and can also be integrated into other systems.

#### TruTip™ technology (Akonni)

Akonni has incorporated SPE of nucleic acids into a filtered pipette tip format (Fig. 10.8B). Using these TruTips™, it is possible to isolate and concentrate PCR-amplifiable nucleic acids in less than 4 minutes. This system has been used in the field for environmental monitoring (Chandler *et al.*, 2010). The TruTip™ technology utilizes a proprietary nucleic acid binding matrix inside the pipette tip, in conjunction with a series of lysis/binding (guanidine thiocyanate), wash (acetone), and elution (Tris) buffers. TruTip™ extraction kits are available for microbial DNA, mycobacteria, human genomic DNA from blood or saliva, microbial RNA, and influenza RNA. Tips of various porosities can be selected depending upon sample viscosity. TruTips™ can be used manually, or can be incorporated into automated dispensing systems.

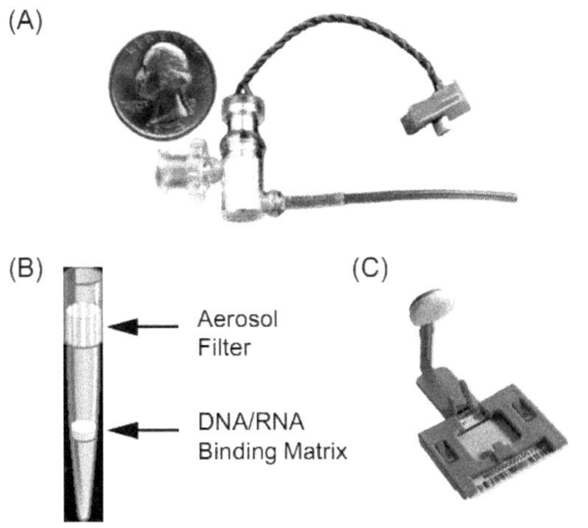


**Figure 10.8** Stand-alone Sample Preparation Devices. (A) OmniLyse™/PureLyse™ cartridge (Claremont BioSolutions) for rapid mechanical pathogens lysis and nucleic acid extraction (Courtesy of Claremont Biosolutions). (B) TruTip™ (Akonni) technology principle for nucleic acid extraction and concentration in filter tips (Adapted from http://www.akonni.com/docs/How_TruTip_Works.pdf). (C) Whatman's EasiCollect™ device for buccal cell sample collection, cell lysis, and nucleic acid extraction via an impregnated FTA filter card (http://www.whatman.com/EasiCollect.aspx).

## EasiCollect FTA™ card (Whatman)

This device (Fig. 10.8C) incorporates a swabbing arm to collect buccal cells from a patient. The swabbing arm is then pressed onto an FTA card incorporated into the device. An indicator dye reflects the area on the card to which the biological sample has been applied. Once the sample has dried, it can be processed immediately or packaged for transportation or storage. Reagents in the FTA card ensure the degradation of proteases, DNases, and RNases, to create a lysed and stabilized sample that presents no biosafety risk. To process the sample, a 2 mm micropunch is removed from the card, and then washed multiple times with different buffers. PCR mastermix reagents can be subsequently added directly to the disc for nucleic acid amplification.

Sample preparation may vary depending upon sample type and target pathogen. For example, due to their greatly reduced size, viral genomes are not effectively trapped in the paper matrix and many are released into the liquid phase during the initial wash stage (Rogers and Burgoyne, 2000). It is recommended to concentrate the viral RNA from the wash buffer, but this increases the complexity of sample processing. Alternatively, the retention of viral genomes within the paper can be promoted by adding high salt concentrations to the wash buffer (Rogers and Burgoyne, 2000).

## Loopamp® PURE DNA extraction kit (Eiken Chemical Company)

In May 2011, Eiken released the Loopamp® Tuberculosis Complex Detection Reagent Kit, which includes a sputum sample preparation and DNA extraction component called the Loopamp® PURE DNA Extraction Kit. The acronym 'PURE' stands for Procedure for Ultra Rapid Extraction. The product and protocol was described by Mitarai *et al.*, who recently performed a laboratory-based evaluation of this kit on clinical samples (Mitarai *et al.*, 2011). The assay targets both *gyrB* and *IS6110*. The extraction kit involves three interlocking components that are sequentially joined together during the extraction process; a heating tube, an adsorbent tube, and injection cap. A 40 μl volume of either raw or processed sputum sample is inserted into the heating tube, where it is heated (in the presence of NaOH) at 90°C for 5 minutes and then cooled to ambient temperature. The heating tube is then inserted into the top of the adsorbent tube and the contents of both tubes are mixed to reduce the pH of the sample and to remove inhibitory compounds that may affect the LAMP assay. The active ingredient in the adsorbent tube contains a zeolite powder (an aluminosilicate). A blue dye in the zeolite powder indicates to the user when the powder is fully mixed. An injection cap is then attached to the bottom of the adsorbent tube. By squeezing the sides of the adsorbent tube, the liquid sample passes through porous material, and the DNA-containing filtrate slowly drips out of the bottom directly into a reaction tube where the cap contains the LAMP TB assay mastermix. The performance of the Loopamp® Tuberculosis Complex Detection Reagent Kit was comparable to PCR-based assays when testing unprocessed sputum. Pretreatment of sputum using *N*-acetyl cysteine and NaOH resulted in poorer performance of the test kit. The authors noted that the volume of specimen placed into the LAMP assay was much less than the other NAT assays used as comparators [the Cobas Amplicor MTB test (Roche) or TRC Rapid MTB test (Tosoh Bioscience)] and therefore samples with fewer target bacteria processed were also less likely to be detected by this method.

### Ustar manual DNA preparation kit

Ustar Biotechnologies are developing manual nucleic extraction systems to purify DNA from clinical specimens. This system consists of disposable, low cost, and simple-to-use components. The first component (Fig. 10.9A) is a stool sample collection and pathogen lysis device, the second component (Fig. 10.9B) is a nucleic acid purification device attached to a syringe barrel. The stool collection and lysis device contains lysis reagents, and has screw caps at either end of the housing cylinder, which prevents leakage of any biohazardous materials once the specimen has been inserted into the lysis reagent. One cap has a small spatula attached that can be used to collect and transfer stool specimen into the device. Once the sample has been added to the lysis reagents, it is mixed by hand and incubated at elevated temperatures to optimize chemical lysis of cells. For the extraction of DNA from less robust pathogens, incubation at ambient temperatures is adequate for sufficient chemical lysis. By unscrewing the cap at the other end of the lysis device and squeezing the device walls, the lysate is aliquoted directly into the DNA purification device. A coarse filter at the orifice prevents the transfer of specimen debris to the purification device. The purification method is based on NA affinity membrane filtration, in conjunction with a syringe unit. The lysed material is passed over the affinity membrane using the syringe plunger. The captured NA is then washed via a series of wash buffers that maintain a low pH and appropriate ionic strength to ensure maximal NA binding, while removing excess salts and detergents. Unlike many other silica-based methodologies, this technology does not use organic solvents in the wash buffer such as chloroform or ethanol as these are either hazardous chemicals or require careful removal prior to the elution of the nucleic acids not to inhibit downstream target amplification. The DNA can be eluted directly into reactions by using a low salt buffer with elevated pH.

To extract TB DNA, the current Ustar methodology requires a centrifuge to pellet and wash cells followed by chemical and heat lysis of cells. An aliquot of cooled lysed material can then be added to the glassified CPA reaction to amplify TB DNA.

(A) (B)

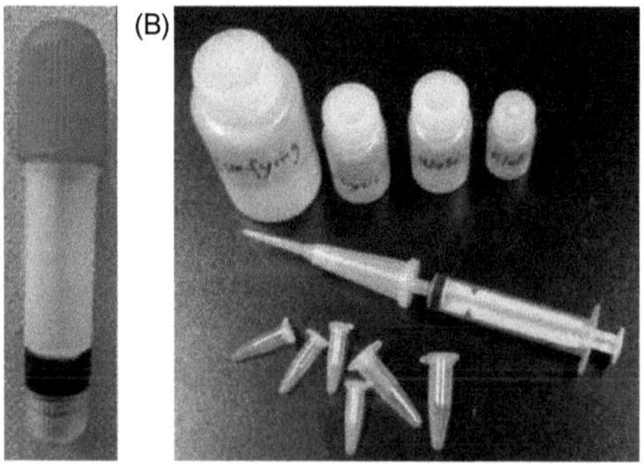

**Figure 10.9** The Ustar manual DNA extraction kit components. (A) The stool sample collection and lysis device houses the lytic reagents. Specimen for lysis can be added via the specimen port capped by a red screw cap containing a small spatula. The other end houses a micromesh filter (black component) to prevent the transfer of large particulates into the purification device. The sample is eluted after lysis into a syringe barrel by opening the black capped port at the bottom of the device and squeezing the tube. The DNA affinity membrane unit, attached to a syringe, is shown (B). The lysis reagents are passed through the membrane using a syringe, as are sequential washes and elution using the same syringe.

### Immiscible phase nucleic acid purification (Northwestern University and others)

Sur *et al.* described an elegant protocol for the extraction and purification of nucleic acids (Sur *et al.*, 2010), which uses chaotropic salts but does not require wash buffers, unlike other extraction methods based upon Boom technology (Boom *et al.*, 1990). This method may represent an improvement over other silica-based SPE methods where there is stringent requirement to wash the beads. Carryover of the chaotropic salts and/or the organic solvent (typically ethanol) used in the wash buffer into the amplification reaction will prevent amplification. Therefore multiple wash step or centrifugation and drying of the silica prior to elution are necessary with other methods. This adds cost, complexity and time to the extraction procedure.

The device consists of a specially designed disposable cartridge with two chambers, one containing a chaotropic salt-based lysis buffer, the other containing a low salt elution buffer, separated by a layer of liquid wax (Fig. 10.10). The specimen is lysed in the chaotropic salt solution and the nucleic acids bind to silica coated paramagnetic beads present in the chamber. Through an externally applied magnetic field, the paramagnetic silica beads are slowly moved from the lysis buffer chamber through the wax phase and finally into the elution chamber. Passing the beads through the liquid wax layer removes inhibitors associated with the specimen and residual lysis buffer from the beads, before the nucleic acids are eluted upon immersion into the low salt buffer chamber. While this technology is not yet commercially available, it has been announced that Quidel Inc. (San Diego, USA) have entered a partnership with Northwestern University to develop a novel fully integrated nucleic acid test format, currently dubbed 'wild cat', that will incorporate this sample preparation method in its overall design (http://www.genomeweb.com/pcrsample-prep/quidel-northwestern-u-developing-automated-sample-result-mdx-platform).

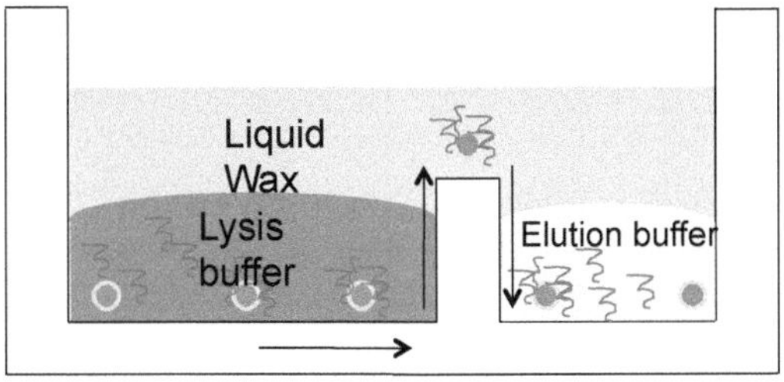


**Figure 10.10** A schematic diagram showing the principal components behind the immiscible phase nucleic acid purification method. Silica coated paramagnetic beads bind nucleic acids in a high salt lysis solution, the beads are captured via a magnetic field and dragged out of the lysis mixture and through an immiscible wax matrix that removes confounding substances. The beads are deposited in the low salt buffer whereby the nucleic acids are eluted from the silica beads and are ready for further manipulation. Adapted from (Sur *et al.*, 2010).

More recently other groups have reported specimen processing systems that expand upon the surface tension principle, and streamline the technology by utilizing linear tubular systems. One embodiment uses an air–liquid immiscible interface (Bordelon *et al.*, 2011) and another a liquid-immiscible interface (Strotman *et al.*, 2012). Both systems use analogous principles: after initial lysis of the sample, the released NAs bind to silica coated magnetic beads. These beads can then be drawn via a magnet through washing stages to remove the lysis materials and buffers to ultimately produce purified NAs in a matter of seconds without the need for centrifugation and addition/aspiration of the wash buffers. Both methods also permit the discrete purification of other specifically targeted cells or molecules as intended for further analysis via appropriately derived capture moieties on the magnetic beads.

### Filtration isolation of nucleic acids (FINA, Northwestern University) and Pro-viral DNA Capture Card (PDCC, Program for Appropriate Technologies in Health (PATH))

The rapid diagnosis of HIV infection is essential for the effective treatment of infants that have contracted HIV from their mothers. Infant HIV diagnosis from whole blood involves detection of pro-viral DNA present in HIV infected T- lymphocytes. However, some components of red blood cells are inhibitory to subsequent detection via PCR (Al-Soud and Radstrom, 2000, 2001; Radstrom *et al.*, 2004), therefore further specimen preparation is required, adding to test complexity, cost, and the time to result. The FINA system, developed at Northwestern University (Jangam *et al.*, 2009), and the PDCC system, developed at PATH, complement the current infant HIV test algorithm that utilizes dried blood spots as a specimen type, while removing many of the downstream specimen processing steps. Both methods describe the use of a thin porous membrane that captures the larger peripheral blood mononuclear cells (PBMCs) while the smaller red blood cells (RBCs) pass through the membrane. Remaining trace amounts of RBCs are further removed with the addition of a wash buffer (Fig. 10.11). These methods have the added advantage of concentrating HIV infected cells, which can potentially improve the sensitivity of HIV infant diagnosis, compared to current methods that purify DNA from either whole blood or dried blood spots. The PBMCs coated membrane can either be immediately transferred to the reaction tube for PCR analysis, or dried for testing at a later date. Both PBMCs capture methods involve very low cost reagents and a simple methodology that may improve upon current specimen collection methods without being disruptive to the current collection protocols using DBS (i.e. extensive retraining of the users is not necessary). Neither tool is currently commercially available, although proof of concept has been demonstrated with both devices.

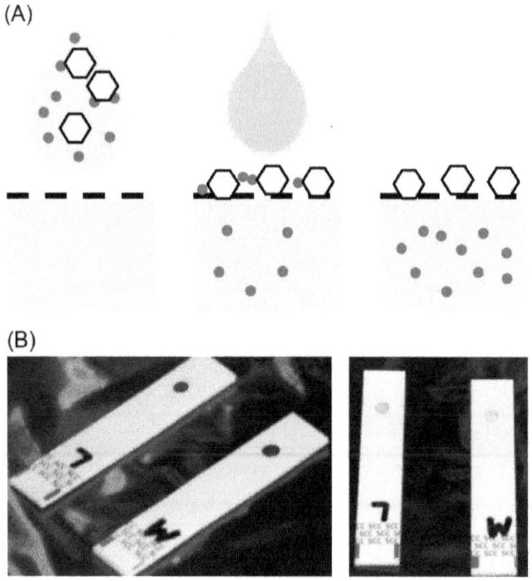

**Figure 10.11** The proviral DNA Capture Card (PDCC) developed at PATH (A) schematic representation of its use: A drop of blood from an infant heel prick is placed on the surface of a membrane. The red blood cells (RBCs) wick though the pad while the peripheral blood mononuclear cells (PBMCs) are captured on the membrane. Subsequent addition of a wash buffer removes remaining RBCs from the surface. The pad can be cored out and placed directly into the PCR tube for amplification. (B) Image of the PDCC, left: after application of a drop of blood; right: after washing. Image produced with permission of PATH.

## Manual extraction of nucleic acids (PATH)

To monitor the effectiveness of anti-retroviral treatment (ART) in suppressing HIV infection, patients are assessed via CD4 T-cell lymphocyte counts and if possible viral load tests (VLT). VLT adds value both to patient management and for public health in terms of detecting early and true ART failure, mitigating drug resistance, and averting new infections (Hamers *et al.*, 2011; Novitsky and Essex, 2012). An investigation into the costs incurred in screening ART recipients revealed that VLT is the most expensive component of ART monitoring and that this cost incrementally increases with greater distance from the test laboratory (Stevens *et al.*, 2012). To address this finding, PATH developed a manual RNA extraction kit that permits RNA purification at the collection site. Processing and stabilizing the RNA before transport eliminates the need for cold chain storage, reduces the cost of transport of specimens to the testing laboratory and may improve test performance. The technology is based on the conventional silica column methods utilizing chaotropic salts for lysis and nucleic acid binding follow by washes and elution of the RNA (Fig. 10.12A). All fluids are introduced to the column via a syringe and all components are supplied as a kit (Fig. 10.12B). This technology supports viral detection at ≥ 500 copies/ml from 200 µl of specimen. This is comparable to commercially available VLTs that require a well-equipped laboratory.

## Simple nucleic acid processing (SNAP)

Blood Cell Storage Inc. (BCSI, Seattle WA) have designed a technology than can quickly, easily, and reliably isolate nucleic acids (DNA and RNA) from blood and a variety of other biological specimens using a variation of the method described by Boom *et al.* (1990). The SNAP technology is based on the principle of silica binding of nucleic acids under high salt conditions. However, unlike other methods that use silica coated magnetic particles or silica membranes, the BCSI technology uses two glass plates aligned in parallel to facilitate capture of nucleic acids when the lysis mixture is passed between them (Fig. 10.13). The material used to position and bond the plates together creates a serpentine channel, through

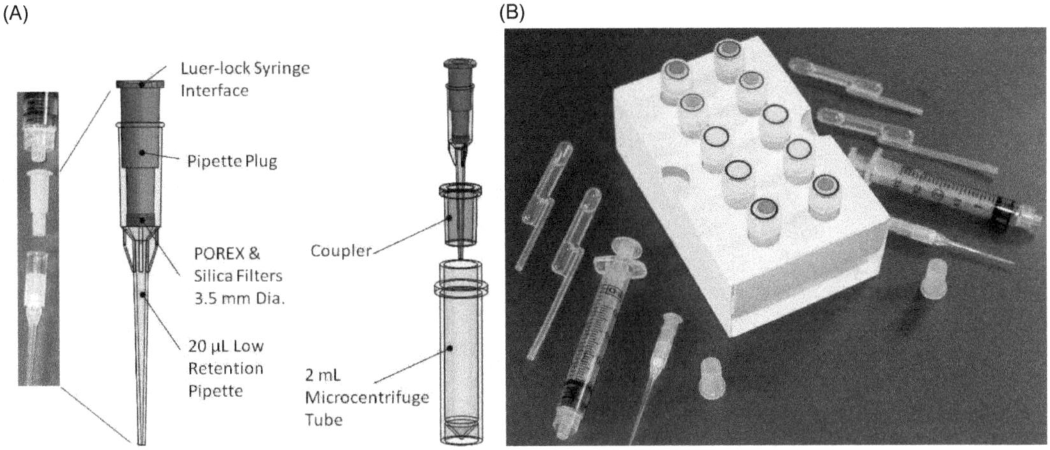


**Figure 10.12** (A) The design of the key components of the PATH field friendly nucleic acid extraction kit. (B) The kits are supplied with all reagents and materials for two specimens. Images produced with permission of G. Domingo, PATH.

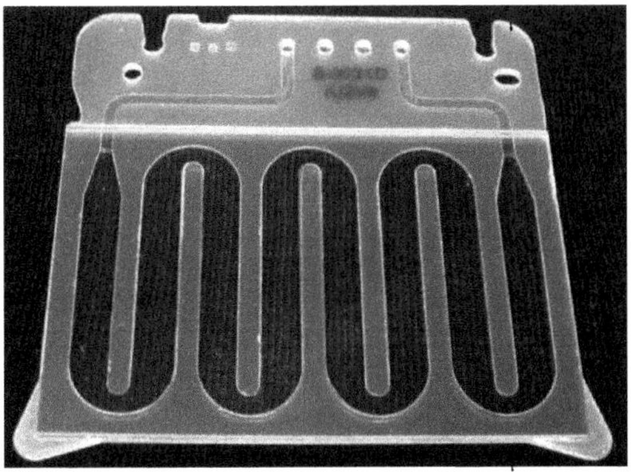

**Figure 10.13** The BCSI SNAP card. The serpentine channel between the glass plates maximizes the surface area available for NA capture while permitting effective washing and removal of processing reagents.

which the reagents flow from an inlet to an outlet, where waste is collected in a separate container. The sample is first lysed in a chaotropic salt buffer and then applied to the device via a syringe to pass between the glass plates. Wash buffers are then applied followed by a low ionic strength elution buffer. The cards have been developed for use with either an automated wash station or more recently for use with manual fluidics applied via a syringe. The effective removal of trace amount of organic solvents from the wash buffer has been demonstrated using compressed air with a three minute drying period. The nucleic acids captured in the cards are stable for several months or may be immediately eluted for further use. The performance of the cards was demonstrated in one study to be similar to an automated magnetic bead-based method using human respiratory specimens spiked with MS2 bacteriophage (Nanassy *et al.*, 2011).

## Molbio sample preparation unit

Molbio Pvt. is a Bangalore-based company that was formed via a collaboration of Bigtec Labs Pvt. (a technology developer) and the Tulip Group (a large diagnostic test manufacturer). The Molbio Uno system consists of reagents, disposables, and pre-calibrated pipettors (Fig. 10.14A–C), with two dedicated instruments, a semi-automated extraction device (the Trueprep MAG™, Fig. 10.14D) and a real-time PCR instrument (the Truelab Uno described later). The extraction instrument is powered via an internal battery or mains electricity and the user chooses the test protocol from a menu of extraction protocols. Specimens are individually processed. Molbio are developing a variety of extraction kits that can be used in conjunction with this device including one for tuberculosis which is currently available. A 1 ml aliquot of the specimen (liquefied if necessary) is pipetted into the lysis tube, which is then inserted into the device. The sample is heated at 85°C for 5 minutes, cooled, and silica coated

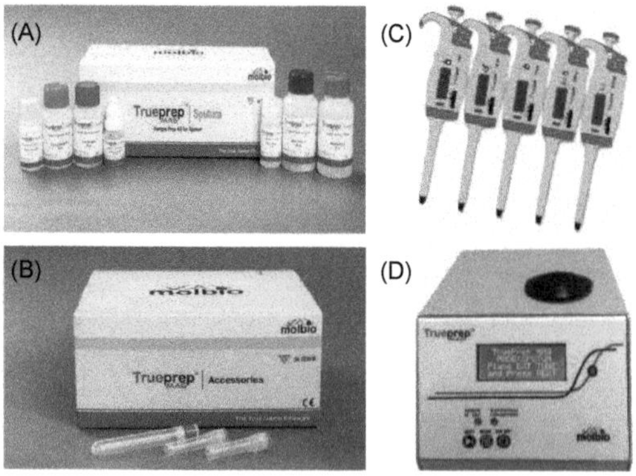


**Figure 10.14** The materials and equipment required to prepare DNA for the Molbio TB assay. (A) All core reagents for liquefaction and DNA purification from sputum are supplied in the sputum kit. (B) The Accessories kit with ancillary tubes for specimen collection, processing and collection of the DNA eluate. (C) Fixed volume pipettors for the transfer of liquid materials and reagents during sample preparation. (D) The Trueprep MAG instrument, with LCD screen user interface. The extraction tube is inserted into the tube holder on top of the device to facilitate sample processing (Text and images reproduced with permission from UNITAID).

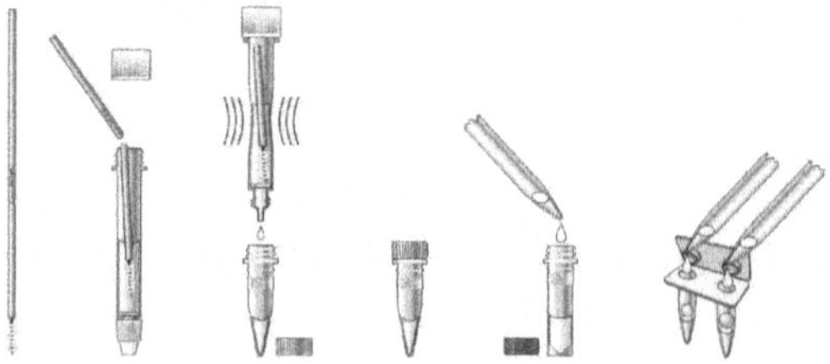

**Figure 10.15** Schematic representation of the Illumigene *Clostridium difficile* assay stool sample preparation process. Adapted from (Meridian, 2011).

magnetic beads are mixed with the sample, leading to capture of nucleic acids on the beads. The operator then sequentially adds/aspirates wash buffers to purify and elute the nucleic acids. The user is prompted to perform each step via an audible alarm in the device. The process takes ~30 minutes to perform and the purified NA can then be used with the PCR instrument.

### Illumigene Stool Sample Preparation System (Meridian)

The FDA cleared, moderate complexity Illumigene *C. difficile* assay involves manual sample preparation of stool samples, as described in this section, followed by isothermal LAMP with turbidimetric detection performed on the IlluMiPro system, as discussed in a later section of this chapter. The sample preparation process is illustrated in Fig. 10.15. A brush, used to collect a stool sample from the patient, is inserted into and broken off inside the so-called Sample Collection Apparatus, which contains sample dilution buffer plus formalin fixed *Staphylococcus aureus* used as internal control. The Sample Collection Apparatus is vortexed to mix the sample with the dilution buffer. After removing the screw cap from the bottom, 5–10 drops of the sample are squeezed through a filter and out of the bottom orifice into a tube, which is then heated to 95°C for 10 minutes to facilitate pathogen lysis. The heat-treated samples are then vortexed again, and a 50-µl aliquot is diluted into the reaction buffer, to further dilute out inhibitors. This reaction mixture is then aliquoted into the 'Test Device', consisting of two amplification tubes, each containing lyophilized LAMP mastermix. Although it has been shown that LAMP can be multiplexed (Liang *et al.*, 2012; Tanner *et al.*, 2012), a duplex LAMP reaction with internal amplification control becomes quite complicated. Therefore, the Illumigene assay uses an external amplification control. Sample prep eluate is split between two reaction tubes, one containing the primers for *C. difficile*, the other for *S. aureus*. The sample preparation process involves seven manual steps, and requires other equipment such as a heat block for the 95 C incubation step, a vortexer, a timer, and precision pipettors. Although this sample preparation process is relatively simple, it may not be appropriate for use in POC settings. This system is designed for use in a moderate complexity centralized laboratory.

## Stand-alone amplification devices

### Non-instrumented nucleic acid amplification (NINA, PATH)

The Non Instrumented Nucleic Acid Amplification (NINA) device under development by PATH is a chemically powered reactor for isothermal assay incubation. This device consists of a thermally insulated body containing calcium oxide, which is activated by adding water (Labarre *et al.*, 2010; Labarre *et al.*, 2011b). A wax-based phase change material (PCM) in contact with the activated calcium oxide modulates then extreme high temperatures generated upon chemical activation through the addition of water. This temperature modulation and the device insulation ensure a stable temperature of 63°C (± 1°C) for over an hour (Fig. 10.16). The tubes containing master-mix are inserted into the region of this phase change material.

This incubation device has been demonstrated to give comparable performance to a conventional PCR machine when using a LAMP assay for the detection of either malarial DNA or HIV-1 RNA (Labarre *et al.*, 2011b; Curtis *et al.*, 2012). A more basic device is described that has similar performance to the NINA heater and which also uses LAMP as the amplification method (Hatano *et al.*, 2010). Although all chemically heated devices described in the literature have used LAMP assays as proof of principle, any isothermal approach may be incubated in this way, provided the incubation temperature of the reaction is greater than the ambient temperature. The incubation temperature of the reaction chamber can be altered by changing the amount or type of exothermic compounds to alter the heat released upon activation of the reaction or by using alternative phase change materials that change the thermal profile after exposure to the heat produced by the exothermic reaction. For example, a suitably adapted NINA heater able to maintain a stable temperature of 55°C has been used successfully to perform isothermal EXPAR coupled NALF detection. Coupled with a simple sample preparation technology, isothermal amplification and visual end-point detection using either NALF, or closed tube turbidimetric/dye-based readout, this device demonstrates that it is possible to perform non-instrumented NAT in the most remote and resource limited settings (Labarre *et al.*, 2011a,b).

## Palm PCR™ (Ahram Biosystems)

Palm PCR™ is a palm-sized portable PCR thermocycler that can amplify up to 2 kb regions within 20 to 39 minutes with high efficiency in a 20 µl reaction volume (Fig. 10.17). The palm-sized unit is powered by a lithium polymer battery, which allows for more than 4 hours of continuous operation, or using line power via an AC/DC adapter. The instrument can process up to 12 samples at once, using disposable PCR tubes specially designed for this system. Although amplification is not monitored in real-time, the system is compatible with other downstream detection methods such as a portable gel device or disposable lateral flow type detectors. The cost of the unit is less

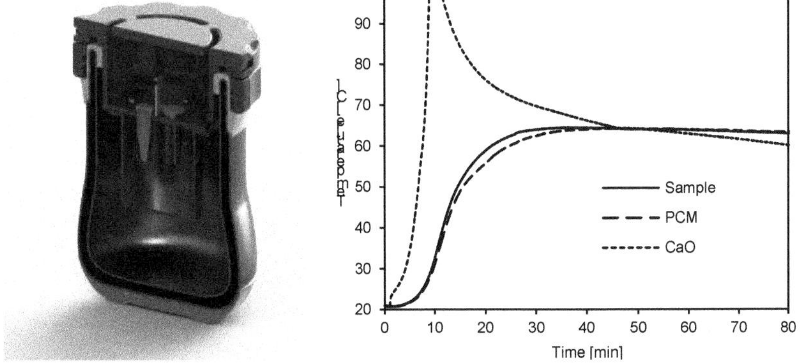

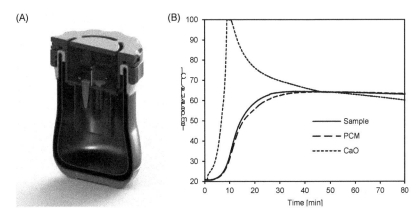


**Figure 10.16** Thermal profiles of the different components in the NINA isothermal incubator upon activation. The activation of the calcium oxide shows a rapid increase in temperature to over 110°C (dotted line). The temperature of the phase change material (dashed line) modulates the temperature in the reaction tube (solid line), which rises from ambient temperature to ~63°C. This temperature is then maintained via the PCM and the insulation of device housing for over 60 minutes. Image courtesy of P. LaBarre, PATH.

**Figure 10.17** PalmPCR™, a portable PCR thermocycler developed by Ahram Biosystems. Up to 12 samples can be amplified simultaneously at the bench or in the field in less than 30 minutes for most standard applications (http://www.ahrambio.com, Image courtesy of Ahram Biosciences).

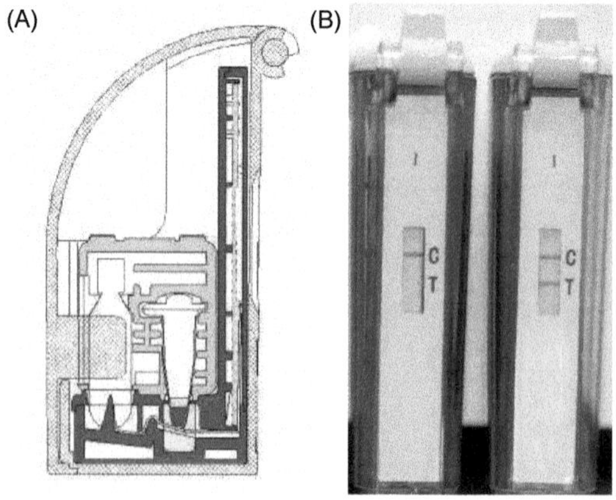


**Figure 10.18** The XCP Nucleic Acid Detection Device (Ustar). (A) schematic cross-section, showing the amplicon cartridge containing a plastic bulb with lateral flow running buffer, and a PCR tube containing amplified mastermix, inserted in the detection chamber holding the lateral flow test strip. (B) The fully enclosed lateral flow cassette for the detection of labelled CPA amplicons. The device on the left shows a negative reaction with the test line (T) remaining blank while the control line (C) indicates appropriate flow and test reagent reactivity for internal quality control of the lateral flow strip. In the device on the right the test line is apparent in addition to the control line indicating a positive CPA test result. Images courtesy of Ustar Biotechnologies and D. Boyle, PATH.

than $6000 with additional expenditures required for PCR reagents and consumables. Palm PCR™ was recently awarded the Instrument Business Outlook (IBO) Silver Design Award of 2012 for innovation in portable analytical instrumentation.

## Stand-alone detection devices

### XCP nucleic acid detection device (Ustar Biotechnologies)/BESt Cassette (BioHelix)

NALF requires the transfer of amplified mastermix to the lateral flow strip, which introduces significant risk of carryover contamination if the amplified mastermix is handled openly. BioHelix (USA) and Ustar Biotechnologies (China) are marketing a fully enclosed disposable device that couples the transfer of amplified mastermix to the lateral flow strip. Ustar Biotechnologies manufacture and market the so-called XCP Nucleic Acid Detection Device in China. The same system is also sold in the USA by BioHelix under the name BioHelix Express Strip (BESt™) Cassette. The system consists of an amplicon cartridge that contains lateral flow running buffer in a plastic bulb, and a detection chamber that contains the antibody-dependent NALF test strip, which detects amplicons dual labelled with biotin and fluorescein (FAM) or with biotin and digoxigenin (DIG) (Fig. 10.18).

For test execution, a standard PCR tube containing amplified mastermix is inserted into the amplicon cartridge next to the running buffer bulb. The amplicon cartridge is closed and inserted into the detection chamber. The detection chamber is then closed using a levered lid, which pushes the amplicon cartridge down into the detection chamber body, where the plastic bulb containing the NALF running buffer and the PCR tube containing amplified mastermix are pierced as the lid snaps shut. The two fluids combine and then run up the lateral flow test strip. After 5–10 minutes, depending on the application, the results can be visually interpreted.

In a preliminary study (Fang *et al.*, 2009), a cross priming amplification (CPA) assay for TB diagnosis coupled with the XCP Nucleic Acid Detection Device was able to detect TB with a sensitivity of 96.9% when compared to smear positive and liquid culture positive samples, which was reduced to 87.5% with smear negative and liquid culture positive samples. With culture negative specimens the specificity was observed to be 98.8%. Screening of 13 non-tuberculosis mycobacteria (NTM) showed the assay had a specificity of 100% for TB. Despite the excellent performance of the amplification and detection method, the authors noted that for this study the DNA extraction process was extensive and required a laboratory for preparation. Ustar Biotechnologies is therefore developing a manual DNA extraction technology, discussed previously, that can complement the CPA TB assay and XCP Nucleic Acid Detection Device, providing a minimally instrumented system for TB NAT. However, sample preparation, amplification and detection are performed manually as separate steps, which make the system relatively labour intensive and prone to error.

In 2011, BioHelix received FDA clearance and moderate complexity designation for the IsoAmp® HSV assay (BioHelix, 2012), which uses the BESt™ Cassette for type-common detection of Herpes Simplex Virus type 1 and 2 from swabs of genital or oral lesions of symptomatic patients. Sample preparation in this case is minimal: the swab is placed into a vial with viral transport medium, and an aliquot of the transport medium is diluted to suitably mitigate inhibition. The diluted sample is then added to a mastermix for isothermal Helicase Dependent Amplification (HDA), incubated for 1Hr at 64°C, followed by visual detection using the BESt Cassette. Despite this ostensibly simple protocol, the IsoAmp® HSV Kit contains 11 different components, and requires the user to provide a heating block, a cooling block, precision pipettors, a timer, a microcentrifuge, a freezer for reagent storage, plus additional disposables. Such an assay may be suitable for moderate complexity centralized laboratories, but would not be appropriate for use at the POC, especially in low resource settings.

#### eSensor (GenMark)

Clinical Micro Sensors (acquired by Motorola, then Osmetech, now called GenMark) have developed a NAT platform called the eSensor to detect and enable genotyping of amplified genetic regions (Pierce and Hodinka, 2012). The system features a modular instrument design, enabling analysis of many different samples in a random access manner. However, sample preparation and amplification have to be performed as separate steps. The eSensor's disposable cartridge contains a fluidic channel that guides amplified mastermix over an array of gold electrodes on a printed circuit board. These gold electrodes are functionalised with a self-assembled monolayer (SAM) that is insulating, but facilitates electron transport to the electrode if a redox active moiety is brought into close proximity to the surface. The target amplicons hybridize to capture probes anchored to the SAM-functionalised electrodes, and to detection probes present in solution. The resulting sandwich complex brings a ferrocene moiety that is attached to the detection probes close to the surface, enabling detection based on AC voltammetry. GenMark recently obtained FDA approval for a respiratory viral panel, and is developing assays for HCV detection and genotyping.

## Systems that integrate amplification and detection

### Systems with end-point detection

#### SAMBA (diagnostics for the real world)

The Simple Amplification Based Assay (SAMBA) is a desktop instrument that automates isothermal nucleic acid amplification coupled to visual lateral flow detection (Lee *et al.*, 2010). SAMBA uses isothermal RNA polymerase mediated amplification, similar to NASBA, coupled with a NALF device that uses a branched polyvalent conjugate approach to enhance sensitivity. The system has been demonstrated for HIV diagnosis. Amplification and detection are executed inside a fully enclosed disposable cartridge to mitigate carryover contamination. The user simply injects extracted nucleic acids into the cartridge, which contains all reagents for amplification and detection, and then inserts the cartridge into the instrument. All remaining steps are performed automatically, and results are available within ~90 minutes. However, sample preparation has to be performed separately and manually, although efforts are under way to automate and integrate this step as well.

## Systems with real-time detection

Many thermocyclers and microfluidic systems to monitor PCR reactions in real-time exist on the market. Most real-time PCR instruments can also be used to incubate and monitor isothermal assays as well. However, due to cost and complexity, most commonly used real-time PCR systems are unsuited for POC applications. The focus here, therefore, is on commercially available (or in late stage development) platforms relevant to IVD NAT that either perform real-time PCR in a more compact less expensive manner, or monitor isothermal reactions in real time. Most of these systems have a small instrument footprint, a reduced need for continuous power, and relatively low throughput, reflecting the needs of POC testing environments in developed and developing countries.

### Real-time PCR: Genedrive (Epistem)

Epistem Ltd. (UK) has developed the Genedrive® real-time PCR instrument (Fig. 10.19A) for use in low resource settings to diagnose a variety of pathogens, including *M. tuberculosis*. The Genedrive™ is a lightweight, portable bench-top instrument that is capable of single test processing. The total assay time, including manual sample preparation, assay incubation and test result generation, is under 45 minutes. The Genedrive™ can be powered via mains electricity or via an internal rechargeable battery. To create a lightweight, low-cost, and effective real-time PCR system, the Genedrive™ involves several innovative technologies, including a

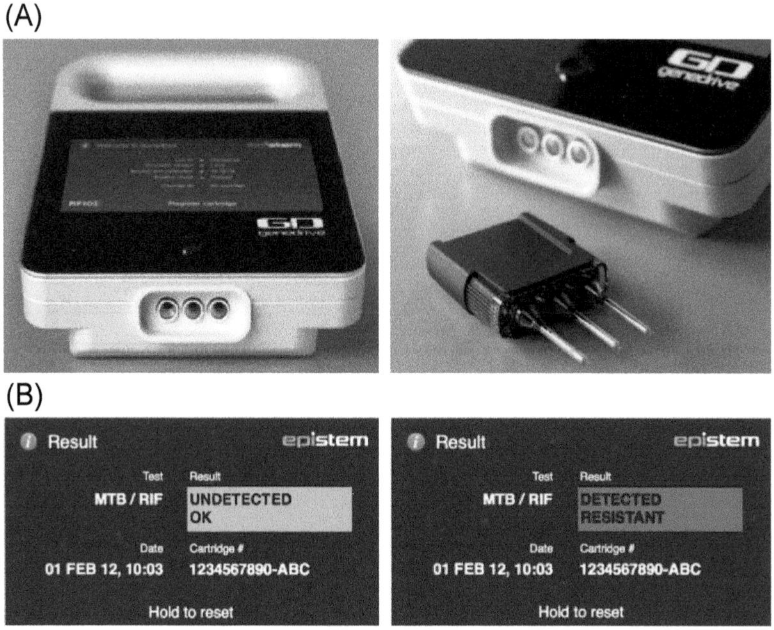


**Figure 10.19** (A) The Epistem Genedrive™ instrument (left) and test cassette (right). The test cassette (containing lyophilized PCR reagents in each tube) is loaded with sample extracts and inserted into the Genedrive™ (porting visible at the front of the Genedrive™) prior to amplification. Source: reproduced with the permission of Epistem, UK. (B) Genedrive™ screen images of test negative and positive results displayed after *M. tuberculosis* analysis. Ease of result interpretation by the user is enhanced by colorimetric differences between a negative (green) and positive test result (red). In addition to MTB identification, rifampin resistance is also indicated. Text and images reproduced with permission form UNITAID.

proprietary heat sink to reduce the power requirements in cooling and heating the PCR reactions samples, and precision optical tubes included in a disposable cassette. The precise technical details of this technology and associated test assays have not yet been fully described. Epistem are developing an assay for *M. tuberculosis* and rifampin resistance diagnosis from either sputum or urine, the Mycobacterium iD® Test-kit. A detailed procedure for how the *M. tuberculosis* test is performed has not been released, and no information is available on the thermal stability of reagents, but test reagents are lyophilized presumably for long term storage at ambient temperature.

The test cassette contains three reaction tubes made of high quality optical material in which the test reactions are performed. Two reaction tubes are used to test for the specific analytes (e.g. *M. tuberculosis* detection and rifampin resistance screening) and the third reaction tube is used to detect the internal process control (IPC). The lyophilized reagents stored in the optical tubes are rehydrated by adding 20 µl of DNA extract to each tube. The target sequences are then amplified via thermal cycling for 30 minutes, and the amplification is monitored in real time in addition to post amplification melt curve analysis of each reaction if required (e.g. detection of alleles conferring rifampin resistance). The raw test data are interpreted by software and the test results are displayed on the Genedrive screen (Fig. 10.19B). Preliminary and unpublished performance data provided by the manufacturer claims a limit of detection of 30 colony forming units/ml using spiked sputum, but to date no peer reviewed performance data are available.

### Real-time PCR: Truelab Uno (MolBio)

The Molbio Truelab™ UNO real-time micro PCR analyser (Fig. 10.20A) is designed for integration with the Trueprep MAG sample processing device described earlier. The core amplification technology utilizes a novel test chip (Fig. 10.20B) which contains lyophilized PCR reagents and serves as the amplification platform to replace a dedicated heating/cooling block as used by other PCR technologies. The chip is a single use disposable, and is designed to prevent reaction evaporation during thermal cycling. Individual chips are automatically processed in the Truelab™ Uno Real-Time micro PCR Analyser. The device can operate from battery or mains power and all reagents and the instrument are designed to be stable at 2–30°C in 10–80% humidity.

The Truelab™ Uno micro PCR Analyser reads fluorescence of two wavelengths in real time to simultaneously detect DNA amplification of the target and the IAC. The IAC ensures appropriate performance of the DNA extraction process, PCR reagents, and equipment. The time to final result is 35 minutes. The host instrument is operated by an embedded microprocessor with Android operating system, plus other related electronics components, with a simple user interface for operation and data input. (Fig. 10.21). This operating system controls PCR execution, analyses

(A) (B)

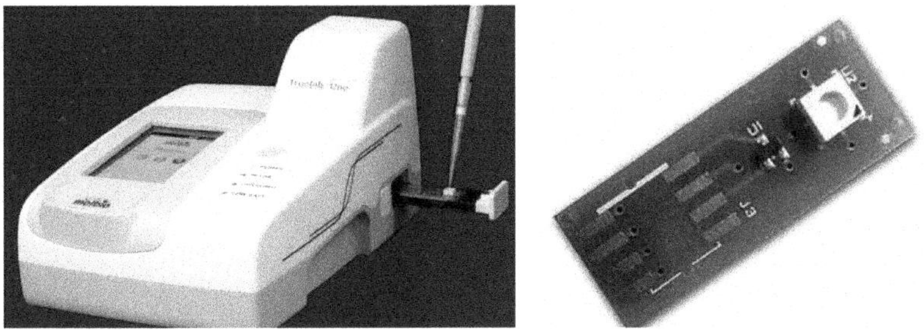

**Figure 10.20** (A) The Truelab™ Uno Real-Time micro PCR Analyser and (B) the Truenat™ micro PCR chip. All key PCR reagents are contained in the square white reaction well on the chip. Eluted DNA is added via pipettor into the reaction well of the chip, which is then inserted into the chip tray, followed by automated real-time PCR analysis. (Text and images reproduced with permission from UNITAID).

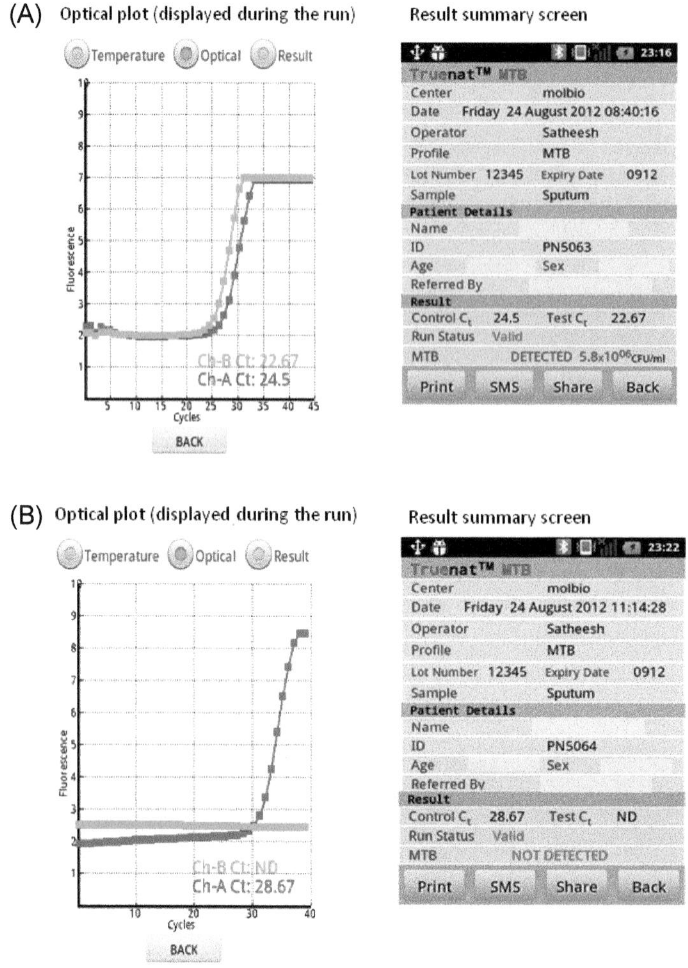


**Figure 10.21** Screen images of test result formats as displayed on the Truelab™ Uno Real-Time micro PCR Analyser for (A) a positive MTB test result, and (B) a negative MTB test result. In both cases, the internal process control amplified, while the target is only amplified in the positive test (Text and images reproduced with permission from UNITAID).

and displays real-time data and the test result, with the added benefits of data storage, wireless data transmission and global positioning capability. Currently there are no peer-reviewed reports describing the performance of the Molbio system, although the test instruments and some diagnostic assays, including one for MTB, are commercially available.

### Real-time isothermal turbidimeters: LA-200 (Eiken) and IllumiPro (Meridian)

The Loopamp Realtime Turbidimeter (LA-200), manufactured by Teramecs and distributed by Eiken, monitors isothermal LAMP reactions in real time based on pyrophosphate precipitation in the presence of $Mn^{2+}$. It is designed for a maximum of four sets of eight-microtube strips and can independently process and measure each set. The device requires line power and an external computer, and is a relatively large bench-top instrument. It has been used for selected applications of clinical relevance, such as for detection of Herpes Virus DNA (Yoshikawa *et al.*, 2004) and malarial parasites (Aonuma *et al.*, 2008).

The IllumiPro-10 incubator/reader, marketed by Meridian as part of the Illumigene molecular diagnostic system, also monitors LAMP reactions in real time based on turbidity changes (FDA, 2011). The system has a relatively small footprint, does not require an external computer, and has an integrated barcode reader for patient ID tracking. The system has two reaction blocks that can be operated independently, each for five reaction tubes. The IllumiPro-10 is currently used in two FDA cleared IVD NATs marketed by Meridian, to detect *C. difficile* from stool samples, and to detect Group B Streptococci from enriched cultures. Sample preparation has to be performed separately (described in a previous section of this chapter).

### Twista™ (TwistDx)

The Twista™ instrument (Fig. 10.22) is marketed by TwistDx Ltd (UK, a wholly owned subsidiary of Alere Inc.) for use with the fluorescence formats of RPA (the TwistAmp™ Exo, TwistAmp™ RT Exo and TwistAmp™ fpg kits). The Twista™ is manufactured by ESE GmbH, an OEM manufacturer acquired by Qiagen, which also sells the instrument in a slightly different configuration as the ESEQuant Tube Scanner, to monitor isothermal amplification reactions other than RPA. The Twista™ can monitor two fluorescence emission signals, typically fluorescein (FAM) and tetramethyl-6-Carboxyrhodamine (TAMRA). The devices have a small footprint and can be easily transported. A rechargeable battery can be used with the devices. The Twista™ is typically operated via a computer to view real-time data. However, the operating programme (with data analysis that indicates positive, negative and indeterminate results) can be downloaded onto the instrument to permit its use independent of an external computer. These results are displayed as '+', '−' and '?' respectively on the liquid crystal display on the device. Each run file can be saved and downloaded later onto a computer from the Twista™.

**Figure 10.22** The TwistDx Twista™ instrument, manufactured by Qiagen, for real-time detection of the isothermal RPA reaction. The battery power pack shown here is housed underneath the instrument. Image courtesy of D. Boyle, PATH.

### BART Bison/PDQ (Lumora)

The Lumora BART (Bioluminescent Assay in Real-Time) couples pyrophosphates generated from nucleic acid synthesis to light emission produced from a thermostable variant of firefly luciferase (Gandelman *et al.*, 2010). As opposed to fluorescence-based detection, BART requires no light source, just a sensitive detector to detect bioluminescence from the reactions. However, this detection format is not amenable to multiplexing as the core technology is based on the release of inorganic phosphate during DNA polymerization, which is not amplicon specific. BART-based detection can, in principle, be coupled with different isothermal amplification reactions, but work to date has been in conjunction with LAMP. The LAMP-BART reaction has been applied for the detection of *Chlamydia trachomatis* (Gandelman *et al.*, 2010). For real-time monitoring of BART, Lumora markets a large instrument, the Bison, which processes 96 well plates and integrates a heating block with a CCD camera. Lumora also provides a compact, battery powered instrument, the PDQ (Fig. 10.23), for reading two strips of eight reaction tubes, based on simple photodiodes for signal detection. The PDQ has a built in user interface, and does not rely on an outside laptop computer.

### Genie (OptiGene)

OptiGene (UK) has an instrument for real-time monitoring of a variety of isothermal assays, the Genie II and a new, smaller instrument, the Genie III (Fig. 10.24). Both instruments can incubate isothermal reactions and read fluorescence or luminescence enabling their use with intercalating

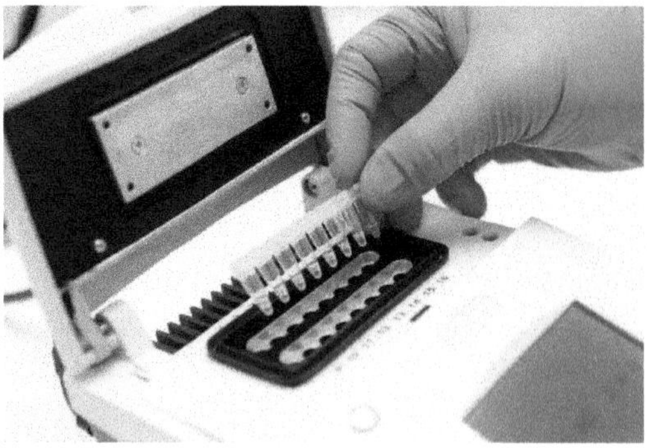

**Figure 10.23** The Lumora BART PDQ system. Image courtesy of Lumora.

(A) (B)

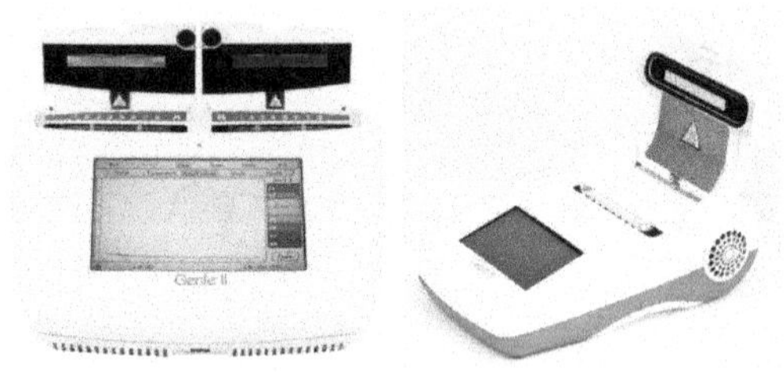

**Figure 10.24** The OptiGene (A) Genie II and (B) Genie III instruments for real-time monitoring of isothermal amplification reactions via fluorescence or luminescence. The Genie II can independently incubate and monitor two sets of up to eight reactions whereas the Genie III can process one set of up to eight reactions. Images with permission from OptiGene.

dyes such as SYBR or Evagreen, fluorescence probes such as FAM in addition to measuring output from BART assays. Both instruments can be powered via an internal rechargeable battery or by external power and can be independently operated via a touch panel display. Both instruments operate across the temperature range used by all isothermal assays (e.g. ambient to 110°C) and have heated lids to prevent changes in reaction volumes during incubation at elevated temperatures. The Genie II utilizes two sets of eight-microtube strips (independently if required) while the Genie III is a similar device that uses only a single strip and therefore is smaller than the Genie II. With the instruments primarily designed for use in the field, the Genie II and Genie III use proprietary 8-microtube strips with more secure lid locking than the traditional PCR strips used with other NATs. This gives greater protection against accidental release of amplicons and contamination of the instrument and/or test site. The OptiGene instruments can perform melt cure analysis after amplification to improve specificity.

## Fully integrated systems

### GeneXpert (Cepheid)

The GeneXpert is an easy to use, fully integrated NAT system with random access modular design (Fig. 10.25). The smaller systems are geared for POC/near patient applications, while large systems such as the 48 module infinity are meant for use in a centralized laboratory.

Nucleic acid sample preparation and PCR amplification with real-time detection are

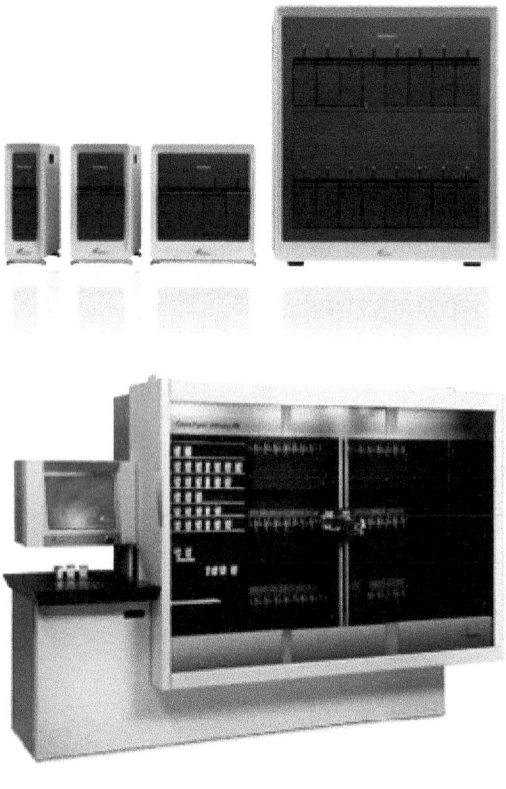

**Figure 10.25** The modular GeneXpert system for fully integrated NAT, including sample preparation coupled with PCR amplification and molecular beacon-based real-time detection. The available platforms (1, 2, 4, 16, or 48 modules) show the trade-off between portability and increased throughput (http://www.cepheid.com/systems-and-software/genexpert-system).

performed inside a disposable, closed system cartridge with all reagents on board. Most of the cartridge body is dedicated to sample preparation and reagent/waste storage. The clinical sample and, in some cases, additional sample preparation reagents are pipetted into the cartridge, which is then closed and inserted into the instrument. A plunger engages with the syringe barrel in the cartridge via a dry interface. Pumping and valving is achieved by moving the plunger up and down, and by rotating the syringe barrel, which at the bottom has a rotating disk valve that connects the different reaction chambers. Depending on the assay, sample preparation either involves solid phase extraction of nucleic acids (e.g. *C. difficile* and Flu assays) or intact pathogen capture from the sample matrix via a filter, followed by washing, and then lysis to liberate genomic material into a buffer (e.g. MTB/RIF assay). Lysis is facilitated through a miniature ultrasonic horn that engages with the cartridge base at the bottom. Sonication is an effective mechanical approach for disrupting lysis-resistant pathogen such as anthrax spores, *C. difficile* and *M. tuberculosis*. The elution buffer containing purified nucleic acids is pumped into a reagent chamber, where it reconstitutes lyophilized mastermix pellets. The complete mastermix is then pumped into a thin square-shaped PCR tube that protrudes from the back of the cartridge. This design facilitates rapid thermocycling due to effective heat transfer through the thin fluid layer. The GeneXpert performs probe-based real-time PCR detection using molecular beacons, and is capable of monitoring six colour channels, using six LEDs and photodiodes that are activated sequentially. In some cases, e.g. for the MTB/RIF assay, the system performs hemi-nested PCR for enhanced specificity. In this case, the cartridge contains separate mastermix pellets with the respective primer pairs. The first round of amplification is performed without fluorescent detection, then the second lyophilized reagent pellet is dissolved in the amplified mastermix, followed by a second round of PCR with real-time detection. The GeneXpert is capable of low levels of multiplexing, which enables simultaneous detection of a limited number of pathogens or drug resistance mutations.

With the GeneXpert, Cepheid has pioneered the move of NAT towards POC for clinical applications such as intrapartum GBS screening in maternity wards (De Tejada *et al.*, 2011) and MRSA screening in hospital wards (Brenwald *et al.*, 2010). The GeneXpert has also significantly advanced the use of NAT in developing countries. Following WHO endorsement in Dec 2010, the Xpert MTB/RIF assay is currently undergoing global roll-out for TB diagnosis. However, the instrument and the cartridges are relative expensive. At the regular list price, a four module system costs $55,000–62,000 and the cost per cartridge for the MTB-Rif assay is 50 euros (~USD72). The Foundation for Innovative New Diagnostics (FIND) negotiated discounted prices for high TB burden countries, reducing the cost per four module instrument plus laptop computer to $17,500 (Foundation for Innovative New Diagnostics, 2012), and the cost per cartridge to $16.86 (Cepheid, 2012). In August 2012, the President's Emergency Plan For AIDS Relief (PEPFAR), UNITAID, USAID and the Bill and Melinda Gates Foundation announced additional substantial subsidies to reduce the cost per cartridge from $16.86 to $9.98 until 2022 (UNITAID, 2012), with the goal of improving market uptake and sustaining the low pricing of the test via market uptake. According to WHO guidelines (WHO, 2011), the GeneXpert has been recommended for use in district and sub-district level laboratories of developing countries, but not for use in more remote settings and primary care facilities. The GeneXpert is relatively expensive and sophisticated, therefore has to be handled with care. Each GeneXpert module has to be calibrated annually, which incurs an additional cost of ~ $1800 at the FIND negotiated reduced price. In the future, this price is expected to drop to $450 through remote on-site calibration. A service contract is typically required to ensure instrument functionality. The GeneXpert requires stable and uninterrupted line power, and ambient operating temperatures ≤ 30°C. Cartridges also have to be stored at 2–28°C, which often means that the system requires an air conditioned facility. The required external computer is susceptible to theft. The cartridges have a limited guaranteed shelf life of six to nine months, which can pose logistical challenges, and are relatively bulky, therefore require substantial storage space. A four-module GeneXpert instrument can process 20 specimens per day, which may not be suitable for laboratories with large patient loads.

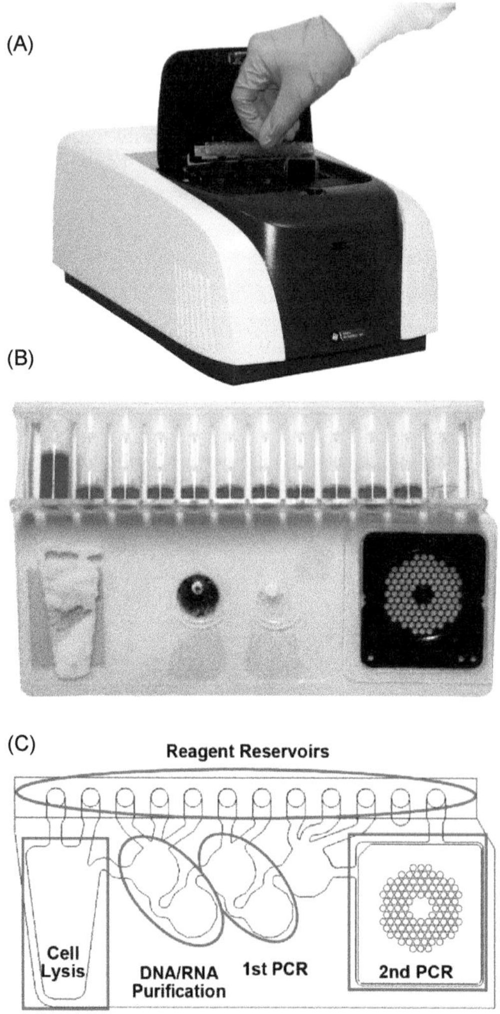


**Figure 10.26** (A) FilmArray instrument by Idaho Technologies. (B) FilmArray cartridge, showing blue fluid in the sample introduction chamber and red fluid distributed throughout the solid reagent chambers after injecting the rehydration solution. Loaded inside the pouch section of the cartridge are ceramic beads, magnetic silica particles, and primers. (C) Schematic representation of processing areas in the cartridge. Modified from (Poritz *et al.*, 2011) with permission.

### FilmArray (Biofire)

The FilmArray (Fig. 10.26A) is another IVD NAT platform that fully integrates nucleic acid sample preparation with real-time PCR. However, as opposed to the GeneXpert, the FilmArray enables much higher levels of multiplexing, and can detect up to 31 separate targets, each in triplicate. Such higher levels of multiplexing are required to detect pathogen panels, or to identify many different drug resistance markers. Biofire recently received FDA clearance and moderate complexity designation for a viral Respiratory Pathogen Panel, and other assays are under development to diagnose sepsis, enteric pathogens, and sexually transmitted infections. Similar to the GeneXpert, the FilmArray was originally developed for biothreat detection. Biofire markets a closely related system under the name Razor, which targets a series of category A, B, and C pathogens.

The FilmArray cartridge (Fig. 10.26B) consists of solid reservoirs for sample introduction and reagent storage along the top, and a heat sealed multi-chamber flexible pouch connected to these reservoirs. Attached to the pouch is an array of ~100 reaction chambers, with a gasket above the reaction chamber wells that acts as a flow restrictor. All reagents are stored in the cartridge in dry lyophilized form under vacuum.

To execute the assay, sample and rehydration buffer are injected through septa into the cartridge. The required volumes are aspirated automatically, drawn in by the vacuum in the cartridge, avoiding the need for precision pipetting. Fluids are moved through the system via pistons inserted into the reservoirs at the top, inflating bladders that press against the flexible pouches, and via blunt pistons that open and close channels within the pouch. The targeted pathogens in the sample, plus an internal control organism that is added to the sample, are mechanically lysed using bead beating with ceramic beads. Liberated nucleic acids are then pushed into the next chamber, where they bind to silica coated magnetic particles. Impurities are removed through three washes from reservoirs in the top. Finally, nucleic acids are eluted, and the eluate is pumped into the next chamber, where it is combined with lyophilized primers and reconstituted mastermix. The system performs nested PCR to increase the sensitivity and specificity of pathogen detection. After the first round of PCR amplification, the mastermix is diluted, then combined with new mastermix reagents and pumped over the array such that the solution is aliquoted into the reaction wells. Each well contains a set of inner primers targeting a specific pathogen or genomic region, therefore multiplexing in this case is based on monitoring a large number of singleplex reactions in parallel. The system uses a intercalating dye-based readout, with LED illumination and a CCD camera to image the wells. The FilmArray can perform melt curve analysis to verify specific amplicon generation. The instrument performs automated data analysis providing readily interpretable results.

Although the FilmArray has many desirable features, it is complex and relatively expensive, with a list price of $49,500. The system requires line power, an external computer, and a number of disposables besides the cartridge. The instrument is not modular, and can only run one sample at a time, therefore has relatively low throughput.

### Liat (IQuum)

The IQuum Lab-In-A-Tube (Liat) system (Fig. 10.27) performs fully integrated closed system nucleic acid sample preparation and PCR amplification with real-time detection (Tanriverdi *et al.*, 2010). The turn-around time is 20 minutes to 1 hour, depending on the assay. Sample preparation, amplification and detection are executed inside a flexible tube, which contains all necessary reagents in segments separated by seals that can be peeled apart if pressure is exerted to a reservoir. Fluids are moved up and down inside the tube through actuators that press against specific segments. The LIAT analyser is a relatively small bench-top system that can be battery operated, does not require an external computer, but can only process one sample at a time. Iquum also has developed the LIAT workstation, which is larger and requires an external computer, but can process eight samples in random access mode.

The LIAT system can handle whole blood, plasma, or swab samples. The sample is added to the top of the tube, which is then inserted into the instrument. Silica-bead solid phase extraction is used for nucleic acid extraction, followed by PCR amplification. Rapid thermocycling is performed by moving the final mastermix up and down to different temperature zones. The system can perform six colour multiplexed target detection. In 2011, Iquum received FDA clearance and moderate complexity designation for the Liat™ Influenza A/B assay and analyser, which is amendable for near-patient settings. Other Liat assays in the pipeline include the Liat™ HIV Quant Assay, Liat™ CMV Quant Assay, and Liat™ Dengue Assay. The LIAT provides clinical sensitivity and specificity comparable to laboratory-based systems (Tanriverdi *et al.*, 2010).

### Enigma ML and FL platforms

Enigma is a UK-based company that was formed out of the Defence Science and Technology Laboratories (Porton Down), which is the UK research centre for the detection of biothreat agents. Enigma has developed an early NAT prototype, called the field laboratory or FL platform, which then evolved into the mini laboratory or ML platform (Fig. 10.28). Both are standalone systems where sample preparation, PCR-based NA

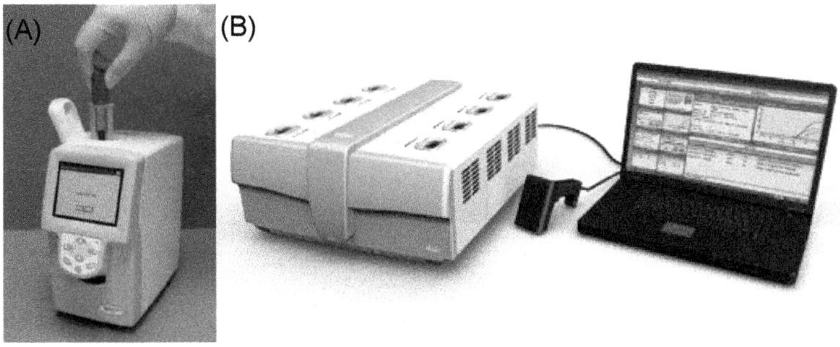


**Figure 10.27** Iquum (A) LIAT analyser, (B) LIAT workstation (http://www.iquum.com).

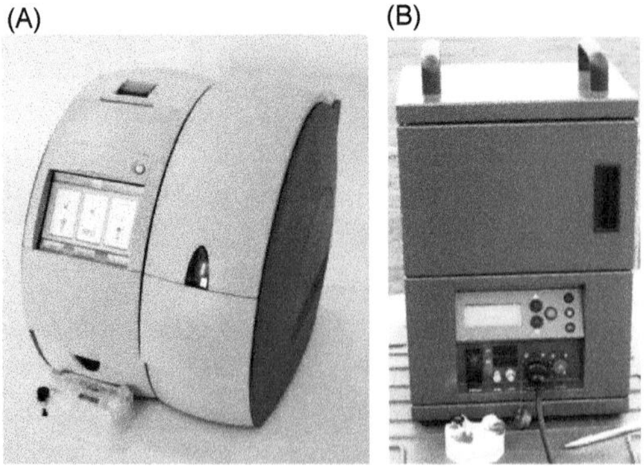

**Figure 10.28** The IVD test instruments and their cartridges produced by Enigma. (A) The Enigma ML system showing the processor and one test unit (right), plus the cartridge used for the ML instrument. (B) The FL platform which has been ruggedised specifically to permit IVD testing outside of the laboratory, and the circular shaped cartridge of the FL instrument. Images courtesy of Enigma.

amplification with real-time detection, data analysis, and result interpretation are fully automated, removing the need for highly trained personnel to perform rapid NAT testing. All reagents for sample processing and subsequent amplification are stabilized, foil sealed and enclosed in the cartridge for storage at ambient temperatures. The ML system is modular and can operate up to six independent test units from one processor. The processor uses a touch screen for data entry and can generate a hard copy of the test result via an integrated printer. No external computer is required. Like most other integrated IVD NAT systems, the ML uses bar-coding of the test cartridge and the specimen collection tube to ensure correct patient identification and test operation. A robotic arm transfers the sample via pipetting between the cartridge components, as opposed to most other fully integrated systems discussed in this section, which use pumping and a valve-based method for fluid handling in the cartridge.

The use of the FL system has been described for detection of bovine viral diarrhoea virus (Wakeley *et al.*, 2010) and foot and mouth disease virus (Madi *et al.*, 2012), both veterinary applications. Extraction and purification of viral RNA is performed via standard SPE, using a lysis buffer and NA capture on magnetic beads, followed by washing and elution. The eluate is combined with the mastermix and ultimately amplified via real-time nested reverse transcription PCR. In a recent study, the FL produced almost identical data to a laboratory-based NAT, but provided results in less than one hour, and testing was performed outdoors (Wakeley *et al.*, 2010).

Although the FL platform has been discontinued, Enigma is performing clinical trials to evaluate an influenza A/B assay using the ML platform, in preparation for a 510k submission to the FDA. This influenza assay uses nasal swabs that are subsequently placed in 1 ml of viral transport medium. The sample is then placed in the cartridge and all steps after the cartridge insertion into the instrument are automated. The time to result is unknown but Enigma claim to have results derived from real-time PCR in under 1 hour. The initial markets for the ML are perceived to be in decentralized laboratories and some POC areas (e.g. physicians' offices). To expand the range of diagnostic tests to be used on the ML platform and it's release into the commercial IVD market, Enigma have recently partnered with GlaxoSmithKline for assay development and the commercialization of the ML instrument.

## Apollo (Biocartis)

Biocartis® (Switzerland) is developing the Apollo, a fully integrated molecular diagnostic system that consists of an instrument, a communication console, and single use disposable cartridges (Fig. 10.29). The system employs mesofluidic processes for specimen extraction and fluid transfer in conjunction with real-time PCR to interrogate DNA and/or RNA-based biomarkers. The system was designed so that multiple instruments can be used in parallel to meet the needs of the user, e.g. a single machine in physicians' offices in comparison to multiple units in larger clinics or hospital laboratories where the test burden is much greater. Like other stand-alone systems, the design focuses on a high performance test device that requires minimal training for use and gives results within 1–2 hours. The primary markets for the Apollo system that have been identified by Biocartis® include infectious diseases and companion diagnostics for drug treatment. Biocartis® are currently partnering with BioMérieux® to develop a suite

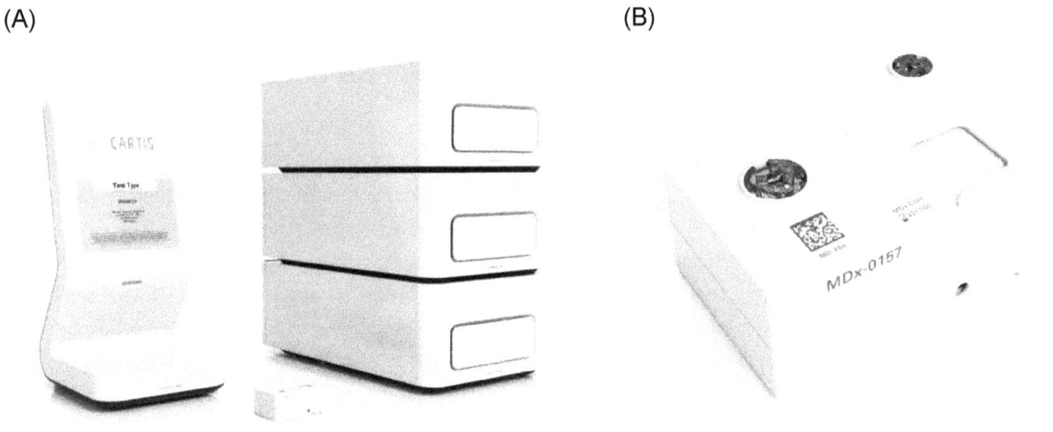


**Figure 10.29** (A) The three primary components of the Biocartis® Apollo system including the controller (left), test cartridge (centre) and three processing units (right). (B) The test cartridge for the Apollo system. The specimen is added into the unit via the port on the lower right hand side of the cartridge (shown closed). A two-dimensional bar-coding system is used for cartridge/test identification upon insertion into the processing unit. Images courtesy of Biocartis®.

of assays for the diagnosis of infectious diseases while Janssen Pharmaceutica (Johnson & Johnson) is their partner for developing assays related to personalized medicine.

The specimen is first added to the cartridge and then this is placed into the machine for NA extraction and purification, NA amplification and finally determination of the test results. Currently, the maximum specimen volume that the cartridge can receive is 1.5 ml but this may be increased to 2.0 ml in future devices. Other specimen types such as swabs can also be added directly into the cartridge. Specimen lysis is performed via a combination of chemical means and sonication. The nucleic acids are captured, washed and eluted through standard solid phase extraction. The eluate is then passed to the amplification component that has five separate reaction chambers each containing lyophilized RT PCR or PCR reagents. The eluate rehydrates the lyophilized reagents in each amplification reaction chamber and the test process can proceed. The optical detection system is designed to detect the spectra of six different fluorescent dyes per chamber, permitting a maximum total of 30 tests per cartridge. Currently it is envisaged that each amplification chamber will have up to five test reactions and one functioning as a quality control test. The Apollo system is anticipated for release in 2013.

### CARD (Rheonix)

Rheonix has developed a system called 'Chemistry And Reagent Device' (CARD), which consists of a microfluidic chip that can be manufactured in a reproducible, inexpensive, and scalable manner by laminating three plastic layers (Zhou *et al.*, 2010a). Chambers and channels are embedded in the top and bottom rigid polystyrene layers. The middle layer is flexible, and can be deflected by applying pressure to the system using pneumatic actuators. This design enables on-chip pneumatic pumping and valving, with intricate and elegant fluidic control (Zhou *et al.*, 2010a). However, although the chip is small, the pneumatic system required to control test execution is relatively bulky. The system has been used for integrated nucleic acid sample preparation, PCR or isothermal amplification, coupled with end-point detection using NALF (Zhou *et al.*, 2010a). Recently, the microfluidic chip has been refined to a so-called quad card that enables simultaneous processing of four clinical samples, multiplexed PCR amplification, and detection using a low density microarray, which uses a principle similar a standard line probe assay (Zhou *et al.*, 2010b). However, reagents are not stored on the cartridge. Rheonix has developed the EncompassMDx Workstation, which uses a conventional liquid handler with positive displacement pipetting to add the necessary reagents to the chip. The workstation controls test execution and detection through the required pneumatic, thermal, and optical systems. Rheonix has performed preliminary clinical demonstration of this system for infectious disease diagnosis of high risk HPV strains, and a panel of sexually transmitted diseases. This comparatively large benchtop instrument is not be suited for POC operation, however Rheonix have developed a portable controller, but this does not automate the reagent addition into the cartridge.

### NAT Analyzer and iNAT system (Alere)

Alere (formerly Inverness Medical) is a key player in the POC IVD industry. While the original company focus was on lateral flow immunoassays, Alere has expanded into other areas over the past years, including nucleic acid testing. During an investor presentation in January 2012, Alere's CEO introduced the NAT analyser and the iNAT system (Fig. 10.30), two platforms that are currently in development and entering clinical trials.

The primary targeted application for the NAT Analyser is near patient HIV viral load monitoring, for which there is a significant need in developing countries such as sub-Saharan Africa. Alere, therefore, plans to launch the system first in high burden developing countries, followed by launch in Europe and the USA. Other assays are in the pipeline, e.g. for hepatitis C (Alere, 2012; Ullrich *et al.*, 2012). According to the January 2012 investor presentation, the NAT analyser is battery operated, robust and portable, does not require an external computer, and does not require external calibration.

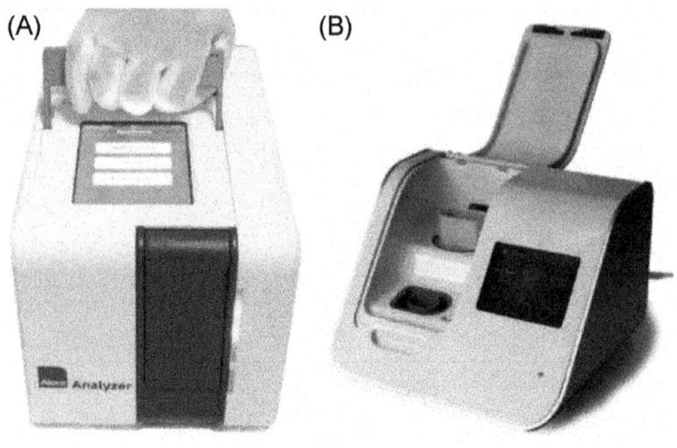


**Figure 10.30** Alere POC molecular diagnostics platforms: (A) PCR-based NAT Analyser; (B) iNAT system, based on isothermal nucleic acid amplification.

For the NAT analyser HIV viral load monitoring assay, amplicon formation is monitored in real time via a multiplexed quantitative microarray-based readout called competitive reporter monitored amplification (CMA). The reaction principle and instrument system for amplification and detection has recently been described (Ullrich *et al.*, 2012).

The Alere iNAT system is based on isothermal nucleic acid amplification. Alere recently acquired TwistDX, the company that developed RPA, and Ionian Technologies, the company that developed the nicking enzyme amplification reaction (NEAR). No clear technical details are available, but the system is claimed to generate results in 15 minutes or less from sample introduction, in a compact and cost-effective format. The first targeted application is influenza, but future assays may target pathogens such as group A streptococcus, RSV, *C. difficile*, *Chlamydia*, *M. tuberculosis* and MRSA.

### Great Basin Portrait Analyzer

Great Basin Corporation has developed a system for integrated sample preparation and isothermal amplification using hotstart blocked primer HDA, with array-based end-point detection (Hicke *et al.*, 2012). The injection moulded cartridge (Fig. 10.31) contains reagents in blister packs, plus channels and chambers for sample extraction, amplification, and detection, with results available in 75 minutes. The recently FDA cleared system enables identification of toxigenic *C. difficile* from stool, with performance comparable to other commercially available *C. difficile* NAT systems (Buchan *et al.*, 2012). Great Basin Corp. has developed a similar assay for *rpoB* genotyping, to identify rifampin resistant *M. tuberculosis* (Ao *et al.*, 2012).

Beyond the systems described herein, isothermal platforms that integrate sample preparation, amplification and detection have been reported in the literature (Lutz *et al.*, 2010; Mahalanabis *et al.*, 2010; Wang *et al.*, 2011; Wu *et al.*, 2011).

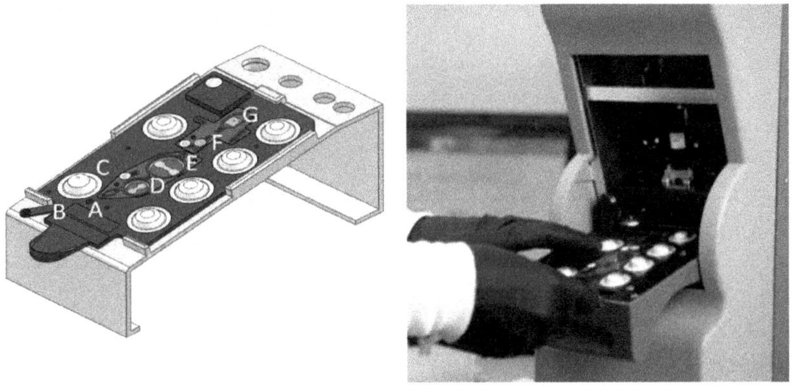


**Figure 10.31** Great Basin Portrait Analyser: cartridge on the loading tray (left) and cartridge inserted into the instrument (right). The cartridge contains a sample port with hinged lid (A,B), blister packs containing liquid reagents (C), chambers for extraction and dilution (D,E), amplification (F), and detection (G). Reproduced from Hicke *et al.* (2012) with permission.

## Conclusions

The development and commercialization of moderate complexity NAT technologies with potential for use at POC for infectious disease diagnosis has gained momentum in the past decade. Cepheid has pioneered this movement, and now offers a broad menu of FDA cleared and/or CE-IVD marked NATs on the GeneXpert platform. Large scale global roll-out of the Xpert MTB/RIF assay has been facilitated through WHO endorsement and major subsidies from various donors. Many other platforms based on PCR and isothermal methods have recently entered the market or are in late stage development. Some offer decreased cost and complexity, others offer advanced features such as the ability to achieve higher levels of multiplexing. Different systems are likely to succeed in different market segments.

Despite these advances, many regulatory hurdles and technical challenges remain. The most significant regulatory hurdle entails obtaining CLIA waived status for NAT, a pre-requisite for use at most POC settings such as physician's office labs. This hurdle will likely be surpassed in the coming years. In terms of technical challenges, instrument complexity and costs are still relatively high for most systems. These challenges have been addressed in part through the shift from PCR to isothermal amplification methods, and by using simpler detection methods such as nucleic acid lateral flow. The remaining technical challenges predominantly revolve around integrating sample preparation with downstream application and detection. Such full sample to answer integration significantly increases the design complexity, but ultimately renders the system more user-friendly and less prone to user error, a pre-requisite for obtaining CLIA waived status. In some cases, such as influenza detection from nasopharyngeal swabs, the sample matrix is comparatively simple, requiring minimal sample preparation. The first CLIA waived NATs are likely to be influenza assays, due to this technical advantage, combined with the clear clinical need to perform rapid testing for highly virulent influenza strains in physician's office labs in the case of a pandemic. Other matrices such as sputum require more involved sample preparation processes. Therefore, sample preparation remains the Achilles' heel for many of the upcoming NAT systems for TB diagnosis in low resource settings.

POC NAT will likely expand beyond the clinical applications listed earlier in this chapter. For example, the Cepheid GeneXpert product pipeline for 2013–2017 includes tests targeted at sexually transmitted diseases and women's health, such as *Chlamydia trachomatis, Neisseria gonorrhoeae*, vaginitis, human papillomavirus, HSV and other tests that are likely to be used in physicians' office labs, such as diagnosis of group A streptococci (causative agents of strep throat), and *Norovirus* (causative agent of diarrhoeal disease). Viral load testing for HCV and HBV is important for the clinical management of patients with hepatitis, and is an important part of the current central laboratory NAT market. A portion of this market may shift to peripheral health settings through assays in development by Cepheid, Alere, and others companies. NAT is becoming more established in developing countries, especially related to TB and HIV diagnosis, drug susceptibility testing, and treatment monitoring. Leading global IVD companies traditionally consider diagnostic testing targeted at developing economies unprofitable, a paradigm that has been challenged by Alere and Cepheid. It is likely that POC NAT systems such as the Alere NAT analyser HIV viral load test will first be validated and launched in high burden developing countries, and eventually penetrate into the market in the developed world. HIV/TB co-infection presents a significant problem, especially in sub-Saharan Africa. Cepheid has already penetrated that market through the GeneXpert MTB/RIF assay, which is likely to facilitate market entry for a GeneXpert HIV viral load assay. Highly multiplexed systems can enable the rapid and differential diagnosis of respiratory diseases, sepsis, and diarrheal disease, which requires panel testing for multiple pathogens, and the rapid identification of many drug resistance mutations, which is important in the diagnosis of, for example, sepsis and MDR/XDR-TB. Novel detection technologies now make it feasible to disseminate highly multiplexed NAT out of central laboratory environments.

NAT has evolved at a very rapid pace in the last two decades, and now forms a significant and growing part of the IVD industry. NAT has already moved from a high complexity to a moderate complexity laboratory environment, and in the near future will likely advance to the point of care in developed and developing countries.

## References


Abe, T., Segawa, Y., Watanabe, H., Yotoriyama, T., Kai, S., Yasuda, A., Shimizu, N., and Tojo, N. (2011). Point-of-care testing system enabling 30 min detection of influenza genes. Lab Chip *11*, 1166–1167.

Abrams, E.J., Simonds, R.J., Modi, S., Rivadeneira, E., Vaz, P., Kankasa, C., Tindyebwa, D., Phelps, B.R., Bowsky, S., Teasdale, C.A., Koumans, E., and Ruff, A.J. (2012). PEPFAR Scale-up of Pediatric HIV Services: Innovations, Achievements, and Challenges. Jaids-J. Acquir. Immune Defic. Syndr. *60*, S105–S112.

Al-Soud, W.A., and Radstrom, P. (2000). Effects of amplification facilitators on diagnostic PCR in the presence of blood, feces, and meat. J. Clin. Microbiol. *38*, 4463–4470.

Al-Soud, W.A., and Radstrom, P. (2001). Purification and characterization of PCR-inhibitory components in blood cells. J. Clin. Microbiol. *39*, 485–493.

Alere (2012). Alere 2011 Anual Report. Available at: http://www.alere.com/content/dam/alere/docs/investor/annualreports/alere-2012ar.pdf (accessed 19 December 2012).

Amaro, A., Duarte, E., Amado, A., Ferronha, H., and Botelho, A. (2008). Comparison of three DNA extraction methods for *Mycobacterium bovis, Mycobacterium tuberculosis* and *Mycobacterium avium* subsp. *avium*. Lett. Appl. Microbiol. *47*, 8–11.

Andresen, D., von Nickisch-Rosenegk, M., and Bier, F.F. (2009). Helicase dependent OnChip-amplification and its use in multiplex pathogen detection. Clin. Chim. Acta *403*, 244–248.

Ao, W.Y., Aldous, S., Woodruff, E., Hicke, B., Rea, L., Kreiswirth, B., and Jenison, R. (2012). Rapid detection of rpoB gene mutations conferring rifampin resistance in Mycobacterium tuberculosis. J. Clin. Microbiol. *50*, 2433–2440.

Aonuma, H., Suzuki, M., Iseki, H., Perera, N., Nelson, B., Igarashi, I., Yagi, T., Kanuka, H., and Fukurnoto, S. (2008). Rapid identification of Plasmodium-carrying mosquitoes using loop-mediated isothermal amplification. Biochem. Biophys. Res. Commun. *376*, 671–676.

Atun, R. (2012). Drug-resistant tuberculosis – current dilemmas, unanswered questions, challenges, and priority needs. J. Infect. Dis. *205 Suppl 2*, S228-S240.

Balasingham, S.V., Davidsen, T., Szpinda, I., Frye, S.A., and Tonjuin, T. (2009). Molecular diagnostics in tuberculosis basis and implications for therapy. Mol. Diagn. Ther. *13*, 137–151.

Barken, K.B., Haagensen, J.A.J., and Tolker-Nielsen, T. (2007). Advances in nucleic acid-based diagnostics of bacterial infections. Clin. Chim. Acta *384*, 1–11.

Batchelor, M., Hopkins, K.L., Liebana, E., Slickers, P., Ehricht, R., Mafura, M., Aarestrup, F., Mevius, D., Clifton-Hadley, F.A., Woodward, M.J., *et al.* (2008). Development of a miniaturised microarray-based assay for the rapid identification of antimicrobial resistance genes in Gram-negative bacteria. Int. J. Antimicrob. Agents *31*, 440–451.

Bauer, M.P., van Dissel, J.T., and Kuijper, E.J. (2009). Clostridium difficile: controversies and approaches to management. Curr. Opin. Infect. Dis. *22*, 517–524.

Berensmeier, S. (2006). Magnetic particles for the separation and purification of nucleic acids. Appl. Microbiol. Biotechnol. 73, 495–504.

BioHelix (2012). IsoAmp HSV Assay Package Insert, Rev 7. Available at: http://www.biohelix.com/pdf/DR-HS01%20IsoAmp%20HSV%20Assay_rev-725Jan2012.pdf (accessed 19 February 2012).

Boehme, C.C., Nabeta, P., Henostroza, G., Raqib, R., Rahim, Z., Gerhardt, M., Sanga, E., Hoelscher, M.,

Notomi, T., Hase, T., *et al.* (2007). Operational feasibility of using loop-mediated isothermal amplification for diagnosis of pulmonary tuberculosis in microscopy centers of developing countries. J. Clin. Microbiol. *45*, 1936–1940.

Boehme, C.C., Nabeta, P., Hillemann, D., Nicol, M.P., Shenai, S., Krapp, F., Allen, J., Tahirli, R., Blakemore, R., Rustomjee, R., *et al.* (2010). Rapid molecular detection of tuberculosis and rifampin resistance. N. Engl. J. Med. *363*, 1005–1015.

de Boer, R., Peters, R., Gierveld, S., Schuurman, T., Kooistra-Smid, M., and Savelkoul, P. (2010). Improved detection of microbial DNA after bead-beating before DNA isolation. J. Microbiol. Methods *80*, 209–211.

Boom, R., Sol, C.J.A., Salimans, M.M.M., Jansen, C.L., Wertheimvandillen, P.M.E., and Vandernoordaa, J. (1990). Rapid and simple method for purification of nucleic acids. J. Clin. Microbiol. *28*, 495–503.

Bordelon, H., Adams, N.M., Klemm, A.S., Russ, P.K., Williams, J.V., Talbot, H.K., Wright, D.W., and Haselton, F.R. (2011). Development of a low-resource RNA extraction cassette based on surface tension valves. ACS Appl. Mater. Interfaces. *3*, 2161–2168.

Brenwald, N.P., Baker, N., and Oppenheim, B. (2010). Feasibility study of a real-time PCR test for meticillin-resistant *Staphylococcus aureus* in a point of care setting. J. Hosp. Infect. *74*, 245–249.

Buchan, B.W., Mackey, T.L.A., Daly, J.A., Alger, G., Denys, G.A., Peterson, L.R., Kehl, S.C., and Ledeboer, N.A. (2012). Multicenter clinical evaluation of the portrait toxigenic *C. difficile* assay for detection of toxigenic *Clostridium difficile* strains in clinical stool specimens. J. Clin. Microbiol. *50*, 3932–3936.

Calmy, A., Ford, N., Hirschel, B., Reynolds, S.J., Lynen, L., Goemaere, E., de la Vega, F.G., Perrin, L., and Rodriguez, W. (2007). HIV viral load monitoring in resource-limited regions: optional or necessary? Clin. Infect. Dis. *44*, 128–134.

Carter, G.P., Douce, G.R., Govind, R., Howarth, P.M., Mackin, K.E., Spencer, J., Buckley, A.M., Antunes, A., Kotsanas, D., Jenkin, G.A., *et al.* (2011). The anti-sigma factor TcdC modulates hypervirulence in an epidemic BI/NAP1/027 clinical isolate of *Clostridium difficile*. PLoS Pathog. *7*, e1002317.

Cepheid (2012). Cepheid cares – Xpert MTB/RIF pricing for high TB burden countries. Available at: http://www.cepheidcares.com/tb/cepheid-vision.html#pricing (accessed 19 December 2012).

Chambers, H.F., and Deleo, F.R. (2009). Waves of resistance: *Staphylococcus aureus* in the antibiotic era. Nat. Rev. Microbiol. *7*, 629–641.

Chandler, D.P., Kukhtin, A., Mokhiber, R., Knickerbocker, C., Ogles, D., Rudy, G., Golova, J., Long, P., and Peacock, A. (2010). Monitoring microbial community structure and dynamics during *in situ* U(VI) bioremediation with a field-portable microarray analysis system. Environ. Sci. Technol. *44*, 5516–5522.

Chang, P.G., McLaughlin, W.A., and Tolin, S.A. (2011). Tissue blot immunoassay and direct RT-PCR of cucumoviruses and potyviruses from the same NitroPure nitrocellulose membrane. J. Virol. Methods *171*, 345–351.

Chantry, C.J., Cooper, E.R., Pelton, S.I., Zorilla, C., Hillyer, G.V., and Diaz, C. (1995). Seroreversion in human immunodeficiency virus-exposed but uninfected infants. Pediatr. Infect. Dis. J. *14*, 382–387.

Ciaranello, A.L., Park, J.E., Ramirez-Avila, L., Freedberg, K.A., Walensky, R.P., and Leroy, V. (2011). Early infant HIV-1 diagnosis programs in resource-limited settings: opportunities for improved outcomes and more cost-effective interventions. BMC Med. *9*, 59.

Clifford, V., Garland, S.M., and Grimwood, K. (2012). Prevention of neonatal group B streptococcus disease in the 21st century. J. Paediatr. Child Health *48*, 808–815.

Cnops, L., Boderie, M., Gillet, P., Van, E.M., and Jacobs, J. (2011). Rapid diagnostic tests as a source of DNA for Plasmodium species-specific real-time PCR. Malar. J. *10*, 67.

Corey, L., and Wald, A. (2009). Current concepts: maternal and neonatal herpes simplex virus infections. N. Engl. J. Med. *361*, 1376–1385.

Corstjens, P., Zuiderwijk, M., Brink, A., Li, S., Feindt, H., Neidbala, R.S., and Tanke, H. (2001). Use of up-converting phosphor reporters in lateral-flow assays to detect specific nucleic acid sequences: a rapid, sensitive DNA test to identify human papillomavirus type 16 infection. Clin. Chem. *47*, 1885–1893.

Corstjens, P., Zuiderwijk, M., Nilsson, M., Feindt, H., Niedbala, R.S., and Tanke, H.J. (2003). Lateral-flow and up-converting phosphor reporters to detect single-stranded nucleic acids in a sandwich-hybridization assay. Anal. Biochem. *312*, 191–200.

Cox, J.T., Lorincz, A.T., Schiffman, M.H., Sherman, M.E., Cullen, A., and Kurman, R.J. (1995). Human papillomavirus testing by hybrid capture appears to be useful in triaging women with a cytologic diagnosis of atypical squamous cells of undetermined significance. Am. J. Obstet. Gynecol. *172*, 946–954.

Curtis, K.A., Rudolph, D.L., and Owen, S.M. (2008). Rapid detection of HIV-1 by reverse-transcription, loop-mediated isothermal amplification (RT-LAMP). J. Virol. Methods *151*, 264–270.

Curtis, K.A., Rudolph, D.L., and Owen, S.M. (2009). Sequence-specific detection method for reverse transcription, loop-mediated isothermal amplification of HIV-1. J. Med. Virol. *81*, 966–972.

Curtis, K.A., Rudolph, D.L., Nejad, I., Singleton, J., Beddoe, A., Weigl, B., Labarre, P., and Owen, S.M. (2012). Isothermal amplification using a chemical heating device for point-of-care detection of HIV-1. PLoS One. *7*, e31432.

Dalovisio, J.R., MontenegroJames, S., Kemmerly, S.A., Genre, C.F., Chambers, R., Greer, D., Pankey, G.A., Failla, D.M., Haydel, K.G., Hutchinson, L., *et al.* (1996). Comparison of the amplified Mycobacterium tuberculosis (MTB) direct test, Amplicor MTB PCR, and IS6110-PCR for detection of MTB in respiratory specimens. Clin. Infect. Dis. 23, 1099–1106.

Dauphin, L.A., Moser, B.D., and Bowen, M.D. (2009). Evaluation of five commercial nucleic acid extraction kits for their ability to inactivate *Bacillus anthracis* spores and comparison of DNA yields from spores and spiked environmental samples. J. Microbiol. Methods 76, 30–37.

Dineva, M.A., Candotti, D., Fletcher-Brown, F., Allain, J.P., and Lee, H. (2005). Simultaneous visual detection of multiple viral amplicons by dipstick assay. J. Clin. Microbiol. 43, 4015–4021.

Dineva, M.A., Mahilum-Tapay, L., and Lee, H. (2007). Sample preparation: a challenge in the development of point-of-care nucleic acid based assays for resource-limited settings. Analyst 132, 1193–1199.

Doebler, R.W., Erwin, B., Hickerson, A., Irvine, B., Woyski, D., Nadim, A., and Sterling, J.D. (2009). Continuous-flow, rapid lysis devices for biodefense nucleic acid diagnostic systems. JALA 14, 119–125.

Enomoto, Y., Yoshikawa, T., Ihira, M., Akimoto, S., Miyake, F., Usui, C., Suga, S., Suzuki, K., Kawana, T., Nishiyama, Y., and Asano, Y. (2005). Rapid diagnosis of herpes simplex virus infection by a loop-mediated isothermal amplification method. J. Clin. Microbiol. 43, 951–955.

Espy, M.J., Uhl, J.R., Sloan, L.M., Buckwalter, S.P., Jones, M.F., Vetter, E.A., Yao, J.D.C., Wengenack, N.L., Rosenblatt, J.E., Cockerill, F.R., *et al.* (2006). Real-time PCR in clinical microbiology: Applications for a routine laboratory testing. Clin. Microbiol. Rev. 19, 165–256.

Fang, R.D., Li, X., Hu, L., You, Q.M., Li, J., Wu, J., Xu, P., Zhong, H.Y., Luo, Y., Mei, J., *et al.* (2009). Cross-priming amplification for rapid detection of Mycobacterium tuberculosis in sputum specimens. J. Clin. Microbiol. 47, 845–847.

FDA (2011). IllumiGene 510k Decision Summary. Available at: http://www.accessdata.fda.gov/cdrh_docs/reviews/K100818.pdf (accessed 19 February 2012).

Fiscus, S.A., Cheng, B., Crowe, S.M., Demeter, L., Jennings, C., Miller, V., Respess, R., and Stevens, W. (2006). HIV-1 viral load assays for resource-limited settings. PLoS Med. 3, 1743–1750.

Foundation for Innovative New Diagnostics (2012). FIND-negotiated prices for Xpert MTB/RIF and country list. Available at: http://www.finddiagnostics.org/about/what_we_do/successes/find-negotiated-prices/xpert_mtb_rif.html (accessed 24 May 2012).

Francois, P., Tangomo, M., Hibbs, J., Bonetti, E.J., Boehme, C.C., Notomi, T., Perkins, M.D., and Schrenzel, J. (2011). Robustness of a loop-mediated isothermal amplification reaction for diagnostic applications. FEMS Immunol. Med Microbiol. 62, 41–48.

Gandelman, O.A., Church, V.L., Moore, C.A., Kiddle, G., Carne, C.A., Parmar, S., Jalal, H., Tisi, L.C., and Murray, J.A. (2010). Novel bioluminescent quantitative detection of nucleic acid amplification in real-time. PLoS One. 5, e14155.

Gardella, C., Huang, M.L., Wald, A., Magaret, A., Selke, S., Morrow, R., and Corey, L. (2010). Rapid polymerase chain reaction assay to detect herpes simplex virus in the genital tract of women in labor. Obstet. Gynecol. 115, 1209–1216.

Gilligan, P., Quirke, M., Winder, S., and Humphreys, H. (2010). Impact of admission screening for methicillin-resistant *Staphylococcus aureus* on the length of stay in an emergency department. J. Hosp. Infect. 75, 99–102.

Goldmeyer, J., Li, H., McCormac, M., Cook, S., Stratton, C., Lemieux, B., Kong, H., Tang, W., and Tang, Y.W. (2008). Identification of *Staphylococcus aureus* and determination of methicillin resistance directly from positive blood cultures by isothermal amplification and a disposable detection device. J. Clin. Microbiol. 46, 1534–1536.

Gracias, K.S., and McKillip, J.L. (2007). Nucleic acid sequence-based amplification (NASBA) in molecular bacteriology: a procedural guide. J. Rapid Methods Automation Microbiol. 15, 295–309.

Gupta, R.K., Hill, A., Sawyer, A.W., Cozzi-Lepri, A., von Wyl, V., Yerly, S., Lima, V.D., Gunthard, H.F., Gilks, C., and Pillay, D. (2009). Virological monitoring and resistance to first-line highly active antiretroviral therapy in adults infected with HIV-1 treated under WHO guidelines: a systematic review and meta-analysis. Lancet Infect. Dis. 9, 409–417.

Halse, T.A., Escuyer, V.E., and Musser, K.A. (2011). Evaluation of a single-tube multiplex real-time PCR for differentiation of members of the Mycobacterium tuberculosis complex in clinical specimens. J. Clin. Microbiol. 49, 2562–2567.

Hamers, R.L., Wallis, C.L., Kityo, C., Siwale, M., Mandaliya, K., Conradie, F., Botes, M.E., Wellington, M., Osibogun, A., Sigaloff, K.C., *et al.* (2011). HIV-1 drug resistance in antiretroviral-naive individuals in sub-Saharan Africa after rollout of antiretroviral therapy: a multicentre observational study. Lancet Infect. Dis. 11, 750–759.

Harries, A.D., and Dye, C. (2006). Tuberculosis. Ann. Trop. Med. Parasitol. 100, 415–431.

Harries, A.D., Zachariah, R., van Oosterhout, J.J., Reid, S.D., Hosseinipour, M.C., Arendt, V., Chirwa, Z., Jahn, A., Schouten, E.J., and Kamoto, K. (2010). Diagnosis and management of antiretroviral-therapy failure in resource-limited settings in sub-Saharan Africa: challenges and perspectives. Lancet Infect. Dis. 10, 60–65.

Hatano, B., Maki, T., Obara, T., Fukumoto, H., Hagisawa, K., Matsushita, Y., Okutani, A., Bazartseren, B., Inoue, S., Sata, T., *et al.* (2010). LAMP using a disposable pocket warmer for anthrax detection, a highly mobile and reliable method for anti-bioterrorism. Jpn. J. Infect. Dis. 63, 36–40.

Helb, D., Jones, M., Story, E., Boehme, C., Wallace, E., Ho, K., Kop, J., Owens, M.R., Rodgers, R., Banada, P., *et al.* (2010). Rapid detection of Mycobacterium tuberculosis and rifampin resistance by use of on-demand, near-patient technology. J. Clin. Microbiol. 48, 229–237.

Hellyer, T.J., and Nadeau, J.G. (2004). Strand displacement amplification: a versatile tool for molecular diagnostics. Expert Rev. Mol. Diagn. 4, 251–261.

Hicke, B., Pasko, C., Groves, B., Ager, E., Corpuz, M., Frech, G., Munns, D., Smith, W., Warcup, A., Denys, G., *et al.* (2012). Automated detection of toxigenic *Clostridium difficile* in clinical samples: isothermal tcdB amplification coupled to array-based detection. J. Clin. Microbiol. 50, 2681–2687.

Higuchi, R., Fockler, C., Dollinger, G., and Watson, R. (1993). Kinetic PCR analysis: real-time monitoring of DNA amplification reactions. Biotechnology (New York) *11*, 1026–1030.

Hofmann, W.P., Dries, V., Herrmann, E., Gartner, B., Zeuzem, S., and Sarrazin, C. (2005). Comparison of transcription mediated amplification (TMA) and reverse transcription polymerase chain reaction (RT-PCR) for detection of hepatitis C virus RNA in liver tissue. J. Clin. Virol. *32*, 289–293.

Inoue, R., Tsukahara, T., Sunaba, C., Itoh, M., and Ushida, K. (2007). Simple and rapid detection of the porcine reproductive and respiratory syndrome virus from pig whole blood using filter paper. J. Virol. Methods *141*, 102–106.

Jangam, S.R., Yamada, D.H., McFall, S.M., and Kelso, D.M. (2009). Rapid, point-of-care extraction of human immunodeficiency virus type 1 proviral DNA from whole blood for detection by real-time PCR. J. Clin. Microbiol. *47*, 2363–2368.

Jeong, Y.J., Park, K., and Kim, D.E. (2009). Isothermal DNA amplification *in vitro:* the helicase-dependent amplification system. Cell Mol. Life Sci. *66*, 3325–3336.

Johannessen, A., Troseid, M., and Calmy, A. (2009). Dried blood spots can expand access to virological monitoring of HIV treatment in resource-limited settings. J. Antimicrob. Chemother. *64*, 1126–1129.

Jung, C., Chung, J.W., Kim, U.O., Kim, M.H., and Park, H.G. (2010). Isothermal target and signaling probe amplification method, based on a combination of an isothermal chain amplification technique and a fluorescence resonance energy transfer cycling probe technology. Anal. Chem. *82*, 5937–5943.

Kaser, M., Ruf, M.T., Hauser, J., Marsollier, L., and Pluschke, G. (2009). Optimized method for preparation of DNA from pathogenic and environmental mycobacteria. Appl. Environ. Microbiol. *75*, 414–418.

Kermekchiev, M.B., Kirilova, L.I., Vail, E.E., and Barnes, W.M. (2009). Mutants of Taq DNA polymerase resistant to PCR inhibitors allow DNA amplification from whole blood and crude soil samples. Nucleic Acids Res. *37*, e40.

Khamadi, S., Okoth, V., Lihana, R., Nabwera, J., Hungu, J., Okoth, F., Lubano, K., and Mwau, M. (2008). Rapid identification of infants for antiretroviral therapy in a resource poor setting: the Kenya experience. J. Trop. Pediatr. *54*, 370–374.

Kim, J., Johnson, M., Hill, P., and Gale, B.K. (2009). Microfluidic sample preparation: cell lysis and nucleic acid purification. Integr. Biol. *1*, 574–586.

Kimberlin, D.W. (2004). Neonatal herpes simplex infection. Clin. Microbiol. Rev. *17*, 1–13.

Kuijper, E.J., Coignard, B., and Tull, P. (2006). Emergence of Clostridium difficile-associated disease in North America and Europe. Clin. Microbiol. Infect. *12*, 2–18.

Kwiatkowski, R.W., Lyamichev, V., de Arruda, M., and Neri, B. (1999). Clinical, genetic, and pharmacogenetic applications of the invader assay. Mol. Diagn. *4*, 353–364.

Labarre, P., Gerlach, J., Wilmoth, J., Beddoe, A., Singleton, J., and Weigl, B. (2010). Non-instrumented nucleic acid amplification (NINA): instrument-free molecular malaria diagnostics for low-resource settings. Conf. Proc. IEEE Eng Med. Biol. Soc. *2010*, 1097–1099.

Labarre, P., Boyle, D., Hawkins, K., and Weigl, B. (2011a). Instrument-free nucleic acid amplification assays for global health settings. Proc. SPIE *8029*, 802902.

Labarre, P., Hawkins, K.R., Gerlach, J., Wilmoth, J., Beddoe, A., Singleton, J., Boyle, D., and Weigl, B. (2011b). A simple, inexpensive device for nucleic acid amplification without electricity-toward instrument-free molecular diagnostics in low-resource settings. PLoS One. *6*, e19738.

Lawn, S.D., and Nicol, M.P. (2011). Xpert (R) MTB/RIF assay: development, evaluation and implementation of a new rapid molecular diagnostic for tuberculosis and rifampicin resistance. Future Microbiol. *6*, 1067–1082.

Lee, H.H., Dineva, M.A., Chua, Y.L., Ritchie, A.V., Ushiro-Lumb, I., and Wisniewski, C.A. (2010). Simple amplification-based assay: a nucleic acid based point-of-care platform for HIV-1 testing. J. Infect. Dis. *201*, S65–S72.

Leelawiwat, W., Young, N.L., Chaowanachan, T., Ou, C.Y., Culnane, M., Vanprapa, N., Waranawat, N., Wasinrapee, P., Mock, P.A., Tappero, J., *et al.* (2008). Dried blood spots for the diagnosis and quantitation of HIV-1: stability studies and evaluation of sensitivity and specificity for the diagnosis of infant HIV-1 infection in Thailand. J. Virol. Methods *155*, 109–117.

Letant, S.E., Ortiz, J.I., tley Tammero, L.F., Birch, J.M., Derlet, R.W., Cohen, S., Manning, D., and McBride, M.T. (2007). Multiplexed reverse transcriptase PCR assay for identification of viral respiratory pathogens at the point of care. J. Clin. Microbiol. *45*, 3498–3505.

Liang, C., Chu, Y.A., Cheng, S.J., Wu, H.P., Kajiyama, T., Kambara, H., and Zhou, G.H. (2012). Multiplex loop-mediated isothermal amplification detection by sequence-based barcodes coupled with nicking endonuclease-mediated pyrosequencing. Anal. Chem. *84*, 3758–3763.

Liu, C., Mauk, M.G., Hart, R., Qiu, X., and Bau, H.H. (2011). A self-heating cartridge for molecular diagnostics. Lab Chip *11*, 2686–2692.

Livak, K.J., Flood, S.J., Marmaro, J., Giusti, W., and Deetz, K. (1995). Oligonucleotides with fluorescent dyes at opposite ends provide a quenched probe system useful for detecting PCR product and nucleic acid hybridization. PCR Methods Appl. *4*, 357–362.

Loeffelholz, M.J., Pong, D.L., Pyles, R.B., Xiong, Y., Miller, A.L., Bufton, K.K., and Chonmaitree, T. (2011). Comparison of the FilmArray respiratory panel and prodesse real-time PCR assays for detection of respiratory pathogens. J. Clin. Microbiol. *49*, 4083–4088.

Lofgren, S.M., Morrissey, A.B., Chevallier, C.C., Malabeja, A.I., Edmonds, S., Amos, B., Sifuna, D.J., von Seidlein, L., Schimana, W., Stevens, W.S., *et al.* (2009). Evaluation of a dried blood spot HIV-1 RNA program for early infant diagnosis and viral load monitoring at rural and remote healthcare facilities. AIDS 23, 2459–2466.

Lutz, S., Weber, P., Focke, M., Faltin, B., Hoffmann, J., Muller, C., Mark, D., Roth, G., Munday, P., Armes, N., *et al.* (2010). Microfluidic lab-on-a-foil for nucleic acid analysis based on isothermal recombinase polymerase amplification (RPA). Lab Chip *10*, 887–893.

McHugh, T.D., Pope, C.F., Ling, C.L., Patel, S., Billington, O.J., Gosling, R.D., Lipman, M.C., and Gillespie, S.H. (2004). Prospective evaluation of BDProbeTec strand displacement amplification (SDA) system for diagnosis of tuberculosis in non-respiratory and respiratory samples. J. Med. Microbiol. *53*, 1215–1219.

McNerney, R., and Daley, P. (2011). Towards a point-of-care test for active tuberculosis: obstacles and opportunities. Nat. Rev. Microbiol. *9*, 204–213.

Madi, M., Hamilton, A., Squirrell, D., Mioulet, V., Evans, P., Lee, M., and King, D.P. (2012). Rapid detection of foot-and-mouth disease virus using a field-portable nucleic acid extraction and real-time PCR amplification platform. Vet. J. *193*, 67–72.

Mahalanabis, M., Do, J., ALMuayad, H., Zhang, J.Y., and Klapperich, C.M. (2010). An integrated disposable device for DNA extraction and helicase dependent amplification. Biomed. Microdevices *12*, 353–359.

Maples, B.K., Holmberg, R.C., Miller, A.P., Provins, J.W., Roth, R.B., and Mandell, J.G. (2009). Nicking and Extension Amplification Reaction for the Exponential Amplification of Nucleic Acids. Ionian Technologies Inc. US2009081670-A1. 14 July 2008. Ref Type: Patent.

Marner, E.S., Wolk, D.M., Carr, J., Hewitt, C., Dominguez, L.L., Kovacs, T., Johnson, D.R., and Hayden, R.T. (2011). Diagnostic accuracy of the Cepheid GeneXpert vanA/vanB assay ver. 1.0 to detect the vanA and vanB vancomycin resistance genes in Enterococcus from perianal specimens. Diagn Microbiol. Infect. Dis. *69*, 382–389.

Mehta, N., Trzmielina, S., Nonyane, B.A.S., Eliot, M.N., Lin, R.H., Foulkes, A.S., McNeal, K., Ammann, A., Eulalievyolo, V., Sullivan, J.L., *et al.* (2009). Low-cost HIV-1 diagnosis and quantification in dried blood spots by real time PCR. PLoS One *4*, e5819.

Melin, P. (2011). Neonatal group B streptococcal disease: from pathogenesis to preventive strategies. Clin. Microbiol. Infect. *17*, 1294–1303.

Mens, P.F., van Amerongen, A., Sawa, P., Kager, P.A., and Schallig, H.D.F.H. (2008). Molecular diagnosis of malaria in the field: development of a novel 1-step nucleic acid lateral flow immunoassay for the detection of all 4 human Plasmodium spp. and its evaluation in Mbita, Kenya. Diagn. Microbiol. Infect. Dis. *61*, 421–427.

Meridian (2011). Illumigene *C. difficile* assay. Available at: http://www.meridianbioscience.com/illumigene (accessed 19 February 2012).

Miller, S., Moayeri, M., Wright, C., Castro, L., and Pandori, M. (2010). Comparison of GeneXpert FluA PCR to direct fluorescent antibody and respiratory viral panel PCR assays for detection of 2009 novel H1N1 influenza virus. J. Clin. Microbiol. *48*, 4684–4685.

Mitani, Y., Lezhava, A., Sakurai, A., Horikawa, A., Nagakura, M., Hayashizaki, Y., and Ishikawa, T. (2009). Rapid and cost-effective SNP detection method: application of SmartAmp2 to pharmacogenomics research. Pharmacogenomics *10*, 1187–1197.

Mitarai, S., Okumura, M., Toyota, E., Yoshiyama, T., Aono, A., Sejimo, A., Azuma, Y., Sugahara, K., Nagasawa, T., Nagayama, N., *et al.* (2011). Evaluation of a simple loop-mediated isothermal amplification test kit for the diagnosis of tuberculosis. Int. J. Tuberc. Lung Dis. *15*, 1211–1217.

Mitarai, S., Karinaga, R., Yamada, H., Mizuno, K., Chikamatsu, K., Aono, A., Sugamoto, T., and Hatano, T. (2012). TRICORE, a novel bead-based specimen concentration method for the culturing of Mycobacterium tuberculosis. J. Microbiol. Methods *90*, 152–155.

Monecke, S., and Ehricht, R. (2005). Rapid genotyping of methicillin-resistant Staphylococcus aureus (MRSA) isolates using miniaturised oligonucleotide arrays. Clin. Microbiol. Infect. *11*, 825–833.

Mori, Y., and Notomi, T. (2009). Loop-mediated isothermal amplification (LAMP): a rapid, accurate, and cost-effective diagnostic method for infectious diseases. J. Infect. Chemother. *15*, 62–69.

Mori, Y., Hirano, T., and Notomi, T. (2006). Sequence specific visual detection of LAMP reactions by addition of cationic polymers. BMC Biotechnol. *6*, 3.

Morrison, T., Hurley, J., Garcia, J., Yoder, K., Katz, A., Roberts, D., Cho, J., Kanigan, T., Ilyin, S.E., Horowitz, D., *et al.* (2006). Nanoliter high throughput quantitative PCR. Nucleic Acids Res. *34*, e123.

Mugasa, C.M., Laurent, T., Schoone, G.J., Kager, P.A., Lubega, G.W., and Schallig, H.D.F.H. (2009). Nucleic acid sequence-based amplification with oligochromatography for detection of trypanosoma brucei in clinical samples. J. Clin. Microbiol. *47*, 630–635.

Nagatani, N., Yamanaka, K., Ushijima, H., Koketsu, R., Sasaki, T., Ikuta, K., Saito, M., Miyahara, T., and Tamiya, E. (2012). Detection of influenza virus using a lateral flow immunoassay for amplified DNA by a microfluidic RT-PCR chip. Analyst *137*, 3422–3426.

Nanassy, O.Z., Haydock, P., Beck, N., Barker, L.M., Hargrave, P., Gestwick, D., Lindsey, W.C., Reed, M.W., and Meschke, J.S. (2011). Extraction of MS2 phage RNA from upper respiratory tract specimens by use of flat glass devices. J. Clin. Microbiol. *49*, 1010–1016.

Nelson, N.C. (1998). Rapid detection of genetic mutations using the chemiluminescent hybridization protection assay (HPA): overview and comparison with other methods. Crit. Rev. Clin. Lab. Sci. *35*, 369–414.

Neumann, G., Noda, T., and Kawaoka, Y. (2009). Emergence and pandemic potential of swine-origin H1N1 influenza virus. Nature *459*, 931–939.

Newell, M.L., Coovadia, H., Cortina-Borja, M., Rollins, N., Gaillard, P., and Dabis, F. (2004). Mortality of infected and uninfected infants born to HIV-infected mothers in Africa: a pooled analysis. Lancet *364*, 1236–1243.

Ngo-Giang-Huong, N., Khamduang, W., Leurent, B., Collins, I., Nantasen, I., Leechanachai, P., Sirirungsi, W., Limtrakul, A., Leusaree, T., Comeau, A.M., *et al.* (2008). Early HIV-1 diagnosis using in-house real-time PCR amplification on dried blood spots for infants in remote and resource-limited settings. J. Acquir. Immune Defic. Syndr. *49*, 465–471.

Niemz, A., and Boyle, D. (2012). Nucleic acid testing for tuberculosis at the point-of-care in high-burden countries. Expert Rev. Mol. Diagn. *12*, 687–701.

Niemz, A., Ferguson, T.M., and Boyle, D.S. (2011). Point-of-care nucleic acid testing for infectious diseases. Trends Biotechnol. *29*, 240–250.

Niwa, T., Kawamura, Y., Katagiri, Y., and Ezaki, T. (2005). Lytic enzyme, labiase for a broad range of Gram-positive bacteria and its application to analyze functional DNA/RNA. J. Microbiol. Methods *61*, 251–260.

Noren, T., Alriksson, I., Andersson, J., Akerlund, T., and Unemo, M. (2011). Rapid and sensitive loop-mediated isothermal amplification test for *Clostridium difficile* detection challenges cytotoxin B cell test and culture as gold standard. J. Clin. Microbiol. *49*, 710–711.

Notomi, T., Okayama, H., Masubuchi, H., Yonekawa, T., Watanabe, K., Amino, N., and Hase, T. (2000). Loop-mediated isothermal amplification of DNA. Nucleic Acids Res. *28*, e63.

Novitsky, V., and Essex, M. (2012). Using HIV viral load to guide treatment-for-prevention interventions. Curr. Opin. HIV. AIDS *7*, 117–124.

Otto, M. (2012). MRSA virulence and spread. Cell. Microbiol. *14*, 1513–1521.

Paule, S.M., Pasquariello, A.C., Hacek, D.M., Fisher, A.G., Thomson, R.B. Jr., Kaul, K.L., and Peterson, L.R. (2004). Direct detection of *Staphylococcus aureus* from adult and neonate nasal swab specimens using real-time polymerase chain reaction. J. Mol. Diagn. *6*, 191–196.

Peiris, J.S.M., de Jong, M.D., and Guan, Y. (2007). Avian influenza virus (H5N1): a threat to human health. Clin. Microbiol. Rev. *20*, 243–267.

Perkins, M.D., and Cunningham, J. (2007). Facing the crisis: improving the diagnosis of tuberculosis in the HIV era. J. Infect. Dis. *196*, S15–S27.

Perreten, V., Vorlet-Fawer, L., Slickers, P., Ehricht, R., Kuhnert, P., and Frey, J. (2005). Microarray-based detection of 90 antibiotic resistance genes of Gram-positive bacteria. J. Clin. Microbiol. 43, 2291–2302.

Piatek, A.S., Tyagi, S., Pol, A.C., Telenti, A., Miller, L.P., Kramer, F.R., and Alland, D. (1998). Molecular beacon sequence analysis for detecting drug resistance in Mycobacterium tuberculosis. Nat. Biotechnol. *16*, 359–363.

Piepenburg, O., Williams, C.H., Stemple, D.L., and Armes, N.A. (2006). DNA detection using recombination proteins. PLoS Biol. *4*, e204.

Pierce, V.M., and Hodinka, R.L. (2012). Comparison of the GenMark diagnostics esensor respiratory viral panel to real-time PCR for detection of respiratory viruses in children. J. Clin. Microbiol. *50*, 3458–3465.

Poritz, M.A., Blaschke, A.J., Byington, C.L., Meyers, L., Nilsson, K., Jones, D.E., Thatcher, S.A., Robbins, T., Lingenfelter, B., Amiott, E., *et al.* (2011). FilmArray, an automated nested multiplex PCR system for multi-pathogen detection: development and application to respiratory tract infection. PLoS One *6*, e26047.

Posthuma-Trumpie, G.A., Korf, J., and van Amerongen, A. (2009). Lateral flow (immuno) assay: its strengths, weaknesses, opportunities and threats. A literature survey. Anal. Bioanal. Chem. 393, 569–582.

Price, C.W., Leslie, D.C., and Landers, J.P. (2009). Nucleic acid extraction techniques and application to the microchip. Lab Chip *9*, 2484–2494.

Puthawibool, T., Senapin, S., Kiatpathomchai, W., and Flegel, T.W. (2009). Detection of shrimp infectious myonecrosis virus by reverse transcription loop-mediated isothermal amplification combined with a lateral flow dipstick. J. Virol. Methods *156*, 27–31.

Radstrom, P., Knutsson, R., Wolffs, P., Lovenklev, M., and Lofstrom, C. (2004). Pre-PCR processing – strategies to generate PCR-compatible samples. Mol. Biotechnol. *26*, 133–146.

Raja, S., Ching, J., Xi, L.Q., Hughes, S.J., Chang, R., Wong, W., McMillan, W., Gooding, W.E., McCarty, K.S., Chestney, M., *et al.* (2005). Technology for automated, rapid, and quantitative PCR or reverse transcription-PCR clinical testing. Clin. Chem. *51*, 882–890.

Rogers, C.D.G., and Burgoyne, L.A. (2000). Reverse transcription of an RNA genome from databasing paper (FTA (R)). Biotechnol. Appl. Biochem. *31*, 219–224.

Rosen, S (2012). The Worldwide Market for In Vitro Diagnostic (IVD) Tests, 8th edn. (Kalorama Information, New York)

Rouet, F., and Rouzioux, C. (2007). The measurement of HIV-1 viral load in resource-limited settings: how and where? Clin. Lab. 53, 135–148.

Rupnik, M., Wilcox, M.H., and Gerding, D.N. (2009). *Clostridium difficile* infection: new developments in epidemiology and pathogenesis. Nat. Rev. Microbiol. *7*, 526–536.

Russell, D.G. (2007). Who puts the tubercle in tuberculosis? Nat. Rev. Microbiol. *5*, 39–47.

Saetiew, C., Limpaiboon, T., Jearanaikoon, P., Daduang, S., Pientong, C., Kerdsin, A., and Daduang, J. (2011). Rapid detection of the most common high-risk human papillomaviruses by loop-mediated isothermal amplification. J. Virol. Methods *178*, 22–30.

Salazar, O., and Asenjo, J.A. (2007). Enzymatic lysis of microbial cells. Biotechnol. Lett. *29*, 985–994.

Schito, M.L., D'Souza, M.P., Owen, S.M., and Busch, M.P. (2010). Challenges for rapid molecular HIV diagnostics. J. Infect. Dis. *201*, S1–S6.

Sollier, E., Rostaing, H., Pouteau, P., Fouillet, Y., and Achard, J.L. (2009). Passive microfluidic devices for plasma extraction from whole human blood. Sensors Actuators B Chem. *141*, 617–624.

Spencer, D.H., Sellenriek, P., and Burnham, C.A. (2011). Validation and implementation of the GeneXpert MRSA/SA blood culture assay in a pediatric setting. Am. J. Clin. Pathol. *136*, 690–694.

Spurgeon, S.L., Jones, R.C., and Ramakrishnan, R. (2008). High throughput gene expression measurement with real time PCR in a microfluidic dynamic array. PLoS One *3*, e1622.

Stevens, D.S., Crudder, C.H., and Domingo, G.J. (2012). Post-extraction stabilization of HIV viral RNA for quantitative molecular tests. J. Virol. Methods *182*, 104–110.

Stevens, W., Sherman, G., Downing, R., Parsons, L.M., Ou, C.Y., Crowley, S., Gershy-Damet, G.M., Fransen, K., Bulterys, M., Lu, L., *et al.* (2008). Role of the laboratory in ensuring global access to ARV treatment for HIV-infected children: consensus statement on the performance of laboratory assays for early infant diagnosis. Open AIDS J. *2*, 17–25.

Stevens, W.S., Scott, L.E., and Crowe, S.M. (2010). Quantifying HIV for monitoring antiretroviral therapy in resource-poor settings. J. Infect. Dis. *201*, S16–S26.

Storhoff, J.J., Marla, S.S., Bao, P., Hagenow, S., Mehta, H., Lucas, A., Garimella, V., Patno, T., Buckingham, W., Cork, W., *et al.* (2004). Gold nanoparticle-based detection of genomic DNA targets on microarrays using a novel optical detection system. Biosens. Bioelectron. *19*, 875–883.

Strotman, L.N., Lin, G., Berry, S.M., Johnson, E.A., and Beebe, D.J. (2012). Facile and rapid DNA extraction and purification from food matrices using IFAST (immiscible filtration assisted by surface tension). Analyst *137*, 4023–4028.

Sur, K., McFall, S.M., Yeh, E.T., Jangam, S.R., Hayden, M.A., Stroupe, S.D., and Kelso, D.M. (2010). Immiscible phase nucleic acid purification eliminates PCR inhibitors with a single pass of paramagnetic particles through a hydrophobic liquid. J. Mol. Diagn. *12*, 620–628.

Tan, E., Wong, J., Nguyen, D., Zhang, Y., Erwin, B., Van Ness, L.K., Baker, S.M., Galas, D.J., and Niemz, A. (2005). Isothermal DNA amplification coupled with DNA nanosphere-based colorimetric detection. Anal. Chem. *77*, 7984–7992.

Tan, E., Erwin, B., Dames, S., Voelkerding, K., and Niemz, A. (2007). Isothermal DNA amplification with gold nanosphere-based visual colorimetric readout for herpes simplex 2 virus detection. Clin. Chem. *53*, 2017–2020.

Tanner, N.A., Zhang, Y., and Evans, T.C. Jr. (2012). Simultaneous multiple target detection in real-time loop-mediated isothermal amplification. Biotechniques *53*, 81–89.

Tanriverdi, S., Chen, L.J., and Chen, S.Q. (2010). A rapid and automated sample-to-result HIV load test for near-patient application. J. Infect. Dis. *201*, S52–S58.

de Tejada, B.M., Pfister, R.E., Renzi, G., Francois, P., Irion, O., Boulvain, M., and Schrenzel, J. (2011). Intrapartum group B Streptococcus detection by rapid polymerase chain reaction assay for the prevention of neonatal sepsis. Clin. Microbiol. Infect. *17*, 1786–1791.

Tomita, N., Mori, Y., Kanda, H., and Notomi, T. (2008). Loop-mediated isothermal amplification (LAMP) of gene sequences and simple visual detection of products. Nat. Protoc. *3*, 877–882.

Tomlinson, J.A., Dickinson, M.J., and Boonham, N. (2010a). Detection of *Botrytis cinerea* by loop-mediated isothermal amplification. Lett. Appl. Microbiol. *51*, 650–657.

Tomlinson, J.A., Dickinson, M.J., and Boonham, N. (2010b). Rapid detection of *Phytophthora ramorum* and *P. kernoviae* by two-minute DNA extraction followed by isothermal amplification and amplicon detection by generic lateral flow device. Phytopathology *100*, 143–149.

Ullrich, T., Ermantraut, E., Schulz, T., and Steinmetzer, K. (2012). Competitive reporter monitored amplification (CMA) – Quantification of molecular targets by real time monitoring of competitive reporter hybridization. PLoS One 7, e35438.

Ulrich, M.P., Christensen, D.R., Coyne, S.R., Craw, P.D., Henchal, E.A., Sakai, S.H., Swenson, D., Tholath, J., Tsai, J., Weir, A.F., *et al.* (2006). Evaluation of the Cepheid GeneXpert system for detecting *Bacillus anthracis*. J. Appl. Microbiol. *100*, 1011–1016.

UNITAID (2012). Public-Private Partnership Announces Immediate 40 Percent Cost Reduction for Rapid TB Test. Available at: http://www.unitaid.eu/index.php?option=com_content&view=article&layout=edit&id=986 (accessed 19 December 2012).

Urdea, M.S. (1993). Synthesis and characterization of branched dna (Bdna) for the direct and quantitative detection of Cmv, Hbv, Hcv, and Hiv. Clin. Chem. *39*, 725–726.

Urdea, M., Penny, L.A., Olmsted, S.S., Giovanni, M.Y., Kaspar, P., Shepherd, A., Wilson, P., Dahl, C.A., Buchsbaum, S., Moeller, G., *et al.* (2006). Requirements for high impact diagnostics in the developing world. Nature *444* (Suppl. 1), 73–79.

Uttayamakul, S., Likanonsakul, S., Sunthornkachit, R., Kuntiranont, K., Louisirirotchanakul, S., Chaovavanich, A., Thiamchai, V., Tanprasertsuk, S., and Sutthent, R. (2005). Usage of dried blood spots for molecular diagnosis and monitoring HIV-1 infection. J. Virol. Methods *128*, 128–134.

Van Ness, J., Van Ness, L.K., and Galas, D.J. (2003). Isothermal reactions for the amplification of oligonucleotides. Proc. Natl. Acad. Sci. U.S.A *100*, 4504–4509.

Vandeventer, P.E., Weigel, K.M., Salazar, J., Erwin, B., Irvine, B., Doebler, R., Nadim, A., Cangelosi, G.A., and Niemz, A. (2011). Mechanical disruption of lysis-resistant bacterial cells by use of a miniature, low-power, disposable device. J. Clin. Microbiol. *49*, 2533–2539.

Vincent, M., Xu, Y., and Kong, H. (2004). Helicase-dependent isothermal DNA amplification. EMBO Rep. *5*, 795–800.

Violari, A., Cotton, M.F., Gibb, D.M., Babiker, A.G., Steyn, J., Madhi, S.A., Jean-Philippe, P., and McIntyre, J.A. (2008). Early antiretroviral therapy and mortality among HIV-infected infants. N. Engl. J. Med. *359*, 2233–2244.

de Vries, J.J.C., Vossen, A.C.T.M., Kroes, A.C.M., and van der Zeijst, B.A.M. (2011). Implementing neonatal screening for congenital cytomegalovirus: addressing the deafness of policy makers. Rev. Med. Virol. *21*, 54–61.

Wakeley, P.R., Errington, J., and Squirrell, D. (2010). Use of a field-enabled nucleic acid extraction and PCR instrument to detect BVDV. Vet. Rec. *166*, 238–239.

Wang, C.H., Lien, K.Y., Wu, J.J., and Lee, G.B. (2011). A magnetic bead-based assay for the rapid detection of methicillin-resistant Staphylococcus aureus by using a microfluidic system with integrated loop-mediated isothermal amplification. Lab Chip *11*, 1521–1531.

Wastling, S.L., Picozzi, K., Kakembo, A.S., and Welburn, S.C. (2010). LAMP for human African trypanosomiasis: a comparative study of detection formats. PLoS. Negl. Trop. Dis. *4*, e865.

WHO (2011). Automated Real-time Nucleic Acid Amplification Technology for Rapid and

Simultaneous Detection of Tuberculosis and Rifampicin Resistance: Xpert MTB/RIF System, Policy Statement. Available at: http://whqlibdoc.who.int/publications/2011/9789241501545_eng.pdf (accessed 24 May 2012).

WHO (2012). Global Tuberculosis Report 2012. Available at: http://apps.who.int/iris/bitstream/10665/75938/1/9789241564502_eng.pdf (accessed 19 December 2012).

Wiedmann, M., Wilson, W.J., Czajka, J., Luo, J.Y., Barany, F., and Batt, C.A. (1994). Ligase chain-reaction (LCR) – overview and applications. PCR Methods Appl. *3*, S51-S64.

Wilson, I.G. (1997). Inhibition and facilitation of nucleic acid amplification. Appl. Environ. Microbiol. *63*, 3741–3751.

Wilson, S., Lane, A., Rosedale, R., and Stanley, C. (2010). Concentration of Mycobacterium tuberculosis from sputum using ligand-coated magnetic beads. Int. J. Tuberc. Lung Dis. *14*, 1164–1168.

Wolk, D.M., Picton, E., Johnson, D., Davis, T., Pancholi, P., Ginocchio, C.C., Finegold, S., Welch, D.F., de Boer, M., Fuller, D., *et al.* (2009). Multicenter evaluation of the Cepheid Xpert methicillin-resistant *Staphylococcus aureus* (MRSA) test as a rapid screening method for detection of MRSA in nares. J. Clin. Microbiol. *47*, 758–764.

Wu, Q., Jin, W., Zhou, C., Han, S., Yang, W., Zhu, Q., Jin, Q., and Mu, Y. (2011). Integrated glass microdevice for nucleic acid purification, loop-mediated isothermal amplification, and online detection. Anal. Chem. *83*, 3336–3342.

Xu, G., Hu, L., Zhong, H., Wang, H., Yusa, S., Weiss, T.C., Romaniuk, P.J., Pickerill, S., and You, Q. (2012). Cross priming amplification: mechanism and optimization for isothermal DNA amplification. Sci. Rep. 2, 246.

Yager, P., Domingo, G.J., and Gerdes, J. (2008). Point-of-care diagnostics for global health. Annu. Rev. Biomed. Eng. *10*, 107–144.

Yoshikawa, T., Ihira, M., Akimoto, S., Usui, C., Miyake, F., Suga, S., Enomoto, Y., Suzuki, R., Nishiyama, Y., and Asano, Y. (2004). Detection of human herpesvirus 7 DNA by loop-mediated isothermal amplification. J. Clin. Microbiol. *42*, 1348–1352.

You, Q., Hu, L., Wang, J., and Zhong, H. (2006). Method for amplifying target nucleic acid sequence by nickase, and kit for amplifying target nucleic acid sequence and its use. Hagzhou Yousida Biotechnology Co Ltd. CN1850981-A; CN100489112-C. 3-10-2006. Ref Type: Patent.

Yulong, Z., Xia, Z., Hongwei, Z., Wei, L., Wenjie, Z., and Xitai, H. (2010). Rapid and sensitive detection of Enterobacter sakazakii by cross-priming amplification combined with immuno-blotting analysis. Mol. Cell. Probes *24*, 396–400.

Zhang, J.L., Guo, Q.Q., Liu, M., and Yang, J. (2008). A lab-on-CD prototype for high-speed blood separation. J. Micromech. Microeng. *18*.

Zhang, Y.H., and Ozdemir, P. (2009). Microfluidic DNA amplification – a review. Anal. Chim. Acta *638*, 115–125.

Zhang, Z., Kermekchiev, M.B., and Barnes, W.M. (2010). Direct DNA amplification from crude clinical samples using a PCR enhancer cocktail and novel mutants of Taq. J. Mol. Diagn. *12*, 152–161.

Zhou, P., Young, L., and Chen, Z.Y. (2010a). Weak solvent based chip lamination and characterization of on-chip valve and pump. Biomed. Microdevices *12*, 821–832.

Zhou, P., Young, L., Spizz, G., Roswech, T., Yasmin, R., Chen, Z., Mouchka, G., Thomas, B., Honey, W., and Montagna, R.A. (2010b). Rheonix CARD (TM) technology a fully automated molecular diagnostic for infectious diseases. J. Mol. Diagn. *12*, 892–893.

Zipper, H., Brunner, H., Bernhagen, J., and Vitzthum, F. (2004). Investigations on DNA intercalation and surface binding by SYBR Green I, its structure determination and methodological implications. Nucleic Acids Res. 32, e103.

Zumla, A., Abubakar, I., Raviglione, M., Hoelscher, M., Ditiu, L., Mchugh, T.D., Squire, S.B., Cox, H., Ford, N., McNerney, R., *et al.* (2012). Drug-resistant tuberculosis–current dilemmas, unanswered questions, challenges, and priority needs. J. Infect. Dis. *205* (Suppl. 2), S228–S240.

# From Bench to Bedside: Development of Polymerase Chain Reaction-integrated Systems in the Regulated Markets

11

Martin Lee, Diane Lee and Phillip Evans


## Abstract

The evolution of the polymerase chain reaction (PCR) as a revolutionary molecular biology tool has been extremely rapid since its inception in 1983. This may be attributed, in part, to the simplicity of the process for the general molecular biologist and also because of the open licensing strategy provided for underpinning products by key stakeholders. The development of commercial products for the applied markets such as food, veterinary, and defence has been greater than for human diagnostics. This sector has been restrictive for most system developers because of the high value of upfront and running royalty licences. In addition, the human diagnostic area is highly regulated in the US (FDA) and the EU (IVDD). The core patents covering the PCR process are approaching their termination dates in key territories, largely negating the requirement for such licences. The opportunity for those developers of Nucleic Acid Tests (NAT) on highly automated laboratory devices and/or simple to use integrated point-of-use PCR devices (iNAT) to 'tap into' the content developed within this broad research sector is enormous. This text is written to provide both technical and commercial advice to those involved in this technology transfer opportunity.

The text is presented as a number of short keynotes in the following areas: (1) PCR formulation; (2) assay stabilization; (3) the sample extraction reagent interface; (4) nucleic acid extraction; (5) signalling and automated analysis; (6) suppliers and GMP; (7) system validation; and (8) the regulatory process. These notes are intended to provide guidance to the key issues rather than a prescriptive manual. The author will direct the reader to the appropriate authority. It will be of most use for individuals and organizations either intending to provide, or to receive, product content. It is based upon the authors' experiences in developing integrated systems for the applied and diagnostics sectors over the last 15 years.


## Introduction

Nucleic acid totally integrated systems (iNAT) have been in development over the last 15 years. The Cepheid GeneXpert® has been at market for some time and several others are at a near market stage. These systems differ from contemporary laboratory integrated systems that tend to be automated using one or more existing solution(s) to achieve a high-throughput analyser, processing many samples of the same type in parallel. Systems such as the GeneXpert® and Enigma®ML address a different requirement. The operation of a complete nucleic acid test by a generalist, rather than specialized operator, and the possibility of a true point-of-need solution remain their primary objective. In doing so they address testing using single discrete patient samples whilst carrying out a mix of assay types that may be introduced into the device in a random access fashion. The development of these iNAT point-of-need devices has therefore required elegant innovative solutions to the key complex integration steps. This has meant that there are important differences between the technological approaches employed in a standard laboratory system and those chosen for an iNAT.

The polymerase chain reaction (PCR) is the primary choice for a robust, rapid and sensitive nucleic acid test. Since PCR's inception it has been applied in too many applications to summarize here. Whilst other amplification methods such as isothermal approaches lend themselves well to iNAT, the main attraction of PCR is the available assay 'content'. From a technical, application specific and geographical perspective there is an unmatched number of existing tests and experienced development teams available. These derive mainly from research and diagnostic service laboratories. However, the trend is for iNAT developers to acquire companies that have existing content and to re-engineer and validate these tests for their specific platform. There are several examples of successful products that have moved from laboratory or research-based platforms into highly integrated automated systems. Acquisitions rather than collaborative approaches occur because knowledge of the transfer process is the key to successful product development. It is far easier for system developers, to acquire the content and address issues themselves, than undertake collaborative programmes of work, since many of the wider technical and commercial issues are less generally understood even by the experienced PCR scientist.

This chapter addresses the key issues for the assay transfer from a standard laboratory test to that of an iNAT. Whilst a PCR focus is maintained the issues are similar for the integration of other amplification methods. Likewise, some of the information is also relevant to contemporary laboratory automated systems. The information here was gathered from the authors combined 20 years' experience of evaluation and development of commercial iNAT PCR devices. The authors present this as a number of key-notes in those areas which are general to the process. These key-notes are therefore not platform specific. They cover issues in technical, regulatory and commercial areas. It is intended that these keynotes are of most use to new technology groups who are seeking to develop an iNAT. It is also intended to guide those who see themselves as the providers of assay content. Whilst specific iNAT platforms differ significantly in their technological approach, the text is designed to be useful in assessing both specific platform and assay transfer readiness, whilst providing an indicative measure of the associated 'value proposition' for such partnerships.

## Background

Any iNAT that is to be developed will be more than a PCR system. Generally speaking, it will have the following additional 'sub-systems':

- *Pre-analytical*: For a true iNAT to achieve the point-of-need requirement it must be free of any ancillary pre-processing or laboratory equipment. Indeed, this is a pre-requisite for achieving CLIA (Clinical Laboratory Improvement Amendments) waiver status (as discussed in key-note 7 below). Pre-analytical processes are normally laborious activities in the high-throughput lab. They include the transfer of the samples from the clinical collection devices into the test consumable. They may also include some pre-processing steps such as heating or vortexing that are not currently automated in existing laboratories.
- *Nucleic Acid Extraction*: Whilst not needed for all NAT, the majority of pathogenic samples require the extraction or removal of the test nucleic acid from an endogenous matrix of cellular material and other nucleic acids. The process ideally both purifies and concentrates the nucleic acid.
- *Stabilized pre-dosed PCR reagents*: iNAT devices have been developed for the point-of-need market. Non-laboratory reagents are transported and stored away from the traditional cold supply chain.
- *Amplification*: Thermal cycling combined with integrated detection.
- *Analysis*: Automated interpretation of data to provide clinically relevant information to the operator.

Throughout the following key-notes the authors make reference to these processes and the interfaces, such as those exemplified in Table 11.1. Understanding these interactions and associated dependencies are the key factors to successful assay transfer.

**Table 11.1** iNAT key interfaces; description, impact and mitigation

| Interface | Description of interface | Process impact and mitigation |
|---|---|---|
| General plastic to chemistry | The type, grade and shape of the plastic material can have an effect on the interaction with some chemical components of the reagents. For example, the plastic may by hydrophobic or hydrophilic. In addition the surface may be porous or pitted. This may have an effect on the retention of material within the container. The abstraction of reagents from the liquid to the plastic may occur | The presence of inhibitory substances such as exogenous metal ions is known to inhibit both the processivity of polymerase enzymes and affect the hybridisation of oligonucleotides (primers and probes). The abstraction of key reagents is similar in that it will also reduce the efficiency of the amplification process. The measurement of these and other effects on the reverse transcription component of a one-tube RT reaction is easier to measure than the effect on PCR component alone because the reverse transcriptase enzyme and process is much more sensitive to inhibition. The PCR reaction amplifies these effects because the process is exponential. Therefore, materials testing of tubes or compensatory reagents is best achieved using a reverse transcriptase PCR at or near to the LoD. This will establish that the reagents are compatible with the tube polymer |
| Reaction lid | The lid or seal of the tubes provides a closure for the PCR vessel. The functions include:<br>• Stopping evaporative loss of reaction solvent(s) during heating steps<br>• A cold point for reagent solvent to condense during cooling steps<br>• To apply an over-pressure to the reaction sample to reduce 'gasing out' which may form gas at the bubble point of the sample | Loss of solvent during the reaction will increase the solute concentration in later cycles. This may change the reaction performance through efficiency differences. This may lead to a reduced level of sensitivity and may also alter specificity. Amplification precision may be poor. It will also affect the thermal loading in the PCR well and therefore any constants associated with that in the thermal control process may be invalid<br><br>Loss of solvent also causes changes in reaction fluorescence. The net increase in fluorophore concentration may appear as an upward drift in the reaction baseline. Refluxing may be the result of excessive evaporation and cause an undulating fluorescent baseline<br><br>Bubbles may cause an optical effect that may have a number of secondary effects:<br>• Fluoroscent baseline drift may be upwards or downwards dependant upon bubble movement within the tube<br>• Steps may appear as bubbles migrate rapidly during heating phase between cycle-to-cycle fluorescent acquisitions<br>• Bubbles may also displace thermal mass from the region of the thermal sensor(s)<br><br>The main mitigation is to provide a passive reference dye into the mix to provide the system with a method to normalise the resulting data for possible volumetric changes<br><br>An imperfect fitting lid may also cause laboratory contamination. It is therefore appropriate to include dUTPs as a minimum contingency plan to use in conjunction with a UNG carry-over prevention approach. Oil overlays may be employed to eliminate evaporative effects in those systems that use epi-illumination. Other optical arrangements may be impeded by such approaches |

**Table 11.1** Continued

| Interface | Description of interface | Process impact and mitigation |
|---|---|---|
| Optical interface | The optical interface provides a means to allow excitation light to enter and fluorescent emission energy to leave the PCR allowing the tube to function as an optical cuvette. A number of factors are known to change performance:<br>• Intrinsic fluorescence of the window material will create an offset signal in acquisitions<br>• The shape of the surfaces will affect the optical transmission efficiency<br>• Scratches will reduce optical coupling<br>• Distortion or stressing of the windows during the moulding processes may cause polarisation effects<br>• Poor surface finish of the manufacturing moulds may nebulise bubble formation on the internal surface | Variability of the intrinsic fluorescence of the optical interface may cause issues with any optical calibration and the algorithmic colour compensation process. This fluorescence may bleach under intense illumination contributing (more so) to reaction signal drift<br>Polarisation effects may cause spectral issues as a result of tube movement during, for example, the heating stages of the PCR<br>Bubbles may cause step fluorescent effects<br>Loss of transmission will reduce the overall gain in the signal and therefore specific tubes may exhibit reduced signal: noise<br>There are few mitigating considerations for a poorly designed optical interface. It may limit the application in terms of the fluorescent approaches that may be employed. It may also affect the overall sensitivity or specificity of the system |
| Thermal contact and homogeneity | Imperfections in the PCR moulding and the resulting thermal performance can have several effects on the performance of the reagents:<br>• By design the tube may not have a high surface area: volume ratio, therefore the intrinsic thermal gradient may be high. The equilibrium time may be long<br>• Tolerance stacking in the mould design may mean that the tube does not make good thermal contact with the heater system. Thermal union is an essential feature of well to well precision and overall performance | Cycle-to-cycle thermal performance of the system drives the reaction specificity. Excessive heating can cause reagent evaporation (see above) and possibly biological reagent degradation<br>Cycle-to cycle thermal equilibrium at the time-point of optical acquisition is critical in achieving a low noise in the assay signal. The effect is a result of the change in fluorescence with temperature which is reciprocal. Excessive cycle-cycle thermal variation is therefore observed (and often wrongly interpreted) as an optical effect. This may lower the overall assay sensitivity. This may also affect any colour compensation or correction process<br>Thermal equilibration times will affect the resolution of nucleic acid probe dissociation analysis. It will also affect High Resolution Melting experiments (HRM)<br>Thermal precision on a cycle-to-cycle basis affects reaction stringency and therefore PCR sensitivity and specificity<br>The authors noted that difference in the thermal design of many systems precludes them from achieving performance of comparators. For example, their intrinsic thermal gradient means that they may never achieve the stringency required for some amplification methods |

## Key-note 1: PCR formulation

### Core formulation

Core PCR formulations differ in their optima in that they are often formulated to a specific platform. These include;

- *The tube format* has specific characteristics that may change the requirement for specific core reaction component(s). For example, glass has been used to great extent in the LightCycler® (Roche Applied Science) and RAPID® (Biofire Diagnostics) family of instruments because of its ideal optical and thermal properties. However, the specific properties of the glass narrow the optima such that bovine serum albumin (BSA) is essential for efficient PCR. BSA reduces the abstraction of specific reagents and analytes that may occur as a result of the binding properties of the glass. It has proven difficult for many labs to get bicine or tricine buffers to work effectively in glass, particularly in combination with manganese ions needed to stimulate some polymerase (*pol*) enzymes for reverse transcriptase activity. The buffering capacity required is generally higher in glass, even for tris-based formulations, e.g. 50 mM tris vs the usual 10 mM. Many of the salts routinely included in PCR to stimulate the processivity of *pol* enzymes in plastic tube formats can actively inhibit the reaction in some glass formats. Other systems may have similar deviations. Not all PCR tubes are made from the same polymer. Whilst polypropylene is the most common, some formats use polycarbonate and others, similar plastics. Some systems may also use the addition of co-polymers or coatings to assist with filling and to reduce adsorption. Therefore, the core reagent requirements may change during the assay transfer process to meet the needs of the specific iNAT tube format.
- *Rapid PCR*: When moving to a faster cycling format from a more conventional lab-based system. The surface area to volume ratio may increase to allow rapid thermal transfer and reduce the thermal gradients for better hybridization stringency. Even when the same polymer is used for the iNAT tube as the originating platform tube that the assay was optimized in, it may be necessary to block the abstraction of active ingredients due to the increased surface area alone. With shorter holds and rapid thermal transitions there is also a requirement for increased mass transfer between the reagents and the analytes because there is less time for the diffusion process and for the various solutes to collide. Shortening the PCR increases the rate at which mass transfer within the solution must occur, if the reaction is to maintain the same efficiency. Therefore, when moving to a rapid format, it is not uncommon to find the bivalent metal ion and polymerase concentration requirement increases to enable rapid cycling. Typically, to achieve rapid PCR for a standard reaction, this changes from 1.5 mM to 3 mM for magnesium ions, and from 0.025 units/μl to 0.04 units/μl for *Taq* polymerase. Other PCR components are in sufficient excess to not require modification with respect to this issue.
- *Hotstart*: Many of the proprietary hotstart enzymes sold for research are supplied with a companion buffer(s). Some hot starts are activated by the shift (drop) in pH in the tris buffer as the reaction is heated. As enzyme activation is favoured by a lower pH, the PCR process is best served by a lower pH, thus reducing the reaction pH to 8.0. Whilst this pH is not optimum for the native *Taq* enzyme, one can achieve a more than satisfactory result, which is significantly enhanced by the polymerase hotstart process. A change in choice of hotstart may require adjustment of this core buffer. This could perturb the overall optima and therefore performance of the assay.

Whilst these items themselves are not major issues and it is fairly common to change and re-optimize conditions between platforms, it may mean that significant re-optimization of the assay is required. Each issue becomes more prominent as the complexity of the assay increases with multiplexing. Specific optima formulations often remain the 'know how' of system developers.

## Probe system

The majority of reported probe-based real-time PCR assays are based upon the 5′ nuclease assay (Roche Molecular Systems) which is retailed for research under the 'TaqMan®' trademark (Roche Molecular Systems and it's licensees). Its success in the research sector is testament to its simplicity of use, the availability of design tools and vast range of suppliers able to support the research community. Most capable instruments today are able to support a multiplex of at least four targets. The 5′ nuclease assay is, therefore, an ideal approach for assays requiring a low level of multiplexing and a quantitative result. The main drawback of the approach is that, whilst it is freely available for research purposes through the purchase of licensed reagents and probes, its application in clinical and other sectors requires specific field-based licences under the core 5′ nuclease assay process patents. Whilst many of these patents are reaching their termination dates, technologies that augment this process will remain subject of obtaining the relevant licences for some time. Examples include specific quenching moieties, modified nucleotides, and binding domains that may increase melting temperature ($T_m$) and facilitate shorter probes. In choosing the 5′ nuclease assay, there may be higher access costs associated with implementation in the diagnostic sector.

Where the platform company has acquired these rights the remaining issues are technological. There are many third party probe methods which are discussed in more detail elsewhere (Lee *et al.*, 2009a). Some provide a greater scope for increased multiplexing whilst reducing the logistics and costs associated with final manufacture. For example, the ability to carry out probe dissociation (using probe technology such as molecular beacons – Chakravorty *et al.* (2011)) increases the theoretical multiplexing capability of a given platform when performed using multiple fluorophores in parallel.

Often it is not an easy process to change the signalling system for a given primer pair. Indeed, some probe methods are intrinsically linked to the primer arrangement. The value of an established test that requires re-engineering for the signalling is questionable, since any associated performance validation data is no longer relevant. It is often much easier to redesign the assay around the preferred signalling system.

## Dye choices

There are many fluorophores available for fluorogenic probes. Many generic fluorophores are available to match the specifications of those well-known proprietary molecules. These may circumvent the patent rights of those suppliers who choose to reserve rights in the diagnostics field at a premium. This is perhaps a minor consideration at an early developmental stage, but the authors highlight this issue because there are, at the moment, few suppliers who provide the full selection of dyes to practice all fluorophore label orientations whilst also being able to manufacture them to meet diagnostic industry standards (see KEY NOTE 8). The dye combinations need to match the optical specification of the integrated system.

## Proprietary reagents

The state-of-the-art of PCR is exemplified through the commercial kits that support life science research. Many of these are in a 'ready-to-go' format. They contain enzymes that may be genetically engineered variants of existing RNA/DNA dependant DNA polymerases, fusions of one more complementary proteins, or novel isolates. These kits make it possible for the end user to quickly develop PCR assays, often negating the need for optimization, and implement their assays in applied and research sectors. However, such kits may represent years of underpinning development. They may also be cocktails of several proprietary reagents and processes that carry a significant licence royalty burden. Some sub-components may not be available as single source items. Examples of such sub components include: hotstart polymerase enzymes; recombinant variants of both DNA dependant DNA polymerases (*pol*) and RNA dependant DNA polymerases (RT); enzymes for carry-over prevention (UNG); specific adjuncts that improve hybridization or reduce GC bias for hybridization performance; binding proteins that enhance the kinetics of hybridization and reaction fidelity.

When an existing assay is formulated around one or more proprietary components then its value may be diminished since it is inevitable that empirical re-optimization will be required to attain an intellectual property freedom to operate position. Knowledge of, and the ability to reformulate is also a requirement for many reagent stabilization processes that add complexity to the chemistry.

## Key-note 2: assay stabilization

iNAT platforms generally try and negate the traditional cold supply chain to allow access into new testing environments such as, for example, accident and emergency wards or the near patient clinic. In doing so, the requirement for ambient stabilization becomes important. There are several technological approaches that fulfil this need; lyophilization is the most commonly known but there are other commercial processes available, for example, micro-encapsulation. Some, if not all, processes are proprietary, may require specific services and possibly attract a licence fee. The intellectual property landscape in this area is very complex and beyond the scope of this chapter.

The lyophilization process requires sublimation of a reaction (or part thereof) in the presence of stabilizing sugars and macromolecules (excipients) that replace the removed water. The sugars, along with other excipients, determine the 'cake' or 'bead' structure, its stability and the effectiveness of its dissolution. The requirement for the physical form (cake or bead) may change between platforms as will the solubility requirement. Some systems may, for example, use a pipette or centrifugal action to help dissolution. Others may use a more fluidic approach. The exact optimal formulation may therefore be very platform specific. An assay in a given stabilized dried format may, therefore, not be readily applied across all integrated systems.

The excipients are solutes that can also affect the solvation of macromolecules. As such, they may affect the core enzymatic and hybridization processes. The stringency of the assay can be significantly affected by the chosen excipients. The extent to which this occurs depends on the assay build and the technology employed. It may be more apparent, for example, in assays that use probe dissociation analysis, because these effects can be visualized by the assay development team. Some may also quench the emission of different fluorophores. Transfer between different lyophilization formulations can result in these different effects. The excipients and stabilization process should be assessed during the assay development stage rather than as an end-process. This is particularly important where the assay is to achieve high level multiplexing and sensitivity is a requirement.

During stabilization there are often losses of specific reagent activities during the sublimation and later drying processes. This includes loss of enzyme(s) and fluorophore functionality. These issues are best addressed through compensation at the formulation stage. Such losses are distinct from those that may occur upon storage. Storage losses need to be modelled using both accelerated and real-time stability studies to ensure the reagents meet their specifications, both in terms of stability, fluorogenic performance and solubility. Accelerated studies use increased temperatures that mimic the molecular collisional events that occur with significantly increased storage times, modelled according to the Arrhenius equation. Such studies are a requisite of regulatory bodies if a claim on product stability is made before real-time data is available.

## Key-note 3: the sample extraction reagent interface

Frozen master mixes are often limited to a 2× formulation. One key advantage of a lyophilized amplification reagent is that it may be re-suspended entirely using sample and this volumetric advantage brings sensitivity improvements for iNAT systems over standard laboratory systems.

The normal process of a 2× dilution of amplification reagent with sample also dilutes inhibitory substances from the extraction. Therefore one has to be sure that the sample processing system for a given assay is optimized such that it removes all inhibitors across a broad range of sample types and operating conditions.

PCR inhibitors may arise, not only from the sample itself, but also from the sample preparation stage. The elution solutions used in many commercial kits are formulated to stabilize the target molecules prior to amplification. For example, they may be buffered or contain EDTA to chelate bivalent metal ions which may be cofactors for co-purified enzymes (nucleases). There may be other solutes associated with the release of nucleic acids from any solid phase if affinity purification is used. These solutes may affect a given PCR's performance. For example EDTA will abstract magnesium and perturb the optimal enzyme-nucleic acid bind conditions (each dNTP has an associated ion). It may, therefore, change the determined $T_m$ in probe dissociation experiments and the stringency of the primers. Under normal use, these solutes may be diluted as part of the normal application of the extracted nucleic acid in any downstream process. For an iNAT, however, it is important that the PCR is optimized using nucleic acids extracted (directly) by the same method that will be employed in the final system or using eluant suspensions of nucleic acids.

## Key-note 4: nucleic acid extraction

Often considered to be of secondary importance to the amplification performance, the extraction process contributes enormously to the overall performance of the platform. It is not only important that the purification and concentration of nucleic acids from the sample is efficient, it is equally important that the process removes contaminating inhibitors. It is also important that the reagents used do not, in some way, contribute to the inhibition of the PCR (see 'Key-note 3').

Many of the iNAT systems have proprietary extraction processes, particularly those based on micro- or mesofluidics, which are far more constrained than those based on column or magnetic bead solid phase extraction process. Systems that can apply the more traditional processes may also be commercially constrained by the choice of specific chemistry that may be used. It is inevitable that the sample extraction process is subject to more change during assay transfer to an integrated system than other assay subsystem.

There are many purification chemistries available, based upon different principles. As most nucleic acid purifications are based upon an affinity binding process, the exact physical form (column/bead/slurry) is perhaps of lesser importance to the reader. Detailing the numerous nucleic acid purification methods available is outside the scope if this chapter. However, we wish to direct the reader to the main characteristics and issues one should be aware of.

One of the main differences between a lab-based process and that required for point-of-need is that the nucleic acid is consumed (in the amplification) within a short period of time. Therefore, the requirement to stabilize the extract by minimizing the effects of nucleases is somewhat reduced. Many of the lengthy protocols that include digestion with broad spectrum proteinases, such as proteinase K, are, therefore, negated.

A significant factor for iNAT systems is that the eluant (containing template) is usually used undiluted in combination with the amplification reagents (see 'Key-note 3'). It is extremely important that the nucleic acid purification process removes inhibitory materials from the sample. Many extraction processes use chaotropic salts as the denaturant for sample lysis. They may also be used in silica-based chemistries to facilitate the binding process. In 'Boom'-based chemistries these two processes (binding and lysis) are combined (Boom *et al.* 1990). Chaotropic agents such as urea and guanidium salts can be very detrimental to the PCR process. Therefore, the volumetric ratio of low salt wash buffer to lysate may need to be significantly higher than for a conventional test. Likewise, where a co-solvent is used in the secondary wash buffer, efficient drying of the solid phase (bead or column) is important, as residual propanol and ethanol may also inhibit the amplification process.

In magnetic bead-based processes, the beads are the solid phase. Unlike gravity or pressure columns, the process requires transfer of the beads using a magnet in combination with a pipette tip and/or 'wand'. Loss of beads in itself can be a significant issue at each transfer step. The mass of beads used determines the assays dynamic range and ultimately the sensitivity. In addition, the mass transfer of nucleic acid on and off the bead surface (converse process for inhibitory substances) is a direct function of the mixing processes, which may need significant optimization to achieve the desired level of detection.

Residual beads that are not removed from the eluant may pose issues for the PCR. They may interfere with the PCR and/or signalling chemistries. This is more prominent in surface chemistries based upon a charge switching process. In addition, beads may interfere with the optical system. Some beads may fluoresce and/or cause light scattering, which may reduce the overall assay signal gain and create a false negative scenario. Residual beads may also cause amplification signalling artefacts such as steps and drift.

## Key-note 5: signalling and automated analysis

The process of automated calling of amplification in iNAT systems, irrespective of the analysis type (amplification and/or melt analysis) is fundamentally different from the more conventional lab-based approaches. Lab-based systems are batch based and the analysis of many samples occur in parallel. There are three fundamental effects of operating a 'single-shot' analysis system:

1 The system cannot have external standards. The requirement for controls is extremely important but this assists mainly with internal validation (see 'Key-note 7'). External standards are generally used in two ways – to provide quantification and, in the case of melt analysis, to provide a calling window or 'bin.'
2 There is no simple way to set a threshold over noise and therefore determine a $C_T$ value or melt peak $T_m$ and area. Many lab-based real-time PCR systems use a threshold value based upon the rise over 'no template' amplification controls. Successful amplification using unknown samples may then be scored against set standards of defined sequence identity. The extent of this problem for an iNAT is dependent upon the platform. Where a colour compensation process is used, the solution is a proportional fluorescence value rather than an absolute one obtained for each dye in a multiplex. When a differential is used in a melt analysis, the rise or peak intensity has no useful dimension. There are solutions to these problems that may require assay specific processes and modifications.
3 Automated systems use a computer-based expert system to provide the test data call. Computers use a systematic, linear, rule-based evaluation to score data performance. They may be tuned to provide the required level of system performance. However, the human brain has evolved to see patterns in data whilst processing many different parameters in parallel. For PCR data analysis, this means that data forms, such as amplification curves and melt peaks, can be rapidly interpreted by the operator. However, their interpretation may often be random, subjective and, therefore, incorrect and inconsistent. The expert system may present a negative result when the human brain suggests a positive one and *vice versa*.

These three observations lead to two important recommendations from the authors:

1 Signalling efficiency is an important metric to quantify a given assays suitability. It is often discrete from the PCR efficiency. The concept of probe efficiency as compared to amplification efficiency has been discussed elsewhere (Lee *et al.*, 2009a). A probe that produces a shallow, linear amplification signal over background noise will never attain a high level of specificity or sensitivity in an automated system. Likewise, melting peaks must be specific and the peaks Gaussian in shape to allow for efficient automated calling.
2 During development and tuning of an algorithm for a specific assay, an orthogonal system must be used to validate the *signalling*. This underpins the assumption that the signal provides the correct output for the PCR i.e. is positive only when there is specific amplicon present. An orthogonal system is one that can analyse the amplification by a process that is alternative to the signalling method used in the iNAT. The ideal solution is separation using, for example, capillary electrophoresis or mass spectrometry. The orthogonal system needs to be more sensitive and ideally exhibit a larger dynamic range than the detector on the iNAT.

The authors would add that, in their experience, the interpretation of amplification data, even by experienced qPCR scientists, is poor when they are analysing single amplification curves. This is particularly important at or near the limit of detection, where the noise level is high in comparison to the *signal* gain of the chemistry. The authors suspect that even on lab-based systems, low copy and no-template samples are often called incorrectly.

## Key-note 6: suppliers and GMP

The supply of components for integrated systems is an area that tends to result in much confusion and uncertainty. At the outset of the development process it is important to understand the delineation of the chain of responsibility. Ultimate responsibility for the product rests with the design owner who, in many cases, is also the marketer of the product.

The best approach to the evaluation of the requirements for the components of integrated systems rests within the design control process. In the case of the *in-vitro* diagnostic devices (IVDD), this is firmly embedded within the scope of ISO13485 and 21CFR820 which are the primary regulatory instruments used to govern and shape medical device development processes.

It is a common mistake to assume that all suppliers of components of the iNAT system must be ISO13485 registered. This is not the case. The iNAT design owner must, however, ensure that critical components of the system are suitably controlled. In doing so, they ensure that there are no unexpected changes to such components and parts and that where changes are proposed, suitable assessment can be made so that any necessary revalidation, as determined by risk assessment, can be performed prior to the change entering the supply chain and the product itself.

The approach, therefore, is to identify the criticality of components early in the development process and establish suitable controls with those suppliers early on. It is critical to start this process early, since the later changes occur in a system, the more expensive and time-consuming they tend to be. Generally, it is assumed that components coming into contact with the sample to be analysed will be critical due to their potential to abstract or add components and ultimately change the result. Other critical components would be those involved in the measurement of the outcome. For example, consider the use of a plastic for housing the extraction reagents in an iNAT system. This may seem relatively trivial but unplanned changes can have dramatic effects on assay outcome. When choosing the plastic, it is important to understand how it may interact with each of the components in the extraction and assay chemistries (that it is exposed to). Trace metals may be abstracted over time and change the performance of the assay. Consideration of addition, catalysis and removal all need to be considered. In most cases, the base system performance is established through the verification and validation process discussed below (see 'Key-note 7'). Inclusion of appropriate challenges at this stage will help identify the criticality of changes in later stages. In most cases it is prudent, even when the risk is deemed to be negligible, to include a form of basic system check to ensure changes do not alter the performance of the system.

Perhaps the biggest challenge in this area is the performance of biologics, since they can be highly variable. This is especially true in the case of enzymes and antibodies, where different batches of notionally similar materials can have dramatically different performance characteristics. It is essential, therefore, that processes are established to ensure that every batch of such biologics used in the system is suitably qualified. In many cases, this will necessitate the adjustment of addition to ensure the correct activity in the final mix. Failure to do so can result in systems that fail to meet the anticipated performance characteristics. This may occur in non-obvious ways, such as enzymes having increased activity delivering reduced stringency. A strong relationship with suppliers is, therefore, fundamental, since they may not appreciate how apparently trivial changes in their process(es) could have catastrophic results in the final system. An example would be a change of the pigment in a polymer that results in considerably more metal ions leaching into the extraction chemistry and unbalancing the PCR reaction. Maintaining a process of notifying and seeking permission from the design owner to make changes is fundamental in avoiding costly knock-on-effects that may not become obvious until the assay has been produced or, worse still, released for distribution and field use.

In practice, it is reasonable for suppliers to be controlled on the basis of risk. Once the assessment has been made and suppliers categorized, it is relatively easy to maintain control. Most iNAT manufacturers will establish some form of regular surveillance of their suppliers, beginning with an audit to establish confidence in the control systems in place. Based on the audit outcome, a quality agreement stating the terms of supply is usually established, which places contractual obligations on the supplier that should ensure supply meets the anticipated requirements. Regular surveillance audits are recommended to ensure that quality does not 'drift'. Where suppliers are deemed less critical, desktop audits and regular contact are sufficient to ensure certification status and ensure that critical changes, that may have an impact on the final product, have not occurred without prior notification. Where generic off-the-shelf reagents are used and suitable direct controls cannot be established, the design owner must establish suitable controls of incoming goods to try to mitigate changes. It is generally best to start at a high level of vigilance and then, through trend analysis of systems such as CAPA and corrective action requests, reduce the controls to an appropriate level over a period of time.

## Key-note 7: system validation

Validation is a fundamental requirement for a medical device system. The discussion here assumes that the requirements of the Quality System Registrar (QSR) and/or ISO13485 have been followed and adhered to throughout the development process. Fundamental to the development is the stage-by-stage verification of components and sub-assemblies as they mature to their final stages.

This key-note considers system validation as opposed to internal validation, which is covered elsewhere (Lee *et al.*, 2009b). Internal validation of tests is usually achieved through a process control. A process control is a molecular mimic or passive target that ideally passes through the entire integrated system and, therefore, provides validation of all subsystems – e.g. nucleic acid extraction, reverse transcription, amplification and analysis.

System validation is a process that needs to be considered from the outset of the development process. Validation is inevitably the (or one of the) last development activities because of its very nature.

The simplest approach to verification is to break down the system into its component parts and ensure that each of them perform at the expected level. Subsequently, the integrated system performance must be verified and subsequently validated, since there is no sure fire way of knowing whether or not the component parts will combine as expected until they are all brought together.

Establishing the performance of the elements of the iNAT on surrogate systems is usually the quickest and most effective way of optimizing the system parts before transferring them to the iNAT system. This allows for a much greater degree of parallel activity, significantly reducing the integrated system development time.

It is valuable in development to consider three parts, sampling/extraction, assay and platform. Prior to any validation or integration experimentation the developer should develop these three elements independently (but with a view to future compatibility). In doing so, the validation testing can be greatly simplified. [refer to FDA 21CFR 820 for validation and verification definition].

For convenience we will categorise the elements of the iNAT process into: sample collection and recovery; pre-analytical processing; nucleic acid extraction; analysis; and reporting.

### Sample collection and recovery

This can often be overlooked but is a fundamental factor in the effectiveness of a test. Where possible, well-established methods of sample collection should be considered, since this will reduce doubt and challenge later. Typical methods of sample collection would be swabs, transit media and buffers, vacutainer collected (and sometimes stabilized) blood and dried samples on paper. There have previously been cases where certain swab types have contained inhibitors that have impaired the performance of the test. If the manufacturer is not going to include the sampling kit, then the labelling should make it clear that the user is responsible for determining how their specific sampling regime may impact the test outcome. Once method(s) have been chosen they must be verified – this normally involves using matrix mimics or spiked samples for the specific biological matrix in question.

## Pre-analytical processing

The period between taking and analysing the sample can often be overlooked but is a critical step that must be considered during any verification and validation work. Consider an analysis for an RNA virus. RNA is notoriously difficult to stabilize over long periods and so the time between taking and analysing the sample may very well affect the outcome. The developer must, therefore, consider how the sample will be handled and retained prior to the next stage of the process. Factors that need to be established may include the temperature at which the sample is held, the maximum time between taking the sample and analysing it, ability to archive the sample for repeat analyses and the presence of potential critical interfering factors. In many developments, a subset of samples is exchanged between trial sites to mimic transport conditions (this is often combined with comparative analysis to determine whether any site specific bias is occurring).

## Nucleic acid extraction

This system goes hand in hand with the analytical stages and should not be underestimated when it comes to criticality and mutual compatibility. There are numerous systems for sample extraction that have been discussed elsewhere. The key point to note here is that the system as a whole should be given due consideration and the trade-off between the analytical steps and the extraction can have a huge bearing on the outcome of the test.

Where possible validation of this step is best approached in stages. Initially, even with well-established kits or processes, it is worth examining and optimizing the individual steps. Experience has shown that off the shelf systems are rarely optimized for specific iNATs and so each step should be challenged. These kits provide generic protocols that often include long lysis times that may not be necessary for the specific iNAT application(s).

The system should be challenged across the range of samples and nucleic acid levels expected in the assay. In particular, the reproducibility of the extraction process at or near the analytical threshold is fundamental and should be challenged rigorously in the presence and absence of common confounding factors known to potentially affect the outcome. Failure to do so often results in regulators refusing tests or requiring labelling statements that are profoundly limiting.

## Analysis

The interpretation of the result is critical in integrated systems. The intent is to remove the user from the process of decision making. Validation here will fall into two processes. The first is where any algorithmic processes are validated first using simulated (computer generated) or model data (a test harness). This is to ensure the embedded program is robust and provides the same result as the development code. The second is where data is generated from real samples on the iNAT to challenge all aspects of the interpretation. In particular it is critical that the reporting procedures are tested robustly around the limits of the system. There are numerous approaches to this but perhaps the most appropriate to base a development system on are those outlined by the Clinical and Laboratory Standards Institute (CLSI).

Ultimately, true validation of the reporting algorithm will only occur during the clinical validation of the integrated system, as it is unlikely, no matter how well designed the in-house verification tests, that the system will have been challenged as well as it can be by true clinical samples. The choice of comparator (see key-note 8) will be fundamental since the credibility of the test rests on the ability of the algorithm to produce a robust and reproducible report. In the case of iNATs this is particularly important because the human element in result interpretation is removed (as much as possible). This is especially true for point-of-care systems where it is unreasonable to expect the user to interpret amplification and/or melt curves.

## Reporting

Reporting of results by iNATs should be clear and unambiguous as with any other test. The key to validation of reporting is to have developed a suitably robust matrix of calls at the outset and then produce surrogate samples to test this matrix at a significant level prior to entering the validation arena. In the case of relatively simple infectious disease tests, this is often a qualified yes/no or positive/negative result providing the extraction and amplification controls have been successful. However, this becomes more complex as the sophistication of the test grows – oncology biomarkers, for instance, do not often provide a black and white answer and often present as increased determinants of risk that can be combined with other markers to inform on clinical management decisions.

## Key-note 8: the regulatory process

NATs do not implicitly require regulatory oversight unless the target market is subject to some form of control. In this section we discuss the requirements of systems produced for some form of clinical application. Research Use Only (RUO) and Application Specific Reagents (ASRs) are outside the scope of this chapter.

CE marking and FDA submissions are the most common regulatory pathways used to qualify medical devices. The primary reason for this is that the two are broadly aligned (although not totally harmonized) and generally form the basis for access to most international markets. Many countries will apply specific local requirements and it is a good idea to secure the aid of a local expert or market access specialist to facilitate qualification and market entry. CE marking of assays is relatively common, and becoming increasingly so, when compared to FDA approval, which is the route that must be followed in the US. This is primarily a function of the differing levels of stringency applied between the approval processes, with CE marking in most cases being viewed as much simpler. This may be contested by some in the industry, but the ability to self-certify certainly reduces time to market and close examination of many current CE marked systems as compared to FDA approved assays suggests this to be true.

The regulatory pathway must be mapped out from early stages of the proposed development to ensure that all elements of the design and development process are recorded and presented in an appropriate manner. Systematic requirements, such as the design control process, are not specifically dealt with here.

At the outset of the development the framework of what is being claimed needs to be established. Commonly, there are two scenarios – a 'me too' scenario where there is a simple and direct comparator, often another iNAT, against which the performance of the assay may be assessed and a novel application scenario for which there is no obvious comparator, such as a novel drug or polymorphism biomarker for predisposition to certain characteristics.

### Simple comparators

In the cases of a simple comparator the process is relatively straightforward, in that the developer has, simply, to establish relative levels of performance against the accredited system. If the comparator is another iNAT/NAT then relative performance can be measured, usually quite easily, by running a comparison of performance trial. Where the comparator is not an iNAT/NAT the process is less straightforward, especially when dealing with conflicting results. This is a particular challenge where the new test is suspected of being more sensitive than the comparator, because the existing gold standard typically takes precedence and results in calls of false positive results against the new test (this also applies where the comparator is another iNAT). Care must be taken when attempting to perform discrepant result analyses since these generally introduce unfair bias in favour of the new test – indeed the FDA actively discourage this approach and will usually reject data reconciled in this manner.

### No existing comparator

Where no system exists to facilitate the comparison the regulatory process is more difficult, usually involving extensive follow up of patients to determine whether, for example, that person goes on to display symptoms of the disease state detected. This can be a costly and time consuming process since symptoms for genetic diseases can take many years to present sufficiently for them to be correlated to the molecular diagnostic result. Many SNPs have been casually associated with disease states over recent years but many have a low predictive value, especially when used individually, and some have been shown to have only a tenuous link. Cystic fibrosis biomarkers are a case in point, where many markers have been reported but no single marker has been found to be wholly or significantly predictive of the disease on its own and many tests are now presented as high-level multiplexes targeting multiple biomarkers. Establishing the performance of such high-level multiplex assays is more complex and expensive since more than one comparator will be required and they are very likely to have unique performance characteristics.

Within CE marking and FDA submissions there are two regulatory pathways that can be followed (although the two pathways for FDA and CE marking are not wholly analogous).

## FDA submissions

### Pre-market approval (PMA)

PMA applies, in the case of IVDDs, to high-risk measurements, such as cancer biomarkers, where there is a high risk to the patient and to devices that have no existing comparator to qualify against. PMA is the most burdensome process and involves a long-term relationship with the regulator over a period of many months or even years. An example of this would be recent approvals to market human papilloma virus assays. Wherever possible it is best to avoid this route since the costs can be prohibitive.

Class III devices are those that support or sustain human life, are of substantial importance in preventing impairment of human health, or which present a potential, unreasonable risk of illness or injury. Owing to the level of risk associated with Class III devices, FDA has determined that general and special controls alone are insufficient to assure the safety and effectiveness of those devices. The must follow the PMA submission process.

The PMA process is a four-step one: (1) a filing review where a limited scientific review is conducted; (2) a substantial regulatory, QS and scientific examination is conducted; (3) an advisory panel review; and (4) a final review and recommendation phase

The basic cost for a PMA is $220,000 with a small business charge of $55,000 (2012 data) – the full scale of charges is dependent upon the situation of the submitters.

### Pre market notification (PMN or 510k)

The more common route for most iNATs is the pre-market notification route (PMN or more commonly 510k). Standard fees charged by the FDA for this service are $4049 with a small business rate of $2204 (2012 data). There are currently three variants of the 510k process: (1) traditional; (2) special; and (3) abbreviated. Once the dossier is submitted the agency has 90 days to respond and issue a clearance to market for the product(s) or to deny access.

Once cleared, there is an on-going commitment from the design owner to maintain the registration and the associated systems. Any changes to the approved product must be assessed and appropriately notified to the agency, at the very least, in a yearly report. In the case of PMAs, proposed changes need to be pre-approved before they are implemented. Any proposed changes must be assessed to determine the potential impact and possible need for revalidation. This assessment must be documented and state whether or not subsequent validation is required.

### CE marking

CE essentially means European Conformity, which when placed on a product is a declaration that that product complies with the appropriate essential requirements applicable to that specific product. It confers freedom of movement within the European free trade area. If a manufacturer is based outside of the European Union it must appoint an EU authorized representative before it can place a product on the market.

IVDDs fall under the auspices of the IVD Directive (98/79/EC) in Europe, if they are intended by the manufacturer to be used for the examination of specimens such as blood, tissue and other body derived fluids and samples for the sole purpose of providing information on a disease state, congenital abnormality, compatibility with a recipient or to monitor a therapy.

CE marking is the European analogue to the FDA process and approval from one system will ensure a development is broadly aligned to being able to register for the other. However, there are subtleties that potential developers and manufacturers should be aware of from the outset to avoid costly additional backfilling work at the end of a development. The full extent of controls applicable will be dependent upon the classification of the device. Within CE marking, this can vary from simple control of production for Class I devices through to implementation of a suitable quality management system (QMS) with a technical file audited by a suitably qualified notified body. It is clear, therefore, that the applicable class needs to be determined as early as possible so that the appropriate framework can be established to support the intended use.

The classifications applied by the FDA and in CE marking are not directly exchangeable and they should not be confused. IVDDs not falling under Annex II of the regulation are subject to self-certification. Certified body notification is required for Annex II listed tests although there are two categories, list A and list B, which carry slightly different approaches, with only List A requiring *design dossier review*. Currently only a limited number of assays will fall into the most burdensome category, Annex II, but this is subject to review and is likely to change in the near future to bring a higher level of consistency to the market. Within the regulation other annexes will apply depending on the approach and these are most easily dealt with by examination of the regulation.

## Summary

The authors have identified key issues that they feel are fundamental considerations for the transfer of assays onto integrated systems within regulated markets. There may be other platform specific considerations. Many research teams who aspire to transfer their assays to these emerging integrated systems will gain the most benefit from these notes that bridge the industry 'mystique' for transitioning from research services to regulated diagnostics systems. The detail of described processes, and attention to risk mitigation, may seem overly complex and burdening, but this is the very substance of the diagnostics industry.

## Key regulatory documents


EURLex (online). Directive 98/79/EC of the European Parliament and of the Council of 27 October 1998 on *in vitro* diagnostic medical devices. Official Journal L 331, 07/12/1998 P. 0001 – 0037. Available at: http://eur-lex.europa.eu/LexUriServ/LexUriServ.do?uri=CELEX:31998L0079:en:HTML (accessed 20 November 2013)

FDA (online). Quality System Regulation QSR Requirements for Medical Device Manufacturers for FDA 21 CFR Part 820 Compliance. FDA 21 CFR Part 820. Available at: http://www.fda.gov/MedicalDevices/ResourcesforYou/Industry/ucm126252.htm (accessed 20 November 2013)

ISO (online). ISO 13485:2003 Medical devices – quality management systems – requirements for regulatory purposes. Available at: http://www.iso.org/iso/iso_catalogue/catalogue_tc/catalogue_detail.htm?csnumber=36786 (accessed 20 November 2013)

MHRA UK (2012). Sale and supply of *in vitro* diagnostics medical devices (IVDs). MHRA UK. September 2012. Available at: http://www.mhra.gov.uk/Howweregulate/Devices/InVitroDiagnosticMedicalDevicesDirective/index.htm


## References


Boom, R., Sol, C.J., Salimans, M.M., Jansen, C.L., Wertheim-van Dillen, P.M., van der Noordaa, J. (1990). Rapid and simple method for purification of nucleic acids. J. Clin. Microbiol. *28*, 495–503.

Chakravorty, S., Aladegbami, B., Thoms, K., Lee, J.S., Lee, E.G., Rajan, V., Cho, E.J., Kim, H., Kwak, H., Kurepina, N., *et al.* (2011). Rapid detection of fluoroquinolone-resistant and heteroresistant Mycobacterium tuberculosis by use of sloppy molecular beacons and dual melting-temperature codes in a real-time PCR assay. J. Clin. Microbiol. *49*, 932–940.

Lee, M.A., Squirrell, D.J., Leslie, D.L., and Brown T. (2009a). Homogeneous fluorescent chemistries for real-time PCR. In Real-time PCR: An Essential Guide, 2nd edn, K. Edward, J. Logan, and N. Saunders, eds (Horizon Bioscience, Norfolk, UK).

Lee, M.A., Leslie D.L., and Squirrell D.J. (2009b). Internal and external controls for reagent validation. In Real-time PCR: An Essential Guide, 2nd edn, K. Edward, J. Logan, and N. Saunders, eds (Horizon Bioscience, Norfolk, UK).

# Part IV

# The Future

# Future of Molecular Diagnostics: The Example of Infectious Diseases

# 12

Eoin Clancy, Kate Reddington, Thomas Barry, Jim F. Huggett and Justin O'Grady

## Abstract

Advances in molecular diagnostics technologies offer clinicians improved infectious diseases tests for use in a hospital setting with regards to specificity, sensitivity and turn-around time to results. These advances have the potential to contribute in a positive way to patient care owing to improved decision making on optimal therapeutic approaches. Such advances may be used independent of traditional culture-based methods, or in some instances complementing culture. In this chapter we review a number of emerging technologies which can be used for both targeted and non-targeted molecular diagnostics in the clinical setting which may play a central role in the future including fully integrated 'sample in – result out' nucleic acid-based tests, digital PCR, mass spectrometry and next-generation sequencing.

## Introduction

In recent years there has been a significant increase in the number of molecular diagnostics technologies and platforms which are likely to become cornerstone in the clinical diagnostics sector. Incorporation of such molecular diagnostics technologies into routine laboratory practice enables better patient management and thus, better patient outcomes (Yang and Rothman, 2004; Mothershed and Whitney, 2006). Some clear advantages are:

- Rapid turn-around time to result – for example, molecular-based approaches can specifically identify an infectious agent in 1.5–2 hours as opposed to traditional culture-based methods which take from 24 hours up to a number of weeks.
- Identification of a specific microorganisms causing infection allows a clinician to make an informed decision as to the optimal therapeutic approach to take for a patient as opposed to unnecessary use of broad spectrum antibiotics.
- The source of outbreaks of infections and transmission dynamics can be identified in a timelier manner.
- The ability to quantify the number of a microorganisms present in a sample allows a clinician to determine colonization versus infection.

In this chapter, we will discuss the current 'state of the art' in molecular diagnostics of infectious diseases and describe some emerging technologies which are likely to be adapted to the clinical sector over the coming years.

## *In vitro* amplification technologies

One of the major advances in nucleic acid diagnostics (NADs) was the invention of the polymerase chain reaction (PCR), described by Kary Mullis in 1985 (Saiki *et al.*, 1985; Mullis and Faloona, 1987). Since then, numerous NADs have been developed and applied in almost every sector including clinical, environmental, food, veterinary etc. In recent years, interest in molecular diagnostics has grown exponentially with the discovery and development of real-time and multiplex real-time PCR/RT-PCR. The main advantages of real-time PCR are: rapid turnaround time; no requirement for post-amplification handling of samples reducing the likelihood of contamination; high specificity and sensitivity; and ability to quantify the amount of target present in a sample (Espy *et al.*, 2006; Anonymous, 2012). A significant disadvantage of real-time PCR is that routinely used instruments are limited to the detection of 5–6 analytes per test.

## Fully automated and integrated molecular diagnostics systems

A growing number of diagnostics companies have developed either fully automated or preferably fully integrated multiplex real-time PCR diagnostic platforms such as *the* Roche COBAS® TaqMan® Analyser (Roche Diagnostics, Basel, Switzerland), BD Max™ System (Becton Dickinson, Franklin Lakes, NJ, USA) and the Cepheid GeneXpert DX system (Cepheid, Sunnyvale, CA, USA), all of which are described in previous chapters.

Molecular diagnostics platforms which have the capacity for multiparametric detection and full integration from sample-in to result-out represent the current state-of-the-art in infectious diseases diagnostics. Such systems offer significant advantages in the clinical sector as: they typically require much less hands on time; they do not require highly skilled personnel for operation; they standardize all steps in the molecular diagnostics workflow; and interpretation of results is less subjective. These systems are based on a number of different technologies including real-time PCR, isothermal amplification and PCR array technology and have been described in detail in previous chapters.

## Recent advances and emerging real-time PCR technologies

In an effort to maintain the advantages of the rapid, closed tube system that real-time PCR offers, a number of recent advances have been described in the literature that increase the multiplexing capability of current real-time PCR systems. Multiplex real-time PCR classically uses a one-colour, one-probe approach to differentiate multiple targets in a sample. As a result, the maximum number of targets that can be detected in a reaction is limited to the number of detection channels on the instrument. Current real-time PCR instruments typically have 4–6 optical channels. Recently, a number of strategies have been developed to overcome the one-colour, one-probe barrier, enabling the detection of multiple targets per channel (Huang *et al.*, 2011; Fu *et al.*, 2012).

### Sloppy molecular beacons

In a recent study by El-Hajj *et al.* (2009), a novel multiplex real-time PCR method based on the amplification of a single conserved target gene detected using a set of four differently labelled 'sloppy molecular beacons' enabled the identification of 27 closely related mycobacterial isolates. 'Sloppy molecular beacons' are unusually long molecular beacons of approximately 40 bp in length. The use of such long probes stabilizes the probes such that they can bind to sequences that are not perfectly matched. The principle of this approach is the amplification of a conserved molecular target that includes a hypervariable region (such as 16S rRNA) that is spanned using the four sloppy beacons each labelled with a different fluorophore. Following the real-time PCR reaction, high resolution melt curve analysis is performed, resulting in different melt peaks in the different channels that are sequence dependant. By using this approach El-Hajj *et al.* (2009) demonstrated that a specific signature could be generated for the majority of mycobacterial species tested, when the combination of $T_m$ values from each of channels was taken into.

A more recent study by Chakravorty *et al.* (2010) used this sloppy molecular beacon approach, incorporating six fluorescently labelled probes, in an effort to develop a universal pathogen identification method in clinical cultures. In this study 270 clinical cultures (including 106 patient blood cultures) were tested and the sloppy molecular beacon method demonstrated 95–97% agreement with conventional identification methods. Another recent study has demonstrated promising results for determining drug resistance using this approach (Chakravorty *et al.*, 2011).

While the use of sloppy molecular beacons looks very promising from a microbial identification and drug susceptibility point of view, the disadvantage of this method is that it can be used only on pure culture isolates.

## Tagging oligonucleotide cleavage and extension (TOCE) technology

Another new concept in multiplex real-time PCR, developed by Seegene (Seoul, Korea), is the tagging oligonucleotide cleavage and extension (TOCE) method. Using this approach dual priming oligonucleotides (DPO – primers with a longer 5′ end and a shorter 3′ end linked with a bridge which is thought to aid specificity (Chun *et al.*, 2007)), pitchers (a tagging oligonucleotide which hybridizes specifically to a target region) and catchers (fluorescently labelled artificial template which has a complimentary tag sequence to part of the pitcher) (Anonymous, 2013a). During primer extension, the pitcher is cleaved by the DNA polymerase and the tagging portion will bind to the perfectly matched catcher sequence (Anonymous, 2013a). During subsequent extension fluorescence is emitted when the fluorescent molecule separates from the quencher, which can be read on a standard real-time thermocycler (Anonymous, 2013a). There are currently seven catchers available with distinct $T_m$ values which can be read in each channel of a real-time instrument (Anonymous, 2013a). As there are typically 4–6 channels on such instruments, it is theoretically possible to multiplex up to 42 analytes using this approach. While information is available on the Seegene website for this technology, it is not available for research use and there has been no articles published to date that detail its performance.

## Isothermal amplification methods

In recent years a number of isothermal *in vitro* amplification techniques have been described in the literature. A major advantage of isothermal amplification is that ramping of temperatures required for PCR is not necessary, therefore expensive machinery is not necessary. Detection of isothermal reactions can often be read by eye, or with the use of electrochemiluminescence which can also reduce the cost associated with equipment. One of the most significant advantages of isothermal amplification methods is in their capacity to be incorporated more easily to near patient or on-chip devices. There are a growing number of isothermal ar
becoming available which
previous chapters.

## Molecular hybridiz

To facilitate higher m
recent advancement in molecular diagnostic
the combination of multiplex PCR followed by hybridization of PCR product to macro/micro arrays. Typically, up to 20 primer pairs can be multiplexed on a standard thermocycler and the resulting products are hybridized to an array containing hundreds to thousands of immobilized capture probes. This type of platform can offer highly multiparametric detection and numerous examples of its utility have been described in the clinical diagnostics sector. Below are descriptions of examples of such recently developed highly multiplexed approaches.

### Prove-it sepsis array

The Prove-it sepsis array (Mobidiag, Helsinki, Finland) is a relatively new commercially available array for the detection of approximately 70 micro-organisms associated with sepsis. The approach taken includes sample preparation, broad range multiplex PCR, followed by hybridization to capture probes immobilized on an array. When using the Prove-it strip array system, automated assay detection and result generation are performed using Prove-IT advisor software. Two recent clinical evaluations demonstrated that sensitivities and specificities for bacterial targets were 94.7% and 98.8% respectively and 99% and 98% for fungal targets (Tissari *et al.*, 2010). This platform represents one of the few diagnostic tests which can identify both bacterial and fungal agents of infection. When automated DNA extraction methods are used, the turnaround time to result is ~4.5 hours and the test covers >90% of micro-organisms associated with sepsis. However a disadvantage of this product is that sample preparation is not integrated.

### Unyvero

The Unyvero (Curetis, Germany) is a fully integrated system which incorporates a universal nucleic acid extraction for bacterial, fungal and gens, followed by amplification and n on a universal test platform (Anony-, 2013b). Currently, the Unyvero platform d one molecular test, the P50 Pneumonia test, are CE-IVD marked (Bissonnette and Bergeron, 2012). This test has the ability to identify approximately 90% of bacterial pathogens associated with pneumonia and a wide range of drug resistances from a variety of patient sample types in approximately 4 hours. Preclinical and CE evaluation studies demonstrate the potential of this test (Anonymous, 2013c); however, there is still a lack of published data on the performance of this test. Curetis are planning to expand their test portfolio, developing cartridges for a broad range of analytes associated with sepsis and wound infections (Anonymous, 2013d).

### Luminex

The X-MAP technology (Luminex, Austin, TX, USA) is a bead-based approach which facilitates high-level multiplexing (up to 200 analytes) in one reaction. Typically, after nucleic acid extraction a highly multiplexed PCR is performed and the products are subsequently hybridized to individual coloured beads. Detection and analysis of results is then performed on the Luminex 100/200 or the MAGPIX instruments. In the context of infectious disease, Luminex have launched the Respiratory Viral Panel (RVP) and the Gastrointestinal Pathogen Panel (GPP), which detect bacterial, viral and parasitic targets. The advantages of this platform are in the high multiplexing capability and sample throughput. However, a significant disadvantage is the need for post-amplification handling of samples which increases the likelihood of contamination.

## Digital PCR

Digital PCR (dPCR) involves the dilution and partitioning of a sample into thousands or even millions of individual reactions (termed partitions), such that each partition contains either '0' or very low copies of the target sequence. Upon amplification, a positive signal will only be detected in those partitions that contain the target sequence. By applying statistical analysis to the proportion of positives, the absolute number of target molecules in the original sample can be determined. This is calculated by applying the equation: $\lambda = -\ln(1-p)$, where $\lambda$ is the average number of target DNA molecules per partition and $p$ is the fraction of positive partitions (Sykes *et al.*, 1992). The term digital PCR was first coined by Vogelstein and Kinzler in 1999 with the demonstrated that 3.9% of the alleles present in the stool of patients with colorectal cancer contained a mutation in the KRAS oncogene (Vogelstein and Kinzler, 1999). More recently, dPCR was shown capable of detecting and quantifying hetero-resistance in drug-resistant *M. tuberculosis* at a ratio of 1:1000, resistant: susceptible *M. tuberculosis* (Pholwat *et al.*, 2013).

In classical real-time PCR, target sequences are quantified by comparing the number of amplification cycles to those of a reference sample. This involves the inclusion of reactions containing known numbers of target template in each experiment. Since dPCR is not reliant on the number of amplification cycles to determine the initial target quantity, calibration standards are less crucial.

A major advantage of dPCR is its ability to differentiate low fold copy number variations. In standard qPCR, a 2-fold difference in template concentration equates to a single threshold cycle (Ct), the ability to discriminate fold differences below a ratio of 1.5 is technically challenging (Weaver *et al.*, 2010; Whale *et al.*, 2012) and is compounded by intra- and interlaboratory qPCR experiment variability. In contrast, dPCR is considerably less variable, permitting the measurement of copy number variation ratios as low as 1.15 are possible (Qin *et al.*, 2008).

In the original paper of Vogelstein and Kinzler, reactions were partitioned into 5 μl volumes. The first dPCR system to become commercially available was launched by Fluidigm in 2006. Since then a number of dPCR platforms have been commercialized and can be broadly classified based on how the reactions are partitioned; chip/plate based and droplet based. Fluidigm and Life Technologies have developed platforms based on reaction chambers within specially designed plates or chips, whilst the Bio-Rad and RainDance platforms are based on the partitioning of reactions in droplets (ddPCR). The individual partition volumes of these platforms varies depending on the system; Fluidigm's 200K prototype chip for example contains 201,786 reaction chambers with each of chamber having a volume of 128 pl whilst the Life Technologies QuantStudio™ 3D platform run 24 chips each containing 20,000 partitions of approximately 1 nl. Bio-Rad's Qx100™ droplet digital™ PCR system is composed of an 8-sample droplet generator and a droplet reader (Hindson *et al.*, 2011). From a 20 μl sample, some 15,000 to 18,000 droplets of the order of 1 nl are generated. The RainDance droplet generator is capable of simultaneously generating 1–10 million ~5 pl sized droplets per sample lane in an 8-lane chip (Kiss *et al.*, 2008).

A number of alternative technologies have been developed that overcome some of the disadvantages of commercially available dPCR systems, such as limited dynamic range and/or the need to dilute the sample prior to amplification. For example, ddPCR requires three separate instruments; a droplet generator, a conventional thermocycler and a droplet reader. The droplet array developed by Hatch *et al.* (2011) combines the benefits of array and ddPCR enabling real-time analysis with increased dynamic range (10–100,000 copies) of over 1 million picolitre-sized droplets packed at high density into a microfluidic chamber (Hatch *et al.*, 2011). The Megapixel dPCR system is a valve-free chip-based array platform that surface tension to enable the partitioning of a sample into more tha 1 million partitions (Heyries *et al.*, 2011). This system was demonstrated to have a dynamic range of seven orders of magnitude.

dPCR typically involves partitioning a sample into pico- or nanolitre-sized partitions such that each partition contains low quantities of the target of interest. Further reducing the volume of each partition has the potential to eliminate the requirement to dilute the sample and therefore the potential for the introduction of contamination and human error. Men *et al.* (2012) have recently demonstrated a biochip consisting of an array of 34 fl partitions packed at a density of 20,000/mm$^2$ for dPCR. Furthermore, reducing the volume/partition enables faster reaction kinetics which permitted the reaction to be performed for just 30 two-step cycles in 35 minutes.

dPCR has been applied in molecular infectious diseases diagnostics including for the absolute quantification of viral load. White *et al.* (2012) demonstrated the use of reverse transcription (RT) dPCR for the absolute quantification of the occult RNA virus GB Virus Type C (GBV-C). They found that RT-dPCR had a lower overall coefficient of variation for viral load testing than RT-qPCR and had a higher overall detection limit. Hayden *et al.* (2013) compared ddPCR to real-time PCR for the detection of cytomegalovirus. In this study, the authors found that ddPCR was less sensitive than real-time PCR, but attributed this to the small total sample volume compared to that for the real-time system used (5 μl for ddPCR compared with 20 μl for real-time PCR). Other examples of dPCR diagnostics assays include a internally controlled ddPCR assay for ocular *Chlamydia trachomatis* infections, a duplex ddPCR for the detection of methicillin-resistant *Staphylococcus aureus* (MRSA) (Kelley *et al.*, 2013). dPCR has been used for single cell analysis and has also be used for the calibration of samples prior to sequencing (Sanchez-Freire *et al.*, 2012; White *et al.*, 2009).

Whilst dPCR has opened up many new avenues for the study of nucleic acids, it is important to note that it does have limitations which must be overcome for it to supersede qPCR in routine analysis. These include cost and limited dynamic range and the fact that smaller may not be always better (Baker, 2012; Henrich *et al.*, 2012).

## Mass spectrometry

Mass spectrometry (MS) was first applied to the identification of bacteria in the 1970s (Anhalt and Fenselau, 1975). Since then, MS has emerged as a powerful tool for the analysis of nucleic acids and proteins and has found wide application in molecular medicine. MS has a number of features that make it an attractive tool for the identification and characterization of microorganisms directly from culture including its broad applicability, its speed (<1 hour (Seng *et al.*, 2009)) and its sensitivity ($\geq 10^3$ CFU; Demirev and Fenselau, 2008; Köhling *et al.*, 2012).

The development of 'soft' ionization techniques such as matrix-assisted laser desorbtion/ionization (MALDI) and electrospray ionization (ESI) has enabled the application of MS to microbial identification. One of the first reports of the application of MALDI TOF to bacterial identification was in 1994, where bacteria were lysed by sonication and analysed (Cain *et al.*, 1994). In subsequent work it was demonstrated that whole intact bacterial cells (3 different species from the genus *pseudomonas*) could be distinguished by MALDI TOF (Holland *et al.*, 1996). Since these pioneering works, MS instruments have become common place in diagnostics laboratories with protocols having been developed that have enabled the identification of fungi (Iriart *et al.*, 2012; Shin *et al.*, 2013) and viruses (Chen *et al.*, 2013).

Two main approaches have been applied to the profiling of bacteria using MALDI TOF; a library-based approach and a bioinformatics-based approach. By far the most common is the library-based strategy. A library-based approach involves comparing the mass spectrum from the unknown sample to that of spectra of known reference bacteria thereby enabling identification. The bioinformatics-based approach involves using mass spectrometry data to identify proteins from primary sequence databases (Perkins *et al.*, 1999). Bioinformatic-based approaches can be subcategorised depending on how the sample is prepared or analysed: (1) intact biomarker identification which primarily uses ribosomal proteins as biomarkers; (2) 'bottom-up' approach that involves digestion of pre-fractionationed proteins prior to MS analysis; and (3) 'top-down' applications that use tandem-MS to fragment intact proteins into smaller sequence-specific peptides (Sandrin *et al.*, 2012).

Whilst MS has become routine in the analysis of cultured bacteria and fungi in many clinical laboratories, the use of MS to type microorganisms based on their nucleic acid sequence has only recently begun to be fully explored. MS analysis of nucleic acids was first demonstrated in the early 2000s and its potential to identify viral and bacterial pathogens soon followed (von Wintzingerode *et al.*, 2002; Hofstadler *et al.*, 2005; Sampath *et al.*, 2005). MS analysis of amplicons generated using multiple pairs of broad-range PCR primers enables non-targeted molecular diagnostics. Not only does this technique permit the sensitive identification of pathogens to the strain level, but also enables the deconvolution of samples containing a mixture of organisms.

Several MS systems have been commercialized for the analysis of nucleic acid sequences. These include the PLEX-ID system from Abbott, the MassCode PCR system from Agilent and the MassARRAY platform from Sequenom. The PLEX-ID platform uses ESI-MS to analyse PCR amplicons generated from (depending on the protocol) 8–36 pairs of broad range PCR primers (Honisch *et al.*, 2007; Baldwin *et al.*, 2009; Sampath *et al.*, 2012). As well as characterizing bacterial markers the PLEX-ID platform has proven useful in the characterization of bacterial drug resistance profiles (Wang *et al.*, 2011).

Agilent's MassCode PCR system involves the reverse transcription-PCR amplification of a sample using primers labelled with small molecules of known molecular weight which result in each amplicon containing dual mass tags. Following PCR, the amplicons are separated by liquid chromatography and the tags are cleaved by UV. Subsequent flow injection into a single quadrupole mass spectrometer equipped with an atmospheric pressure chemical ionization (APCI) source enables detection of the tags. The MassCode platform has been successfully used to identify viruses (Briese *et al.*, 2005) and for the typing of bacteria (Richmond *et al.*, 2011).

Sequenom's MassARRAY iPLEX platform uses MALDI-TOF MS coupled with single-base extension PCR for high-throughput multiplex SNP detection (Oeth *et al.*, 2008). The MassARRAY iPLEX platform has been used to genotype methicillin-resistant *Staphylococcus aureus* (Syrmis *et al.*, 2011). Another MassARRAY application, iSEQ is based on the *in vitro* transcription of PCR amplicons which are then base-specifically cleaved, with the resultant fragments subjected to analysis by MALDI-TOF MS. The iSEQ platform has been applied to the Multilocus Sequence Typing of *Streptococcus pneumoniae* (Dunne *et al.*, 2011) and *Neisseria meningitidis* (Honisch *et al.*, 2007).

## Sequencing

Since the transition from Sanger-based sequencing to 'next generation' sequencing (NGS) in early 2008, there has been a profound reduction in sequencing costs. The full economic cost of sequencing a typical bacterial genome now stands at less than £50 (Köser *et al.*, 2012a; Priest *et al.*, 2012).

The avalanche of sequence information from NGS initiatives is profoundly impacting the field of molecular diagnostics. At the time of writing (May 2013), The Genomes Online Database (GOLD, http://www.genomesonline.org/) contained 4329 complete (or permanent draft) bacterial genome sequences (Pagani *et al.*, 2012). Other online databases have seen corresponding exponential increases in sequence information. For example, the latest release (10.31, Accessed May 2013) of the Ribosomal Database Project (RDB, http://rdp.cme.msu.edu/) contains 2,639,157 rRNA sequences. In September 2008 (release 10.3), the RDB database contained 676,998 rRNA sequences (Cole *et al.*, 2009). One example of how whole genome sequencing (WGS) is impacting molecular diagnostics is in the discovery of orthologous biomarkers in previously difficult to differentiate microorganisms (Jordan *et al.*, 2011).

Whilst the growing quantity and quality of sequence information available is enabling the design of more accurate diagnostics assays, the routine use of NGS in molecular diagnostics currently remains both cost and time prohibitive (Köser *et al.*, 2012a). In some instances, however, molecular drug susceptibility testing is warranted, such as in the case of slow growing microorganisms or species known to carry multi-drug resistance where the turn-around time for standard phenotypic testing can take weeks. In these cases and for epidemiological typing, sequencing could have an important role to play in the control of the disease. One such example is that of the *Mycobacterium tuberculosis* complex (MTC), which, as well as being slow growing, also have the ability to evolve their fitness to become highly resistant strains (Daum *et al.*, 2012; Sun *et al.*, 2012).

In clinical microbiology, NGS has been most widely applied to the study of outbreaks and to characterize transmission events. Examples of this include the use of WGS to characterize the German enterohaemorrhagic *E. coli* O104:H4 outbreak of 2011 (Mellmann *et al.*, 2011), to monitor a *Legionella* outbreak (Reuter *et al.*, 2013), the investigation of a MRSA outbreak (Köser *et al.*, 2012b), the tracking of a *K. pneumoniae* outbreak (Snitkin *et al.*, 2012) and tracing the origin of the 2010 Haitian *Vibrio cholerae* outbreak strain (Chin *et al.*, 2011). Developments in sample preparation are leading to streamlining of sequencing workflows. For example Seth-Smith *et al.*, using immunomagnetic separation and multiple displacement amplification were able to produce the whole bacterial genome sequence of *Chlamydia trachomatis* directly from clinical samples without culture (Seth-Smith *et al.*, 2013). WGS has also recently been shown capable of resolving mixed cultures, even when using short (50 bp) read lengths enabling a sequencing run to be completed in 5 hours (Long *et al.*, 2013).

The ultimate application for NGS in the clinical microbiology laboratory is non-targeted culture independent identification and antibiotic resistance profiling of bacterial, viral and fungal pathogens in clinical specimens. This would be particularly useful in diseases with complex aetiologies such as sepsis, pneumonia and gastrointestinal infections leading to improved patient management and antibiotic stewardship. There remain a number of technical hurdles that must be overcome before NGS is routinely used in clinical microbiology (Köser *et al.*, 2012a,b). Advances in sample preparation from diverse clinical samples, some of which contain very few pathogens of interest in a very high human or bacterial background, is required. Also, whilst the capability to generate sequence information quickly enough to affect patient management is undoubtedly in place, streamlining analysis to provide actionable information from the large quantities of sequence data to healthcare workers with no knowledge of genome sequencing must be overcome.

## Concluding remarks

The field of infectious diseases diagnostics is currently undergoing a revolution as a result of developments in PCR, mass spectrometry and next generation sequencing. These advances are driving the development of faster, more sensitive and more specific molecular diagnostics, helping to usher in the era of personalized medicine. With advances in microfluidics and the ability to integrate these technologies on miniaturized, automated devices, capable of performing complex multiparametric assays, the stage is set for the movement of molecular diagnostics testing from large centralized clinical laboratories to their deployment at the point-of-need. Non-targeted molecular technologies such as sequencing and MS PCR which provide 'a molecular agar plate' have the potential to change the way we diagnose infectious diseases by replacing, rather than complementing, culture.

## References

Anhalt, J.P., and Fenselau, C. (1975). Identification of bacteria using mass spectrometry. Anal. Chem. 47, 219–225.

Anonymous (2012). Dorak MT: Real-Time PCR. Available at: http://www.dorak.info/genetics/realtime.html (accessed 11 September 2013).

Anonymous (2013a). Seegene. Available at: http://www.seegene.com/en/research/core_040.php (accessed 11 September 2013).

Anonymous (2013b). Cuertis. Available at: http://www.curetis.com/en/technology/curetis-technology/ (accessed 11 September 2013).

Anonymous (2013c). Curetis P50 pneumonia product insert. Available at: http://www.curetis.com/fileadmin/curetis/media/struktur/Products/Instructions/00118_Unyvero_Pneumonia_Application_Guide_V2_0.pdf (accessed 11 September 2013).

Anonymous (2013d). Curetis product pipeline. Available at: http://www.curetis.com/en/products/unyverotm-cartridge-pipeline/ (accessed 11 September 2013).

Baker, M. (2012). Digital PCR hits its stride. Nat. Methods *9*, 541–544.

Baldwin, C.D., Howe, G.B., Sampath, R., Blyn, L.B., Matthews, H., Harpin, V., Hall, T.A., Drader, J.J., Hofstadler, S.A., Eshoo, M.W., *et al.* (2009). Usefulness of multilocus polymerase chain reaction followed by electrospray ionization mass spectrometry to identify a diverse panel of bacterial isolates. Diagn. Microbiol. Infect. Dis. *63*, 403–408.

Bissonnette, L., and Bergeron, M.G. (2012). Multiparametric technologies for the diagnosis of syndromic infections. Clin. Microbiol. Newslett. *34*, 159–168.

Briese, T., Palacios, G., Kokoris, M., Jabado, O., Liu, Z.Q., Renwick, N., Kapoor, V., Casas, I., Pozo, F., Limberger, R., *et al.* (2005). Diagnostic system for rapid and sensitive differential detection of pathogens. Emerg. Infect. Dis. *11*, 310–313.

Cain, T.C., Lubman, D.M., Weber, W.J., and Vertes, A. (1994). Differentiation of bacteria using protein profiles from matrix-assisted laser desorption/ionization time-of-flight mass spectrometry. Rapid Commun. Mass Spectrom. *8*, 1026–1030.

Chakravorty, S., Aladegbami, B., Burday, M., Levi, M., Marras, S.A.E., Shah, D., El-Hajj, H.H., Kramer, F.R., and Alland, D. (2010). Rapid universal identification of bacterial pathogens from clinical cultures by using a novel sloppy molecular beacon melting temperature signature technique. J. Clin. Microbiol. *48*, 258–267.

Chakravorty, S., Aladegbami, B., Thoms, K., Lee, J.S., Lee, E.G., Rajan, V., Cho, E.-J., Kim, H., Kwak, H., Kurepina, N., *et al.* (2011). Rapid detection of fluoroquinolone-resistant and heteroresistant Mycobacterium tuberculosis by use of sloppy molecular beacons and dual melting-temperature codes in a real-time PCR assay. J. Clin. Microbiol. *49*, 932–940.

Chen, W.-H., Hsu, I.H., Sun, Y.-C., Wang, Y.-K., and Wu, T.-K. (2013). Immunocapture couples with matrix-assisted laser desorption/ionization time-of-flight mass spectrometry for rapid detection of type 1 dengue virus. J. Chromatogr. A *1288*, 21–27.

Chin, C.-S., Sorenson, J., Harris, J.B., Robins, W.P., Charles, R.C., Jean-Charles, R.R., Bullard, J., Webster, D.R., Kasarskis, A., Peluso, P., *et al.* (2011). The origin of the haitian cholera outbreak strain. N. Engl. J. Med. *364*, 33–42.

Chun, J.-Y., Kim, K.-J., Hwang, I.-T., Kim, Y.-J., Lee, D.-H., Lee, I.-K., and Kim, J.-K. (2007). Dual priming oligonucleotide system for the multiplex detection of respiratory viruses and SNP genotyping of CYP2C19 gene. Nucleic Acids Res. *35*, e40.

Cole, J.R., Wang, Q., Cardenas, E., Fish, J., Chai, B., Farris, R.J., Kulam-Syed-Mohideen, A.S., McGarrell, D.M., Marsh, T., Garrity, G.M., *et al.* (2009). The Ribosomal Database Project: improved alignments and new tools for rRNA analysis. Nucleic Acids Res. *37*, D141–D145.

Daum, L.T., Rodriguez, J.D., Worthy, S.A., Ismail, N.A., Omar, S.V., Dreyer, A.W., Fourie, P.B., Hoosen, A.A., Chambers, J.P., and Fischer, G.W. (2012). Next-generation ion torrent sequencing of drug resistance mutations in Mycobacterium tuberculosis strains. J. Clin. Microbiol. *50*, 3831–3837.

Demirev, P.A., and Fenselau, C. (2008). Mass spectrometry for rapid characterization of microorganisms. Annu. Rev. Anal. Chem. *1*, 71–93.

Dunne, E.M., Ong, E.K., Moser, R.J., Siba, P.M., Phuanukoonnon, S., Greenhill, A.R., Robins-Browne, R.M., Mulholland, E.K., and Satzke, C. (2011). Multilocus sequence typing of *Streptococcus pneumoniae* by use of mass spectrometry. J. Clin. Microbiol. *49*, 3756–3760.

El-Hajj, H.H., Marras, S.A.E., Tyagi, S., Shashkina, E., Kamboj, M., Kiehn, T.E., Glickman, M.S., Kramer, F.R., and Alland, D. (2009). Use of sloppy molecular beacon probes for identification of mycobacterial species. J. Clin. Microbiol. *47*, 1190–1198.

Espy, M.J., Uhl, J.R., Sloan, L.M., Buckwalter, S.P., Jones, M.F., Vetter, E.A., Yao, J.D.C., Wengenack, N.L., Rosenblatt, J.E., Cockerill, F.R., *et al.* (2006). Real-time PCR in clinical microbiology: applications for routine laboratory testing. Clin. Microbiol. Rev. *19*, 165–256.

Fu, G., Miles, A., and Alphey, L. (2012). Multiplex detection and SNP genotyping in a single fluorescence channel. PLoS One *7*, e30340.

Hatch, A.C., Fisher, J.S., Tovar, A.R., Hsieh, A.T., Lin, R., Pentoney, S.L., Yang, D.L., and Lee, A.P. (2011). 1-Million droplet array with wide-field fluorescence imaging for digital PCR. Lab Chip *11*, 3838–3845.

Hayden, R.T., Gu, Z., Ingersoll, J., Abdul-Ali, D., Shi, L., Pounds, S., and Caliendo, A.M. (2013). Comparison of droplet digital PCR to real-time PCR for quantitative detection of cytomegalovirus. J. Clin. Microbiol. *51*, 540–546.

Henrich, T.J., Gallien, S., Li, J.Z., Pereyra, F., and Kuritzkes, D.R. (2012). Low-level detection and quantitation of cellular HIV-1 DNA and 2-LTR circles using droplet digital PCR. J. Virol. Methods *186*, 68–72.

Heyries, K.A., Tropini, C., VanInsberghe, M., Doolin, C., Petriv, O.I., Singhal, A., Leung, K., Hughesman, C.B., and Hansen, C.L. (2011). Megapixel digital PCR. Nat. Methods *8*, 649-U664.

Hindson, B.J., Ness, K.D., Masquelier, D.A., Belgrader, P., Heredia, N.J., Makarewicz, A.J., Bright, I.J., Lucero, M.Y., Hiddessen, A.L., Legler, T.C., *et al.* (2011). High-throughput droplet digital PCR system for absolute quantitation of DNA copy number. Anal. Chem. *83*, 8604–8610.

Hofstadler, S.A., Sampath, R., Blyn, L.B., Eshoo, M.W., Hall, T.A., Jiang, Y., Drader, J.J., Hannis, J.C., Sannes-Lowery, K.A., Cummins, L.L., *et al.* (2005). TIGER: the universal biosensor. Int. J. Mass Spectrom. *242*, 23–41.

Holland, R.D., Wilkes, J.G., Rafii, F., Sutherland, J.B., Persons, C.C., Voorhees, K.J., and Lay, J.O. (1996). Rapid identification of intact whole bacteria based on spectral patterns using matrix-assisted laser desorption/ionization with time-of-flight mass spectrometry. Rapid Commun. Mass Spectrom. *10*, 1227–1232.

Honisch, C., Chen, Y., Mortimer, C., Arnold, C., Schmidt, O., van den Boom, D., Cantor, C.R., Shah, H.N., and Gharbia, S.E. (2007). Automated comparative sequence analysis by base-specific cleavage and mass spectrometry for nucleic acid-based microbial typing. Proc. Natl. Acad. Sci. U.S.A. *104*, 10649–10654.

Huang, Q., Zheng, L., Zhu, Y., Zhang, J., Wen, H., Huang, J., Niu, J., Zhao, X., and Li, Q. (2011). Multicolor combinatorial probe coding for real-time PCR. PLoS One *6*, e16033.

Iriart, X., Lavergne, R.-A., Fillaux, J., Valentin, A., Magnaval, J.-F., Berry, A., and Cassaing, S. (2012). Routine identification of medical fungi by the new Vitek MS matrix-assisted laser desorption ionization–time of flight system with a new time-effective strategy. J. Clin. Microbiol. *50*, 2107–2110.

Jordan, I.K., Conley, A.B., Antonov, I.V., Arthur, R.A., Cook, E.D., Cooper, G.P., Jones, B.L., Knipe, K.M., Lee, K.J., Liu, X., *et al.* (2011). Genome sequences for five strains of the emerging pathogen *Haemophilus haemolyticus*. J. Bacteriol. *193*, 5879–5880.

Kelley, K., Cosman, A., Belgrader, P., Chapman, B., and Sullivan, D.C. (2013). Detection of methicillin-resistant *Staphylococcus aureus* by a duplex droplet digital polymerase chain reaction. J. Clin. Microbiol. *51*, 2033–2039.

Kiss, M.M., Ortoleva-Donnelly, L., Beer, N.R., Warner, J., Bailey, C.G., Colston, B.W., Rothberg, J.M., Link, D.R., and Leamon, J.H. (2008). High-throughput quantitative polymerase chain reaction in picoliter droplets. Anal. Chem. *80*, 8975–8981.

Köhling, H.L., Bittner, A., Müller, K.-D., Buer, J., Becker, M., Rübben, H., Rettenmeier, A.W., and Mosel, F. (2012). Direct identification of bacteria in urine samples by matrix-assisted laser desorption/ionization time-of-flight mass spectrometry and relevance of defensins as interfering factors. J. Med. Microbiol. *61*, 339–344.

Köser, C.U., Ellington, M.J., Cartwright, E.J.P., Gillespie, S.H., Brown, N.M., Farrington, M., Holden, M.T.G., Dougan, G., Bentley, S.D., Parkhill, J., *et al.* (2012a). Routine use of microbial whole genome sequencing in diagnostic and public health microbiology. PLoS Pathog. *8*, e1002824.

Köser, C.U., Holden, M.T.G., Ellington, M.J., Cartwright, E.J.P., Brown, N.M., Ogilvy-Stuart, A.L., Hsu, L.Y., Chewapreecha, C., Croucher, N.J., Harris, S.R., *et al.* (2012b). Rapid whole-genome sequencing for investigation of a neonatal MRSA outbreak. N. Engl. J. Med. *366*, 2267–2275.

Long, S.W., Williams, D., Valson, C., Cantu, C.C., Cernoch, P., Musser, J.M., and Olsen, R.J. (2013). A genomic day in the life of a clinical microbiology laboratory. J. Clin. Microbiol. *51*, 1272–1277.

Mellmann, A., Harmsen, D., Cummings, C.A., Zentz, E.B., Leopold, S.R., Rico, A., Prior, K., Szczepanowski, R., Ji, Y., Zhang, W., *et al.* (2011). Prospective genomic characterization of the german enterohemorrhagic *Escherichia coli* O104:H4 outbreak by rapid next generation sequencing technology. PLoS One *6*, e22751.

Men, Y., Fu, Y., Chen, Z., Sims, P.A., Greenleaf, W.J., and Huang, Y. (2012). Digital polymerase chain reaction in an array of femtoliter polydimethylsiloxane microreactors. Anal. Chem. *84*, 4262–4266.

Mothershed, E.A., and Whitney, A.M. (2006). Nucleic acid-based methods for the detection of bacterial pathogens: present and future considerations for the clinical laboratory. Clin. Chim. Acta *363*, 206–220.

Mullis, K.B., and Faloona, F.A. (1987). [21] Specific synthesis of DNA *in vitro* via a polymerase-catalyzed chain reaction. In Methods in Enzymology, W. Ray, ed. (Academic Press, San Diego, CA), pp. 335–350.

Oeth, P., Mistro, G.d., Marnellos, G., Shi, T., and Boom, D. (2008). Qualitative and quantitative genotyping using single base primer extension coupled with matrix-assisted laser desorption/ionization time-of-flight mass spectrometry (MassARRAY®). Methods Mol. Biol. *578*, 307–343.

Pagani, I., Liolios, K., Jansson, J., Chen, I.M.A., Smirnova, T., Nosrat, B., Markowitz, V.M., and Kyrpides, N.C. (2012). The Genomes OnLine

Database (GOLD) v.4: status of genomic and metagenomic projects and their associated metadata. Nucleic Acids Res. *40*, D571-D579.

Perkins, D.N., Pappin, D.J.C., Creasy, D.M., and Cottrell, J.S. (1999). Probability-based protein identification by searching sequence databases using mass spectrometry data. Electrophoresis *20*, 3551–3567.

Pholwat, S., Stroup, S., Foongladda, S., and Houpt, E. (2013). Digital PCR to detect and quantify heteroresistance in drug resistant *Mycobacterium tuberculosis*. PLoS One *8*, e57238.

Priest, N.K., Rudkin, J.K., Feil, E.J., van den Elsen, J.M.H., Cheung, A., Peacock, S.J., Laabei, M., Lucks, D.A., Recker, M., and Massey, R.C. (2012). From genotype to phenotype: can systems biology be used to predict *Staphylococcus aureus* virulence? Nat. Rev. Microbiol. *10*, 791–797.

Qin, J., Jones, R.C., and Ramakrishnan, R. (2008). Studying copy number variations using a nanofluidic platform. Nucleic Acids Res. *36*, e116.

Reuter, S., Harrison, T.G., Köser, C.U., Ellington, M.J., Smith, G.P., Parkhill, J., Peacock, S.J., Bentley, S.D., and Török, M.E. (2013). A pilot study of rapid whole-genome sequencing for the investigation of a *Legionella* outbreak. BMJ Open 3, e002175.

Richmond, G.S., Khine, H., Zhou, T.T., Ryan, D.E., Brand, T., McBride, M.T., and Killeen, K. (2011). MassCode liquid arrays as a tool for multiplexed high-throughput genetic profiling. PLoS One *6*, e18967.

Saiki, R.K., Scharf, S., Faloona, F., Mullis, K.B., Horn, G.T., Erlich, H.A., and Arnheim, N. (1985). Enzymatic amplification of beta-globin genomic sequences and restriction site analysis for diagnosis of sickle cell anemia. Science *230*, 1350–1354.

Sampath, R., Hofstadler, S.A., Blyn, L.B., Eshoo, M.W., Hall, T.A., Massire, C., Levene, H.M., Hannis, J.C., Harrell, P.M., Neuman, B., *et al.* (2005). Rapid identification of emerging pathogens: coronavirus. Emerg. Infect. Dis. *11*, 373–379.

Sampath, R., Mulholland, N., Blyn, L.B., Massire, C., Whitehouse, C.A., Waybright, N., Harter, C., Bogan, J., Miranda, M.S., Smith, D., *et al.* (2012). Comprehensive biothreat cluster identification by PCR/electrospray-ionization mass spectrometry. PLoS One 7, e36528.

Sanchez-Freire, V., Ebert, A.D., Kalisky, T., Quake, S.R., and Wu, J.C. (2012). Microfluidic single-cell real-time PCR for comparative analysis of gene expression patterns. Nat. Protoc. 7, 829–838.

Sandrin, T.R., Goldstein, J.E., and Schumaker, S. (2012). MALDI TOF MS profiling of bacteria at the strain level: a review. Mass Spectrom. Rev. 32, 188–217.

Seng, P., Drancourt, M., Gouriet, F.d.r., La Scola, B., Fournier, P.-E., Rolain, J.M., and Raoult, D. (2009). Ongoing revolution in bacteriology: routine identification of bacteria by matrix-assisted laser desorption ionization time-of-flight mass spectrometry. Clin. Infect. Dis. *49*, 543–551.

Seth-Smith, H.M.B., Harris, S.R., Skilton, R.J., Radebe, F.M., Golparian, D., Shipitsyna, E., Duy, P.T., Scott, P., Cutcliffe, L.T., O'Neill, C., *et al.* (2013). Whole-genome sequences of *Chlamydia trachomatis* directly from clinical samples without culture. Genome Res. *23*, 855–866.

Shin, J.H., Ranken, R., Sefers, S.E., Lovari, R., Quinn, C.D., Meng, S., Carolan, H.E., Toleno, D., Li, H., Lee, J.N., *et al.* (2013). Detection, identification, and distribution of fungi in bronchoalveolar lavage specimens by use of multilocus PCR coupled with electrospray ionization/mass spectrometry. J. Clin. Microbiol. *51*, 136–141.

Snitkin, E.S., Zelazny, A.M., Thomas, P.J., Stock, F., Henderson, D.K., Palmore, T.N., and Segre, J.A. (2012). Tracking a hospital outbreak of carbapenem-resistant *Klebsiella pneumoniae* with whole-genome sequencing. Sci. Transl. Med. *4*, 148ra116.

Sun, G., Luo, T., Yang, C., Dong, X., Li, J., Zhu, Y., Zheng, H., Tian, W., Wang, S., Barry, C.E., *et al.* (2012). Dynamic population changes in mycobacterium tuberculosis during acquisition and fixation of drug resistance in patients. J. Infect. Dis *206*, 1724–1733.

Sykes, P.J., Neoh, S.H., Brisco, M.J., Hughes, E., Condon, J., and Morley, A.A. (1992). Quantitation of targets for PCR by use of limiting dilution. Biotechniques *13*, 444–449.

Syrmis, M.W., Moser, R.J., Whiley, D.M., Vaska, V., Coombs, G.W., Nissen, M.D., Sloots, T.P., and Nimmo, G.R. (2011). Comparison of a multiplexed MassARRAY system with real-time allele-specific PCR technology for genotyping of methicillin-resistant Staphylococcus aureus. Clin. Microbiol. Infect. *17*, 1804–1810.

Tissari, P., Zumla, A., Tarkka, E., Mero, S., Savolainen, L., Vaara, M., Aittakorpi, A., Laakso, S., Lindfors, M., Piiparinen, H., *et al.* (2010). Accurate and rapid identification of bacterial species from positive blood cultures with a DNA-based microarray platform: an observational study. Lancet 375, 224–230.

Vogelstein, B., and Kinzler, K.W. (1999). Digital PCR. Proc. Natl. Acad. Sci. U.S.A. *96*, 9236–9241.

Wang, F., Massire, C., Li, H., Cummins, L.L., Li, F., Jin, J., Fan, X., Wang, S., Shao, L., Zhang, S., *et al.* (2011). Molecular characterization of drug-resistant Mycobacterium tuberculosis isolates circulating in china by multilocus PCR and electrospray ionization mass spectrometry. J. Clin. Microbiol. *49*, 2719–2721.

Weaver, S., Dube, S., Mir, A., Qin, J., Sun, G., Ramakrishnan, R., Jones, R.C., and Livak, K.J. (2010). Taking qPCR to a higher level: analysis of CNV reveals the power of high throughput qPCR to enhance quantitative resolution. Methods *50*, 271–276.

Whale, A.S., Huggett, J.F., Cowen, S., Speirs, V., Shaw, J., Ellison, S., Foy, C.A., and Scott, D.J. (2012). Comparison of microfluidic digital PCR and conventional quantitative PCR for measuring copy number variation. Nucleic Acids Res. *40*, e82.

White, R., Blainey, P., Fan, H.C., and Quake, S. (2009). Digital PCR provides sensitive and absolute

calibration for high throughput sequencing. BMC Genomics *10*, 116.
White, R.A., Quake, S.R., and Curr, K. (2012). Digital PCR provides absolute quantitation of viral load for an occult RNA virus. J. Virol. Methods *179*, 45–50.
von Wintzingerode, F., Böcker, S., Schlötelburg, C., Chiu, N.H.L., Storm, N., Jurinke, C., Cantor, C.R., Göbel, U.B., and van den Boom, D. (2002). Base-specific fragmentation of amplified 16S rRNA genes analyzed by mass spectrometry: a tool for rapid bacterial identification. Proc. Natl. Acad. Sci. U.S.A. *99*, 7039–7044.
Yang, S., and Rothman, R.E. (2004). PCR-based diagnostics for infectious diseases: uses, limitations, and future applications in acute-care settings. Lancet Infect. Dis. *4*, 337–348.

# Index

## S



## T



## U



## V



## W



## Z